FARADAY DISCUSSIONS OF THE CHEMICAL SOCIETY
NO. 69 1980

Phase Transitions in Molecular Solids

THE FARADAY DIVISION
CHEMICAL SOCIETY
LONDON

A GENERAL DISCUSSION

ON

Phase Transitions in Molecular Solids

26th, 27th and 28th March, 1980

A General Discussion on Phase Transitions in Molecular Solids was held at the University of Exeter on 26th, 27th and 28th March, 1980. The President of the Faraday Division, Professor J. S. Rowlinson, F.R.S., was in the chair: about 130 Fellows of the Faraday Division and visitors from overseas attended the meeting.
Among the overseas guests were:

Mr. H. Arntz, *West Germany*
Dr. G. Baranovic, *Yugoslavia*
Dr. J. L. Baudour, *France*
Dr. P. Blanpain, *Belgium*
Prof. R. Blinc, *Yugoslavia*
Mr. W. J. Briels, *The Netherlands*
Dr. J. Bruneaux, *France*
Mr. H. Cailleau, *France*
Prof. S. Califano, *Italy*
Dr. L. L. Chapoy, *Denmark*
Prof. G. Chapuis, *Switzerland*
Mr. M. E. Cosines, *West Germany*
Prof. R. M. J. Cotterill, *Denmark*
Dr. G. Coulon, *France*
Prof. L. A. De Graaf, *The Netherlands*
Dr. N. Descamps, *France*
Dr. R. Destro, *Italy*
Dr. J. P. Dumas, *Morocco*
Dr. A. Dworkin, *France*
Mr. C. Ecolivet, *France*
Dr. M. Eldrup, *Denmark*
Dr. B. Ewen, *West Germany*
Prof. H. Fontaine, *France*
Prof. R. Fouret, *France*
Dr. J. Fousek, *Czechoslovakia*
Dr. A. Fuchs, *France*
Mr. J. I. Geijsel, *The Netherlands*
Prof. M. Ghelfenstein, *France*
Mr. A. Gierard, *France*
Dr. A. Hüller, *West Germany*
Dr. Y. Koga, *Canada*
Mrs. N. Le Calvé, *France*
Dr. T. Luty, *Poland*
Dr. J. F. Maguire, *The Netherlands*
Prof. S. Masimo, *Italy*
Mr. R. J. Meier, *The Netherlands*
Prof. J. Meinnel, *France*
Dr. J. C. Messager, *France*
Prof. K. H. Michel, *Belgium*
Dr. M. More, *France*
Prof. J. A. Morrison, *Canada*
Dr. F. Moussa, *France*
Dr. A. Novak, *France*
Dr. J. C. A. Offringa, *The Netherlands*
Prof. R. Pick, *France*
Dr. K. O. Prins, *The Netherlands*
Dr. M. Rey, *France*
Prof. J. Rohleder, *Poland*
Dr. S. Sarbach, *West Germany*
Mr. R. C. F. Schaake, *The Netherlands*
Mr. K. J. Schenk, *Switzerland*
Prof. G. M. Schneider, *West Germany*
Dr. M. Schott, *France*
Dr. J. A. Schoten, *The Netherlands*
Prof. A. Schuyff, *The Netherlands*
Mr. P. Seiler, *Switzerland*
Prof. S. Seki, *Japan*
Mr. M. Sprik, *The Netherlands*
Prof. G. R. Strobl, *West Germany*
Prof. H. Suga, *Japan*
Dr. H. Szwarc, *France*
Mr. D. Van Der Putten, *The Netherlands*
Dr. J. Virlet, *France*
Dr. R. D. Weir, *Canada*
Prof. E. F. Westrum, *U.S.A.*
Mr. H. W. Wieldraaijer, *The Netherlands*
Dr. A. Würflinger, *West Germany*
Dr. M. Zabkowska, *Poland*

Organising Committee

Prof. A. J. Leadbetter (*Chairman*)
Dr. N. Boden
Mrs. Y. A. Fish
Dr. R. W. Munn
Dr. N. G. Parsonage
Dr. G. S. Pawley
Prof. J. N. Sherwood
Dr. L. A. K. Staveley
Dr. D. A. Young

ISBN: 0-85186-748-0
ISSN: 0301-7249

Printed in Great Britain by Fletcher & Son Ltd, Norwich

CONTENTS

Double-well Potentials and Structural Phase Transitions in Polyphenyls

By Hervé Cailleau, Jean-Louis Baudour and Jean Meinnel

Groupe de Physique Cristalline E.R.A. au C.N.R.S. no. 015,
Université de Rennes, Campus de Beaulieu, 35042 Rennes Cedex, France

and Ary Dworkin

Laboratoire de Chimie Physique des Materiaux Amorphes, L.A. au C.N.R.S. no. 75,
Université Paris Sud, Centre Scientifique, 91405 Orsay Cedex, France

and Fernande Moussa

Laboratoire Leon Brillouin, Cen Saclay, B.P. 2,
91190 Gif Sur Yvette, France

and Claude M. E. Zeyen

Institut Max von Lauë-Paul Langevin,
B.P. 156 X, 38042 Grenoble Cedex, France

Received 20*th December*, 1979

In their crystalilne phases, the conformation of non-rigid polyphenyl molecules results from a delicate balance between competing intramolecular and intermolecular forces. At low temperature, the polyphenyl crystals undergo structural phase transitions associated with stabilization of a non-planar conformation, with respect to a torsional angle between the phenyl rings. An interesting feature is the different natures of these phase transitions: displacive in biphenyl and order–disorder in *p*-terphenyl. In order to illustrate this point we consider the following features: the thermal motion in the high-temperature phases, the influence of hydrostatic pressure on the phase transition in *p*-terphenyl, the transitional heat capacities, the observation of the critical slowing down of fluctuations in *p*-terphenyl and the existence of incommensurate phases in biphenyl.

In a molecular crystal, the intramolecular potentials are generally weakly perturbed by the intermolecular interactions so that the internal vibrations are similar to those of the isolated molecule. In the case of a double-well intramolecular potential, however, very special features may arise if a delicate balance between competing intramolecular and intermolecular forces exists: in particular, the system is able to undergo a structural phase transition. This fact is well illustrated by materials such as non-rigid polyphenyl molecules: biphenyl, *p*-terphenyl, *p*-quaterphenyl . . . (fig. 1).

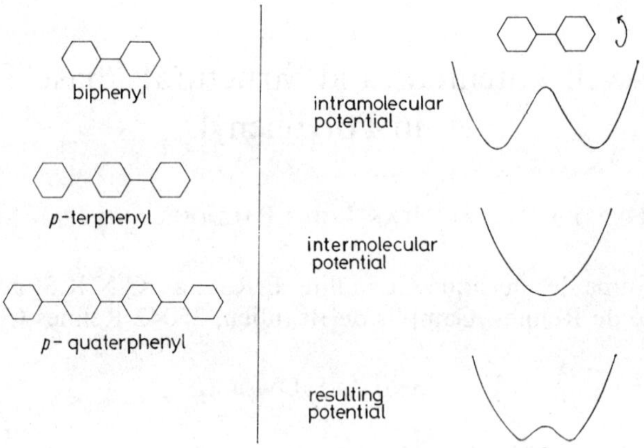

FIG. 1.—Schematic drawing of intramolecular, intermolecular and resulting torsional potential in polyphenyls.

NON-RIGID POLYPHENYL MOLECULES

The isolated molecule is non-planar, as a torsional angle exists between the planes of the phenyl rings:[1] the conjugation energy between phenyl rings is not sufficient to overcome the *ortho*-hydrogen repulsion, and this situation induces the existence of a double-well intramolecular potential. The conformation of these molecules strongly depends on the environment: the torsional angle in biphenyl is $\approx 40°$ in the gas phase and is reduced to 20° in the liquid phase. In the solid state, the different polyphenyls are apparently planar[3-5] at room temperature and their crystals are isostructural within the usual space group for flat molecules ($P2_1/a$). At low temperature, the polyphenyl crystals undergo antiferrodistorsive structural phase transitions (table 1) associated with a stabilization of the molecules in a non-planar configuration.[6-8] This behaviour is an indication of the delicate balance between intramolecular and intermolecular forces. Resulting from crystal forces which tend to place the phenyl rings parallel,[9] the intermolecular part of the potential

TABLE 1.—TRANSITION TEMPERATURE IN POLYPHENYL CRYSTALLINE PHASES

	biphenyl	*p*-terphenyl	*p*-quaterphenyl
hydrogenated	40 K[24]	191 K[16]	243 K[23]
deuterated	38 K[28]	179 K[16]	

has a single-well shape and is antagonistic to the intramolecular barrier. A double-well potential, corresponding to the instability of the planar configuration, subsists with a smaller barrier height between two close minima than in the isolated molecule (fig. 1): the molecules are no longer rigid with respect to the torsional angle.

An interesting feature is that different polyphenyls undergo phase transitions of different natures: displacive in biphenyl, it becomes order–disorder in *p*-terphenyl. The purpose of this paper is to illustrate the difference between the mechanisms of these two types of phase transitions in polyphenyls.

DOUBLE-WELL POTENTIAL AND STRUCTURAL PHASE TRANSITION

As indicated above in polyphenyl crystalline phases, the double-well potential results essentially from the competition between the *ortho*-hydrogen repulsion and crystal packing forces. In *p*-terphenyl the *ortho*-hydrogen repulsion forces are about twice as large as in biphenyl, hence the distance between the two minima is larger and the barrier height is higher than in biphenyl (fig. 2). Consequently, the torsional angle in the low-temperature phase is about 21° in *p*-terphenyl [7] and only 10° in biphenyl.[6]

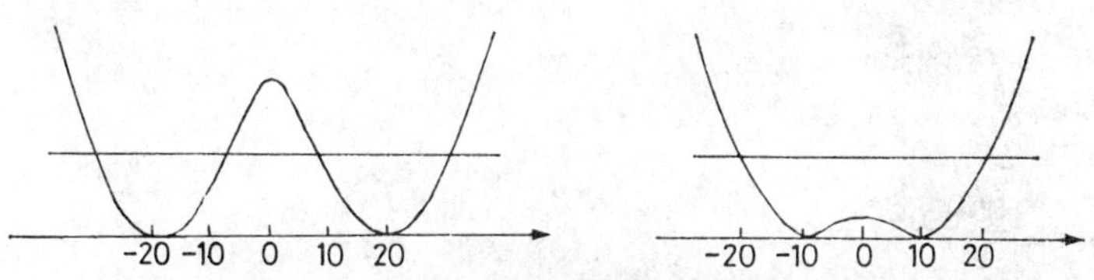

FIG. 2.—Schematic drawing of the crystalline torsional potential in (*a*) *p*-terphenyl (order–disorder transition) and (*b*) biphenyl (displacive transition).

In the high-temperature phase of *p*-terphenyl each molecule is well localized in one of the two bottoms of its double-well (fig. 2), and the disorder consists of an equal orientational distribution over the two wells in the crystal. The jump from one well to the other one is a single-molecule relaxation process governed by an activation energy. Between two jumps, vibrations also exist within the well and they are not strongly temperature dependent. On approaching the phase transition from above, pretransitional effects appear with the growth of short-range clusters.

The mechanism of the displacive phase transition in biphenyl is different. At room temperature the barrier energy is negligible compared with the thermal energy (fig. 2) and large-amplitude torsional vibration occurs. On lowering the temperature, the influence of the barrier height becomes more and more important and the frequency of the torsional mode is lowered. This soft mode condenses in a superlattice reflection at the transition.

The mechanisms of structural phase transitions presented here are oversimplified and more elaborated models may be needed.[10,11]

THERMAL MOTION IN THE HIGH-TEMPERATURE PHASE

Evidence for differences between the behaviour of biphenyl and *p*-terphenyl can be obtained *via* the determination of libration tensor from diffraction data. Table 2 shows the mean-square librational amplitudes of phenyl rings around the long molecular axis obtained with an harmonic single-well model at different temperatures. In biphenyl this amplitude decreases with temperature and is smaller than in *p*-terphenyl for which this amplitude is almost temperature independent. In *p*-terphenyl the picture of a double-well potential is rather suitable. Such a description is simply obtained by halving the atomic occupations outside the molecular axis on either side

TABLE 2.—MEAN-SQUARE LIBRATIONAL AMPLITUDES OF THE PHENYL RING AROUND THE LONG MOLECULAR AXIS WITH AN HARMONIC MODEL

temp./K	biphenyl	*p*-terphenyl (central rings)	*p*-quaterphenyl (central rings)
300	105.9(deg^2)	260.3(deg^2)	178.3(deg^2)
200		248.0	
110	45.7	19.3	

of the mean plane. The reliability factor values, compared with those for single-well models, are thus significantly improved.[4] The resulting double-peaked structure in the probability density function is characteristic of disorder (fig. 3).

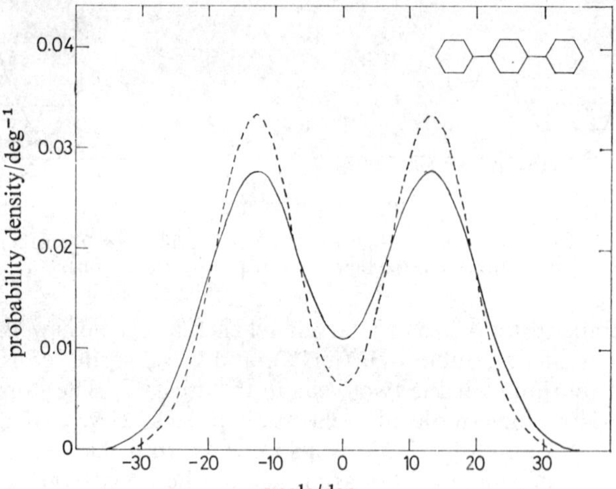

FIG. 3.—Probability density function for the *p*-terphenyl central ring librations around the long molecular axis from neutron data at 200 K (broken line) and X-ray data at 300 K (full line).

INFLUENCE OF HYDROSTATIC PRESSURE ON THE PHASE TRANSITION IN *p*-TERPHENYL

The compressibility of aromatic molecular crystals is very high compared with that of inorganic materials, and the investigations of phase transitions in polyphenyls under high hydrostatic pressure may be very fruitful. When pressure is increased, the polyphenyl intermolecular potential becomes steeper, whereas the intramolecular one remains unchanged: thence in the resulting double-well potential, the barrier height becomes smaller between two closer minima. Therefore the transition temperature is expected to decrease with increasing pressure. This behaviour is different from that observed for most inorganic materials in which high pressure affects the balance between competing short-range and long-range forces.[12]

An experiment has been performed on a neutron diffractometer at the Institut Lauë-Langevin. A deuterated *p*-terphenyl crystal was mounted within an helium high-pressure cell[13] and a standard cryostat. At high pressure, the intensity of a superlattice reflection presents a discontinuity step at the transition, so the determination of the transition temperature is easy. As expected, the transition is shifted to lower temperature when pressure is increased (fig. 4). This shift is very large and is not linear with pressure. In addition, at low temperature a new phase appears and a triple point exists at ≈70 K and 3.6 kbar. This new phase is associated with an important change in the unit cell, but its structure has not yet been determined.

The magnitude in the discontinuity step observed at high pressure increases quickly as the pressure increases. At atmospheric pressure the intensity variation with temperature of a superlattice reflection was found to be continuous or very close to continuous.[3] Moreover, the pseudo critical exponent β obtained by a plot of intensity data according to $I \propto (T_c - T)^{2\beta}$ is small: $\beta \simeq 0.15$. This small value can be associated with a classical Landau expansion of the free energy for a first-order transition, the square of the order parameter having a similar temperature dependence near the transition.[15]

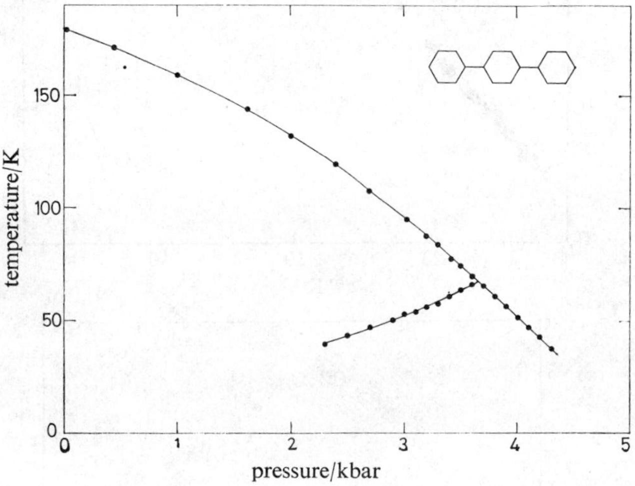

FIG. 4.—Phase diagram of deuterated *p*-terphenyl.

TRANSITIONAL HEAT CAPACITY

The heat capacities of hydrogenated biphenyl and *p*-terphenyl were measured in the vicinity of the transition temperature (fig. 5). In *p*-terphenyl a thermal anomaly extending over 20 K was found[16] with a maximum of C_p located at 193.33 K. The overall transition enthalpy was found as $\Delta H_t = 86.0$ J mol^{-1} and the transition entropy (obtained by graphical integration of a C_p/T against T curve) was $\Delta S_t =$ 0.49 J K^{-1} mol^{-1}. This transition seemed to be continuous at the temperature resolution used (≈0.8 K) and independent work performed by Chang[17] with a better resolution (0.01 K) confirmed this point. No thermal anomaly was found in biphenyl[18] either near 40 K or near 15 K where Cullick and Gerkin discovered another phase transition by e.p.r. measurements.[19] An upper limit to the transition enthalpy may be given as 10 J mol^{-1}.

The fact that ΔH_t is hardly noticeable in biphenyl whereas it is easily measured in *p*-terphenyl appears as a confirmation of the different nature of the phase transitions in these two compounds. In the displacive case the phase transition takes place between two relatively ordered structures and one may expect a smaller change of entropy than in order–disorder case. In this instance however, it has to be remarked

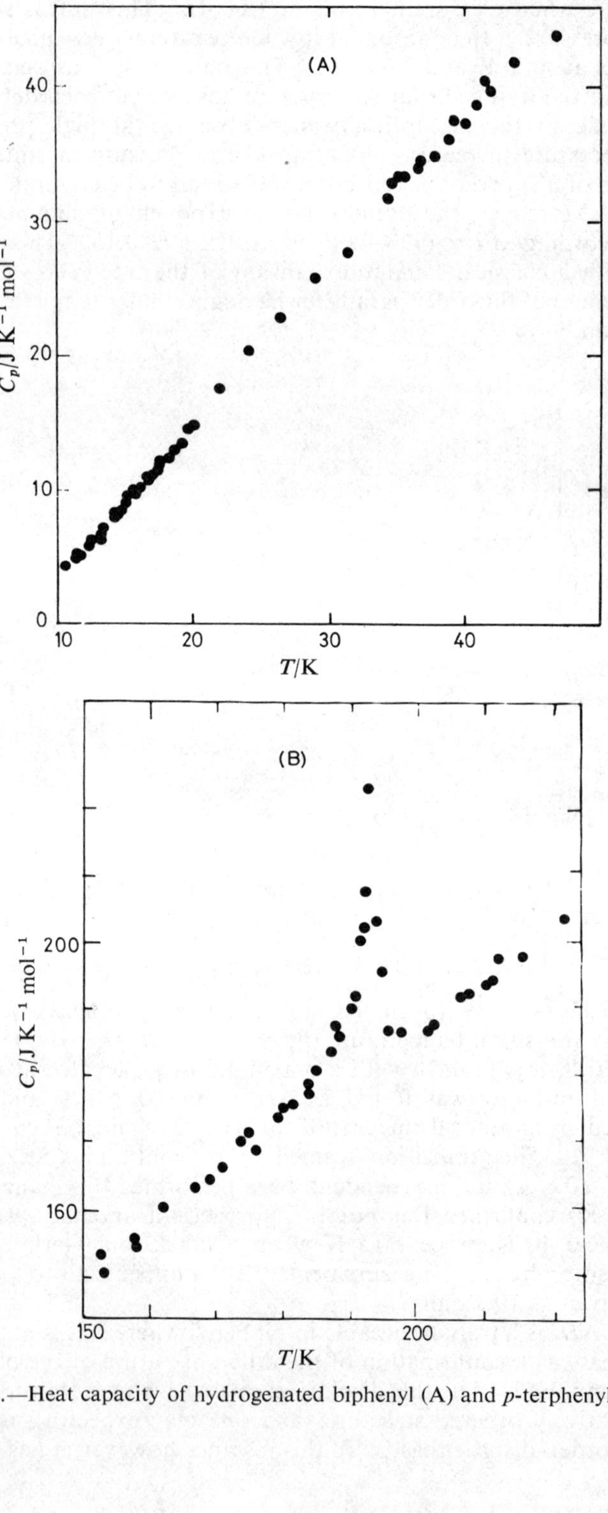

FIG. 5.—Heat capacity of hydrogenated biphenyl (A) and *p*-terphenyl (B).

that ΔS_t of *p*-terphenyl is very small compared with $R \ln 2 = 5.76$ J K^{-1} mol^{-1} which would happen in the case of a pure order–disorder transition. This behaviour can be understood by assuming the existence of strong short-range correlations between disordered molecules.

CRITICAL SLOWING DOWN IN *p*-TERPHENYL

In *p*-terphenyl, as in biphenyl, the pretransitional short-range correlation effects are located close to a Brillouin zone boundary point, and neutron scattering investigations are particularly suitable. A triple-axis experiment on deuterated *p*-terphenyl has shown the presence of critical quasielastic scattering close to the C($\frac{1}{2}$,$\frac{1}{2}$,0) point.[10] This quasi-elastic component is due to the formation of clusters of the ordered structure within the disordered phase. The energy width is proportional to the inverse of the cluster relaxation time τ_c: this lifetime is about 2×10^{-11} s at room temperature. The energy width could not be followed down to the phase transition because of the

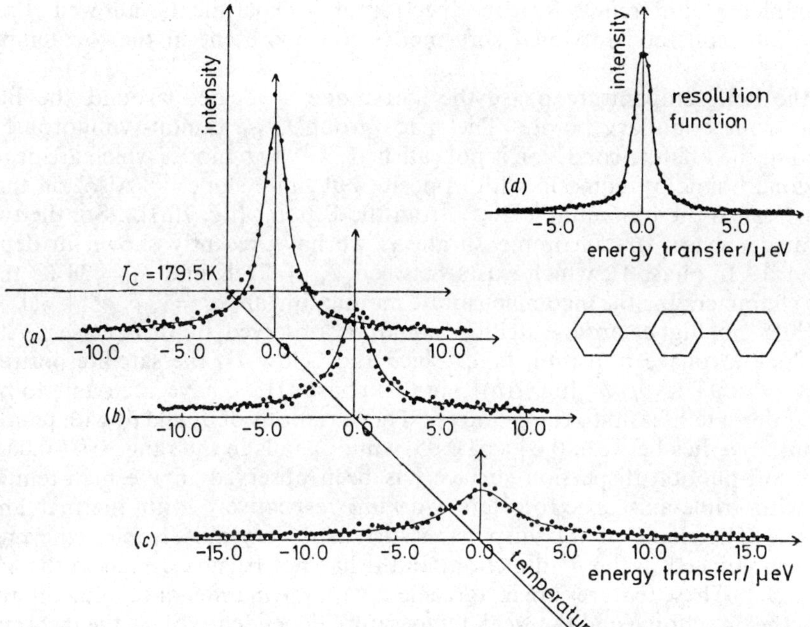

FIG. 6.—Energy spectra of neutrons scattered in deuterated *p*-terphenyl at the superlattice point ($\frac{5}{2}$, $\frac{1}{2}$, 0). The full curves represent a fit of a convolution of a single Lorentzian scattering law with the resolution function of the spectrometer. The fitted values of the h.w.h.m. obtained are 0.36, 0.73 and 1.67 μeV at, respectively, $T_c + 0.3$, $T_c + 0.8$ and $T_c + 1.6$ K, which yield the cluster lifetimes given in the text.

limited energy resolution. Also two very high resolution experiments were performed at the Institut Lauë-Langevin, the first one on the backscattering spectrometer[21] IN 10 and a second one, very recently, on the spin-echo machine IN 11. Fig. 6 shows the intensity of scattered neutrons, in the first experiment, as a function of energy and temperature. As for magnetic phase transitions, a critical slowing down of fluctuations is clearly observed: the intensity diverges and the energy width decreases on approaching the phase transition. The incoherent scattering, which at room temperature is of the same order of magnitude as the coherent scattering, is negligible close to the transition temperature T_c. Resolution corrections have been performed using a single Lorentzian scattering law. Values of the lifetime τ_c of 2.5×10^{-9}, 5.7×10^{-9} and 11.5×10^{-9} s were obtained at, respectively, $T_c + 1.6$, $T_c + 0.8$ and $T_c + 0.3$ K. The neutron spin-echo data are not yet completely analysed, but they confirm the preceding results. Furthermore, an indication of the existence of a central elastic peak was found close to the transition temperature.

On the other hand, it is interesting to notice that, as expected in an order–disorder phase transition, the frequencies of torsional internal modes do not depend very much on temperature, as shown by Raman scattering studies.[22,23]

INCOMMENSURATE PHASES IN BIPHENYL

The dynamical properties associated with the displacive transition of biphenyl are completely different. Raman spectroscopy experiments allowed Bree and Edelson[24] to conclude torsional soft modes to be present in the low-temperature phases.

In the high-temperature phase the soft mode is located around the B(0, $\frac{1}{2}$, 0) Brillouin zone boundary point. The space group $P2_1/a$ is non-symmorphic and at the B point the Lifshitz condition is not satisfied:[25,26] two modes which are degenerate at the zone boundary come in with opposite but finite slope.[27] Also, on the lower phonon branch the minimum is away from the B point [fig. 7(*a*)]. So, the two low-temperature phases are incommensurate as we have recently shown in deuterated biphenyl.[28] In phase II, which exists between $T_{II} = 21$ K and $T_I = 38$ K, the wave vectors characterizing the incommensurate modulation are $\boldsymbol{q}_S = \pm\delta_a \boldsymbol{a}^* \pm \frac{1}{2}(1 - \delta_b)\boldsymbol{b}^*$ [fig. 7(*b*)]; no higher-order satellites could be observed in this phase. At T_{II}, a partial lock-in phase transition takes place and below T_{II} the satellite positions become $\boldsymbol{q}_S = \pm\frac{1}{2}(1 - \delta_b)\,\boldsymbol{b}^*$ [fig. 7(*b*)]; in this phase III we have been able to measure higher-order satellites up to third-order. The variations of δ_a and δ_b with temperature are small;[28] δ_a lies between 0.04 and 0.05, while δ_b falls in the range 0.07-0.085.

The soft-phonon dispersion surface has been observed in the high-temperature phase with triple-axis spectrometers working respectively with thermal and cold neutrons. The shape of this dispersion surface is very anisotropic: the minimum is well pronounced in the $\boldsymbol{b}^*$ direction, and it has not been observed in the $\boldsymbol{a}^*$ direction (fig. 8). These features are in agreement with the maintenance of incommensurability in the direction $\boldsymbol{b}$ and the weak temperature dependence of δ_b, the incommensurate wave vector remaining close to that corresponding to the deep minimum of the soft phonon branch in this direction.

In the high-temperature phase, the mode which corresponds to the minimum of the appropriate dispersion surface is easily resolved in the temperature range 50-200 K and shows a pronounced softening [fig. 9(*b*)]. Above 200 K the observation of this mode is difficult, the damping being important as expected in aromatic molecular crystals; close to the transition temperature, the mode becomes overdamped [fig. 9(*a*)].

On the other hand, we have very recently observed a new excitation superimposed on the overdamped mode in the incommensurate phases. The experimental results are discussed and reported elsewhere.[29] The dispersion of this excitation is found to follow a linear law originating at the satellite reflection with a slope very much lower than that of the lowest acoustic mode. This new excitation branch

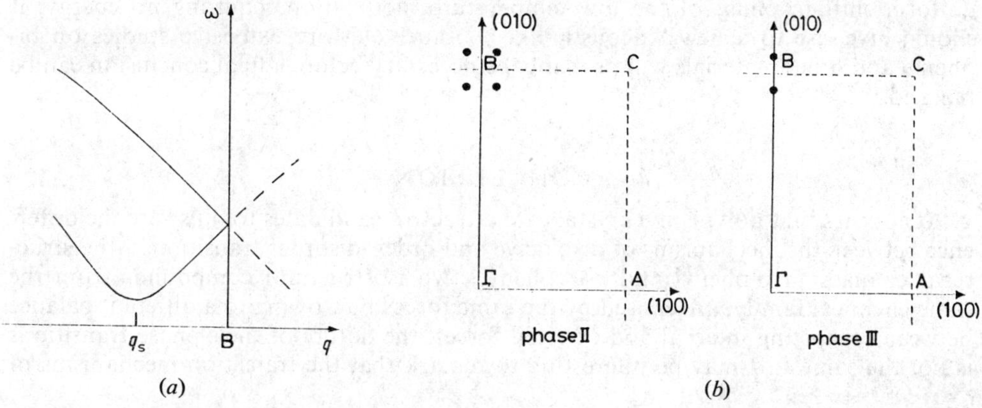

FIG. 7. (*a*) Schematic drawing of the torsional mode dispersion curves close to the B (0, ½, 0) point in biphenyl. (*b*) Locations of satellite reflections in the (*h k* 0) scattering plane for phase II and phase III. Dotted lines correspond to the limits of the first Brillouin zone.

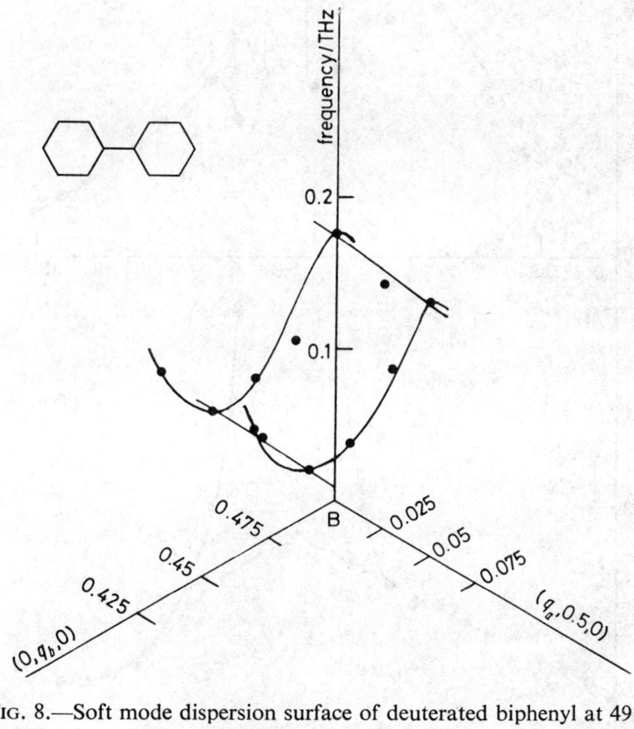

FIG. 8.—Soft mode dispersion surface of deuterated biphenyl at 49 K.

could be the phase-mode one. A phase mode, called a phason by Overhauser,[30] is a Goldstone mode corresponding to the continuous broken phase symmetry: a uniform shift in phase of the low temperature modulation requiring no energy, it should give rise to a new " acoustic-like " branch. More extensive studies on biphenyl and other materials will probably be necessary before a final conclusion can be reached.

CONCLUSION

It appears that polyphenyl crystals are attractive candidates to illustrate the difference between the mechanisms of displacive and order–disorder transitions: the structural changes in biphenyl and *p*-terphenyl, two isostructural compounds from the same chemical family, are caused by the same forces but, owing to a different balance between competing internal and external forces, the nature of their phase transitions is not the same. It may be interesting to remark that the transition mechanisms of

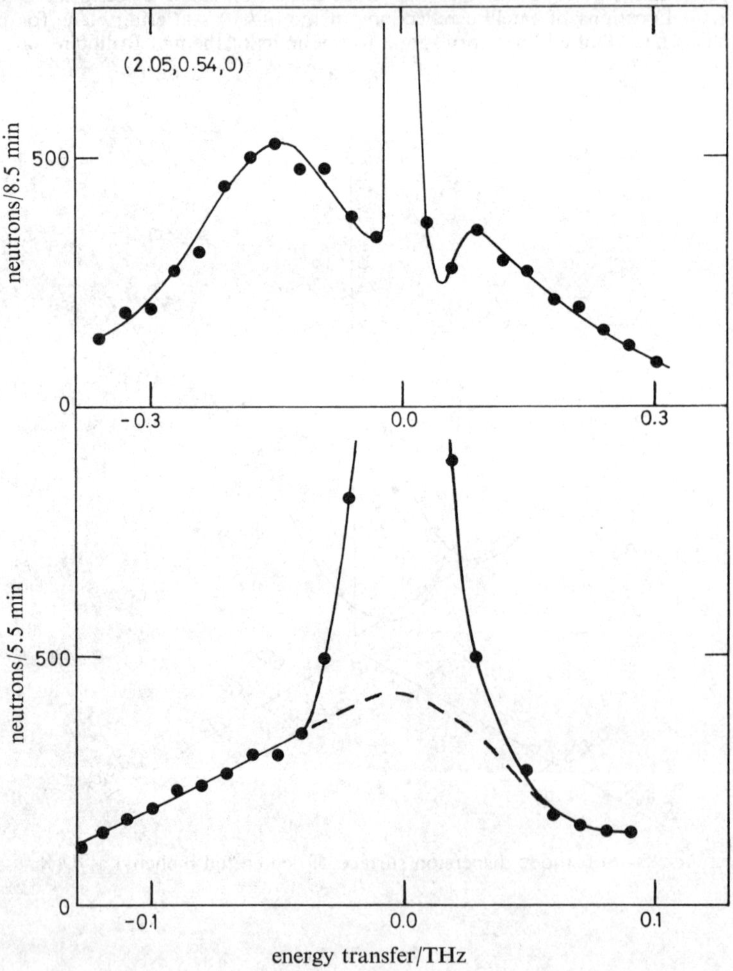

(*a*)

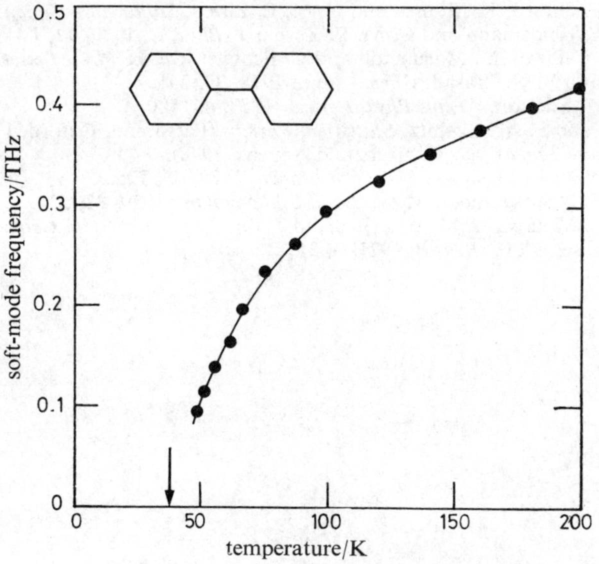

(*b*)

FIG. 9. (*a*) Constant-*Q* scans in deuterated biphenyl at two different temperatures (55 K above and 38 K below) at a point where a satellite reflection appears below 38 K. The broken line represents roughly the overdamped mode subtracted from the elastic part. The full lines are guides to the eye. (*b*) Soft-mode frequency (non-fitted) as a function of temperature in deuterated biphenyl.

these molecules are different from the orientational ordering in an angle dependent periodic potential; more particularly in this last case, the soft-mode behaviour does not exist.

The authors are grateful to J. Bouillot, A. Heidemann, F. Mezei and C. Vettier of I.L.L.; also to A. Girard of Groupe de Physique Cristalline for helpful discussions and assistance during measurements.

[1] O. Bastiansen, *Acta Chem. Scand.*, 1949, **3**, 408.
[2] H. Suzuki, *Bull. Chem. Soc. Japan*, 1959, **32**, 1340.
[3] G. P. Charbonneau and Y. Delugeard, *Acta Cryst. B*, 1977, **33**, 1586.
[4] J. L. Baudour, H. Cailleau and W. B. Yelon, *Acta Cryst. B*, 1977, **33**, 1773.
[5] Y. Delugeard, J. Desuche and J. L. Baudour, *Acta Cryst. B*, 1976, **32**, 702.
[6] H. Cailleau, J. L. Baudour and C. M. E. Zeyen, *Acta Cryst. B*, 1979, **35**, 426,
[7] J. L. Baudour, Y. Delugeard and H. Cailleau, *Acta Cryst. B*, 1976, **32**, 150.
[8] J. L. Baudour, Y. Delugeard and P. Rivet, *Acta Cryst. B*, 1978, **34**, 625.
[9] J. C. Messager, M. Sanquer, J. L. Baudour and J. Meinnel, 1st European Crystal Meeting, Bordeaux, 1972.
[10] S. Aubry, *J. Chem. Phys.*, 1975, **62**, 3217.
[11] T. Schneider and E. Stoll, *Phys. Rev. B*, 1976, **13**, 1216.
[12] G. A. Samara, T. Sakudo and K. Yoshimitsu, *Phys. Rev. Letters*, 1975, **35**, 26.
[13] J. Paureau and C. Vettier, *Rev. Sci. Instr.*, 1975, **46**, 963.
[14] H. Cailleau, J. L. Baudour, A. Girard and W. B. Yelon, *Solid State Comm.*, 1976, **20**, 577.
[15] J. P. Bachheimer and G. Dolino, *Phys. Rev. B*, 1975, **11**, 3195.
[16] H. Cailleau and A. Dworkin, *Mol. Cryst. Liq. Cryst.*, 1979, **50**, 217.
[17] S. S. Chang, 7th ASME Symp. Thermophys. Prop., 1977.
[18] A. Dworkin and H. Cailleau, to be published.
[19] A. Cullick and R. E. Gerkin, *Chem. Phys.*, 1977, **23**, 217.

[20] H. Cailleau, A. Girard, F. Moussa and C. M. E. Zeyen, *Solid State Comm.*, 1979, **29**, 259.
[21] H. Cailleau, A. Heidemann and C. M. E. Zeyen, *J. Phys. C*, 1979, **12**, L411.
[22] A. Girard, H. Cailleau, Y. Marqueton and C. Ecolivet, *Chem. Phys. Letters*, 1978, **54**, 479.
[23] B. A. Bolton and P. N. Prasad, *Chem. Phys.*, 1978, **35**, 331.
[24] A. Bree and M. Edelson, *Chem. Phys. Letters*, 1977, **46**, 500.
[25] L. D. Landau and E. M. Lifshitz, *Statistical Physics* (Pergamon, Oxford, 1968).
[26] A. Guyon, *Thèse 3ème Cycle* (Université d'Amiens, 1978).
[27] V. Dvorak and Y. Ishibashi, *J. Phys. Soc. Japan*, 1978, **45**, 775.
[28] H. Cailleau, F. Moussa and J. Mons, *Solid State Comm.*, 1979, **31**, 521.
[29] H. Cailleau, F. Moussa, C. M. E. Zeyen and J. Bouillot, *Solid State Comm.*, in press.
[30] A. W. Overhauser, *Phys. Rev. B*, 1971, **3**, 3173.

Phase Transitions in Crystals of Chain Molecules

Relation between Defect Structures and Molecular Motion in the Four Modifications of n-$C_{33}H_{68}$

By Bernd Ewen and Gert R. Strobl

Institut für Physikalische Chemie, Universität Mainz,
Jakob-Welder-Weg 15, 6500 Mainz, W. Germany

and Dieter Richter

Institut für Festkörperforschung, KFA Jülich,
5170 Jülich, W. Germany

Received 17*th December*, 1979

n-Tritriacontane (n-$C_{33}H_{68}$) exhibits three solid–solid phase transitions before melting. Applying a variety of experimental techniques, including X-ray scattering, Raman spectroscopy, n.m.r., dielectric relaxation and quasielastic neutron scattering, it was possible to specify for all four modifications defect structures and the corresponding molecular motions.

Each phase transition is accompanied by a step-like decrease in the degree of order resulting from the successive onset of rotational jumps, translational jumps in chain direction and the creation and diffusion of intrachain defects.

Molecular crystals may exhibit one or more solid–solid phase transitions before melting.[1] These transitions in most cases are of the first order and accompanied by the introduction of defects. They can generally be regarded as reactions of the crystal lattice to the onset of specific types of molecular motion.

In the special case of oligomeric chain molecules, which crystallize as lamellar systems, defect structures may concern the orientational order of chains, the lamellar surface or the internal chain conformation.

This paper gives an example and presents detailed studies on the solid-phase behaviour of n-tritriacontane (n-$C_{33}H_{68}$). This crystalline n-alkane passes through four stable modifications between room temperature and the melting point. A variety of experimental techniques was applied in order to characterize in detail the molecular motions and defect structures in the four solid phases with the general aim of gaining a better understanding of phase transitions in chain molecular crystals.

SOLID PHASE BEHAVIOUR OF n-$C_{33}H_{68}$

n-Tritriacontane (n-$C_{33}H_{68}$), which was synthesized in a manner[2] that excluded the presence of neighbouring homologues, exhibits three solid–solid phase transitions before melting (T_m = 71.8 °C) as can be seen from the d.s.c. diagram (see fig. 1). The four crystalline modifications which appear with increasing temperature are denoted in the following as A, B, C, D. Modification A is the orthorhombic low-temperature form found with all odd-numbered paraffins. Modification D is usually called the " rotator phase ", expressing a view introduced first by Müller[3] according

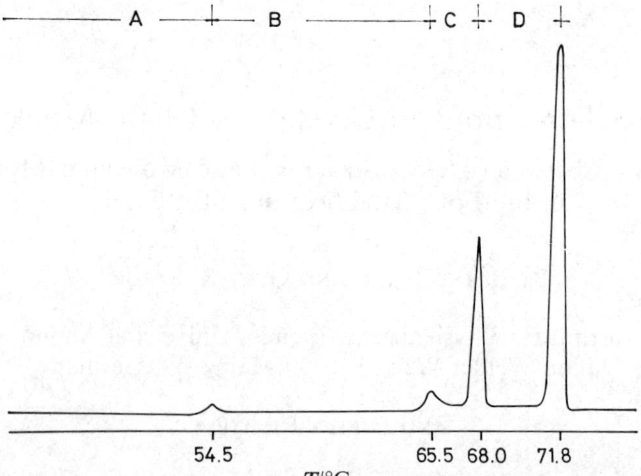

FIG. 1.—D.s.c. thermogram of solution crystallized n-$C_{33}H_{68}$, showing three solid–solid phase transitions before melting. A, B, C and D denote the four solid modifications.

to which the chains should rotate quasi-freely about their long axes. Hitherto no attention has been directed to the modifications intermediate between A and D.

Wide-angle X-ray scattering experiments[4] were used to determine the structures in these four modifications. For the highly disturbed crystal lattice in modification D it was not possible to specify the unit cell parameters. However, essential parameters like the angle ψ between the chain axes and the normal to the lamellar surface,

TABLE 1.—n-$C_{33}H_{68}$, RESULTS OF WIDE-ANGLE X-RAY SCATTERING EXPERIMENTS

η_c: electron density in the interior of the lamellae; ψ: angle between chain axis and surface normal.

modification	subcell	crystal structure	η_c	ψ	remarks
A(25 °C)	orthorhombic	orthorhombic $a =$ 7.44 Å $b =$ 4.96 Å $c =$ 87.65 Å	0.341	0	space group *Pcam* layer surface parallel to $(001)_s$ plane of the sublattice
B(60 °C)	orthorhombic	monoclinic $a =$ 7.57 Å $b =$ 4.976 Å $c =$ 87.94 Å $\beta =$ 88.8 or 91.2°	0.334	0	layer surface parallel to $(001)_s$ plane of the sublattice twin structure
C(66 °C)	orthorhombic	monoclinic $a =$ 8.02 Å $b =$ 4.985 Å $c =$ 88.0 Å $\beta =$ 73.5 or 106.5°	0.332	18.5	layer surface parallel to $(101)_s$ and $(10\bar{1})_s$ planes of the sublattice twin structure
D(69 °C)	nearly, but definitely not completely, hexagonal		0.318	19.5	multiple twin structure

the electron density η_c in the interior of the lamellae and the cross-section S per chain were obtained with sufficient accuracy. The results of these investigations are summarized in table 1.

RESULTS OF SMALL-ANGLE X-RAY SCATTERING AND RAMAN SPECTROSCOPY

For perfect n-alkane lamellae all CH_3-end groups of the molecules are situated in the planar lamellar surfaces. This situation is expected to change if motions of extended chains with a component in the chain direction become active or if interchain defects, which may diffuse along the chains, lead to a chain shortening. In both cases the originally planar structure of the lamellar surface is perturbed. Perturbations of this kind can be examined by measuring the absolute intensity of the small-angle X-ray scattering (SAXS). Since these investigations play a central role for the following discussion, the principles of evaluation are sketched briefly here. A detailed explanation of the procedures is given in ref. (5).

As can be seen from fig. 2(*a*), in a perfect, well ordered n-alkane crystal adjacent

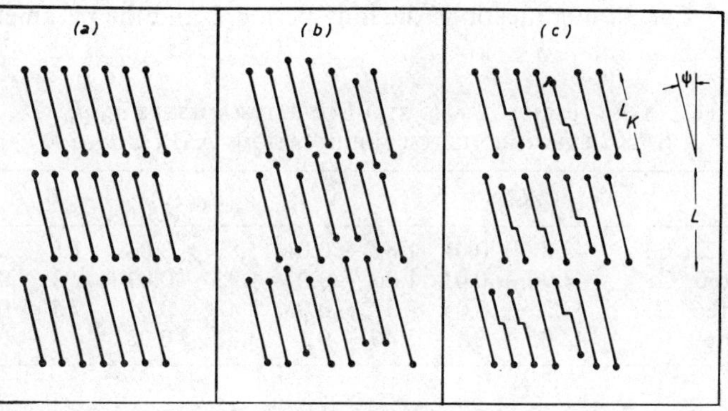

FIG. 2.—Examination of the interfacial order in n-alkane crystals with the aid of small angle X-ray scattering experiments. Deviations from the equalities $d_s = d_{av}$, $d_{av} = D_t$ (*a*), arising from longitudinal shifts ($d_s = d_{av}$, $d_{av} < D_t$) (*b*) and from intrachain defects ($d_s < d_{av}$, $d_{av} < D_t$) (*c*).

lamellae are separated by a plane layer of voids. Its thickness d_s is given by the difference of the long period L and the projection of the extended chain on the normal to the lamellar surface

$$d_s = L - 1.27 \text{ Å } n \cos \psi \tag{1}$$

(n is the number of C atoms in the backbone and ψ the angle between chain direction and surface normal).

This lamellar structure is reflected in a SAXS experiment by a series of (00*l*) reflections, their intensities being solely determined by the electron density profile of the interfacial region. Any disorder in the lamellar interfaces arising from the introduction of defects changes this profile and hence the intensities in a characteristic manner. Two important structure parameters can be directly derived.

First, the extrapolated intensity of the zeroth-order reflection I_0 is proportional to the square of d_{av}, the average thickness of the voids between adjacent end groups.

$$I_0 \sim d_{av}^2. \tag{2}$$

Secondly D_t, the overall thickness of the void layer, is related to the curvature at the origin

$$\left.\frac{d^2 I_l}{dl}\right]_{l=0} \sim D_t^2. \tag{3}$$

D_t is connected with the second moment of the electron density profile, σ^2, by

$$D_t = (12\sigma^2)^{1/2}. \tag{4}$$

Using eqn (2) and (3) one can decide about the occurrence of defects and distinguish between different types. For a well ordered lamellar system all three quantities d_s, d_{av} and D_t should be equal [see fig. 2(*a*)]. If the chains are randomly shifted in chain direction but remain stretched [see fig. 2(*b*)], D_t is increased with respect to d_s, the averaged void thickness d_{av} remaining unchanged. In the event that intrachain defects as kinks[6] or Reneker defects[7] are introduced [see fig. 2(*c*)], additional vacancies are created leading to an increase in both d_{av} and D_t, with respect to d_s. Thus, by measuring the absolute intensities of SAXS reflections and comparing D_t and d_{av} with d_s, it becomes possible to characterize the lamellar surfaces in a quantitative manner.

In table 2 the experimental results of the SAXS investigations on n-$C_{33}H_{68}$ are summarized. For all modifications the long period L and the parameters d_s, d_{av} and D_t are listed.

TABLE 2.—$C_{33}H_{68}$: LONG PERIOD L AND STRUCTURE PARAMETERS d_s, d_{av}, D_t, DETERMINED BY SAXS EXPERIMENTS FOR MODIFICATIONS A, B, C AND D

	L/Å	d_s/Å	d_{av}/Å	D_t/Å
A(25 °C)	43.8 + 0.01	1.83 ± 0.01	1.82 ± 0.03	1.8 ± 0.8
B(60 °C)	43.96 ± 0.01	1.96 ± 0.01	1.95 ± 0.03	2.3 ± 0.8
C(66 °C)	42.2 ± 0.05	2.37 ± 0.05	2.42 ± 0.05	7.5 ± 0.6
D(69 °C)	42.05 ± 0.1	2.46 ± 0.1	3.32 ± 0.08	10.1 ± 0.6

The data of modification A prove that the crystals, which were obtained from a dilute solution, are perfect at room temperature. The well ordered lamellar surface remains nearly unchanged when passing to modification B. In modification C one still finds coincidence between d_{av} and d_s; however, there is a significant increase in D_t. This result clearly signals the occurrence of longitudinal disorder, the chains remaining in the extended all-*trans* conformation. Finally, in modification D the SAXS data are indicative of additional intrachain defects.

Their occurrence could be confirmed by the i.r. spectrum,[8] where bands appeared (at 1310 and 1350 cm^{-1}), which are considered to represent local modes attached to a *gauche–trans–gauche* sequence in an otherwise stretched paraffin chain.[9,10]

Additional evidence for the occurrence of intrachain defects follows from the low-frequency Raman spectrum. It can be observed that the accordion mode, which is assigned to the all-*trans* conformation,[11] shows an intensity decrease upon passing to modification D. Fig. 3 shows the accordion modes in the modifications A and D, both being normalized to the integral intensity of the CH_2-twisting band (1295 cm^{-1}), which can be used as an internal scattering standard.[12] The result indicates the occurrence of intrachain defects for $\approx$40% of the chains.

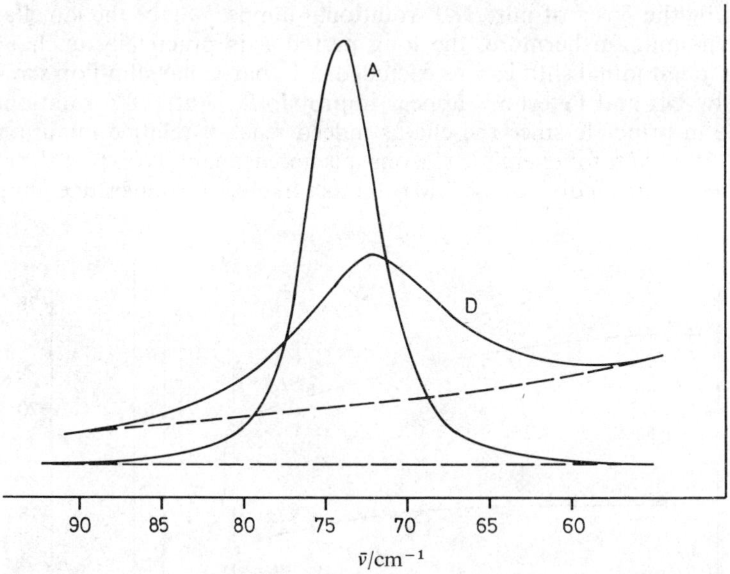

FIG. 3.—Accordion vibration bands in the low frequency Raman spectrum of n-$C_{33}H_{68}$, obtained for modifications A and D after normalization with regard to the CH_2 twisting intensity.

RESULTS OF WIDE-LINE N.M.R.

In contrast to SAXS, where only small changes of the characteristic parameters occur when passing from A to B (see table 2), major effects can be seen in the wide line n.m.r. spectra.[8] These are shown in fig. 4, where the line width and the second moment are plotted as a function of temperature between −150 °C and the melting point. Step-like decreases are observed at the transitions A → B and C → D, whereas the transition B → C is passed continuously.

A quite similar decrease was found by Olf and Peterlin[13] in n.m.r. experiments on the neighbouring homologue n-$C_{32}H_{66}$ using a uniaxially oriented sample. On the basis of model calculations[14] these authors concluded from the orientational dependence of the second moment that the motion becoming active at the first step should involve 180° rotational jumps of the extended chains about their long axis. As a mechanism of motion they suggested flip-flop screw jumps, *i.e.*, 180° rotational jumps coupled with a simultaneous shift in chain direction over one CH_2 unit. By such a kind of motion the packing of the methylene groups in the interior of the lamella is unchanged.

Similar measurements on n-$C_{33}H_{68}$ led to the same experimental result.[8]

At the transition from C → D the angular dependence of the second moment suggests a quasi rotation-like motion of the CH_2 groups about the long axis. The experimentally determined second moment, 4.8 G^2,[8] is smaller than the value of 5.8 G^2 expected for rigid-rod-like rotating chains as was initially assumed by Müller.

DEFECT STRUCTURES IN THE FOUR MODIFICATIONS OF n-$C_{33}H_{68}$

From the experiments described a molecular picture evolves as sketched in fig. 5.

At room temperature the crystals are perfect with the possible exception of a few single orientational defects [see fig. 5(A)]. The transition to modification B is accompanied by the onset of pure 180°-rotational jumps. Since the lamellar surface remains planar and, furthermore, the long period L is practically unchanged, any simultaneous longitudinal shift can be excluded. In particular, flip-flop screw jumps, as proposed by Olf and Peterlin,[13] appear improbable. Pure 180°-rotational jumps seem possible in principle since the chains indeed reach a relative minimum after a 180° turn.[15] However, for energetic reasons it is unreasonable to expect the molecules to perform these jumps individually. Much more likely is a cooperative jump process

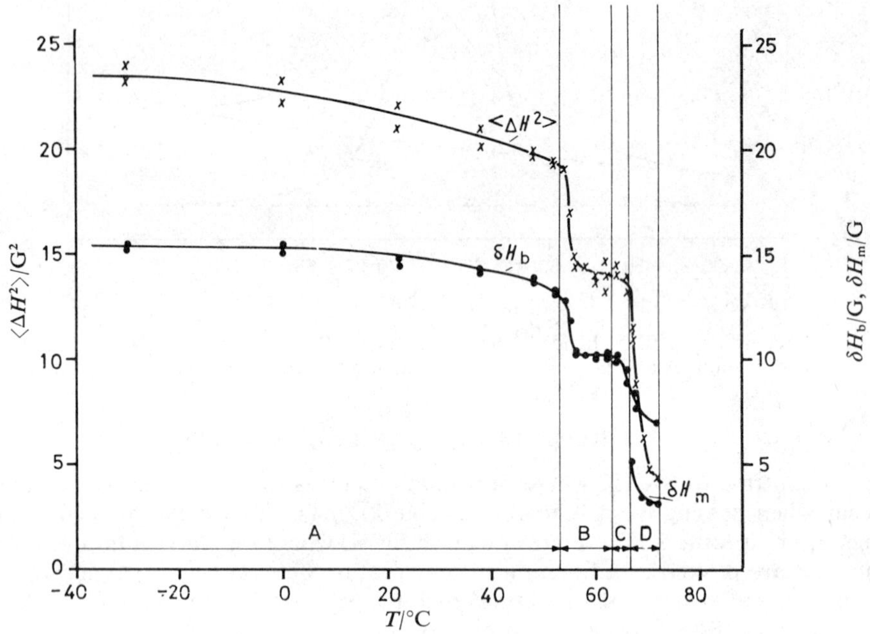

FIG. 4.—Second moment $\langle \Delta H^2 \rangle$ and linewidth δH_b of the wide-line n.m.r. spectra of an isotropic n-$C_{33}H_{68}$ sample. In modification D a second component with linewidth δH_m appears.

which preserves an orientational short range order [see fig. 5(B)]. The orientational long range order, present within modification A, is broken down.

With the transition to modification C [see fig. 5(C)] the originally planar lamellar surfaces are destroyed. This result indicates that longitudinal disorder is present, resulting from a motion with a component parallel to the chain direction. One possibility would be a change from the pure rotational-jumps of modification B to a flip-flop screw motion. In this case the motional components parallel and perpendicular to the chain axes would be strongly correlated. Alternatively one can also imagine a motional behaviour where the 180° rotational jumps, active in modification B, are superposed by independent translational jumps in the chain direction. This question cannot be decided from the experimental data presented so far. The results of quasielastic neutron scattering experiments[16,17] described in the next section will permit a decision.

The wide-angle X-ray pattern obtained for modification D[4] indicates that the crystal lattice is highly distorted, the regular CH_2 group packing being largely broken down [see fig. 5(D)]. This is supported by the low value of the n.m.r. second moment, which suggests a considerable degree of chain twisting. SAXS intensities, i.r. and

Raman spectroscopic observations provide evidence for intrachain defects. Estimates based on the Raman intensity of the accordion mode and, assuming simple kink defects (g^+–t–g^- sequences), on the SAXS parameter d_{av}, give a concentration of defected chains of the order 0.4-0.7.[5] As reflected by the value measured for D_t, there is an additional increase in the motional components parallel to the chain axes.

As a consequence the simple rotator model commonly used for describing the high temperature modification of n-alkanes should be modified with two respects: First, one has to include intrachain motions like high amplitude torsional oscillations

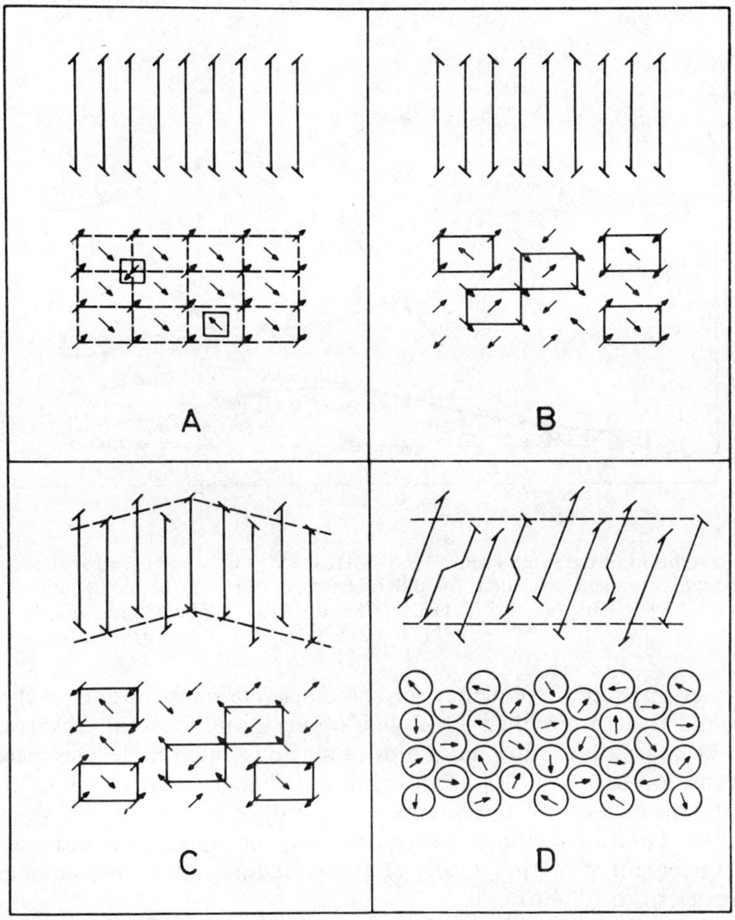

FIG. 5.—Defect structures in the four modifications of n-$C_{33}H_{68}$, in views perpendicular and parallel to the chain direction. The orientation of the zig-zag planes is indicated by arrows. A, single orientational defects; B, coupled 180° jumps, orientational short-range order; C, superposition of restricted longitudinal motion; D, hindered rotation, spacial limited diffusion in chain direction, chain twisting, kink defects.

and defect diffusion. Secondly, there is an additional component of motion parallel to the chain axis involving the molecules as a whole.

The experiments described so far do not permit a determination of the time scales to be attributed to the different types of motion. For this purpose two other experimental methods, namely dielectric relaxation measurements[18] and quasielastic neutron scattering[16,17] were applied. Since the quasielastic neutron scattering is simultaneously sensitive to the motional behaviour in time as well as in space, this method provides additional more detailed information.

DIELECTRIC RELAXATION MEASUREMENTS

The dielectric relaxation experiments were performed on a sample of n-$C_{33}H_{68}$ to which a small amount (5%) of a symmetric ketone ($C_{33}H_{66}O$) with the same chain length was added.[18] The results are shown in fig. 6, where the dielectric loss tangent

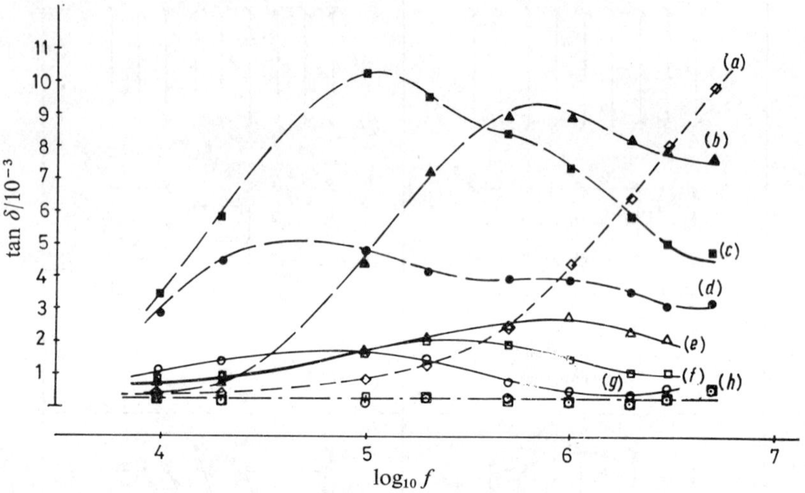

FIG. 6.—Solid solution of the symmetric ketone $C_{33}H_{66}O$ (5%) in n-$C_{33}H_{68}$. Frequency dependence of the dielectric loss tangent, measured for different temperatures in the modifications A, B and C. (*a*) 66 °C (C), (*b*) 64 °C (B), (*c*) 57.5 °C (B), (*d*) 55 °C (A/B), (*e*) 50 °C (A), (*f*) 40 °C, (*g*) 30 °C, (*h*) 68, 69 °C (D).

is plotted as a function of frequency for some discrete temperatures in the modifications A, B and C. In modification A one observes only a weak absorption signal, which may be caused by rotational jumps of single isolated molecules [see fig. 5(A)]. With the transition to modification B a significant change occurs, leading to a complex lineshape and an increase of the intensity. The observed relaxation strength corresponds to 180° rotational jumps between equally occupied orientational positions. This is in agreement with the results of the wide-line n.m.r. measurements.[8] The jump frequencies are 10^5 - 10^6 s^{-1}.

A further discontinuous change is observed at the transition to modification C. Since the maximum of tan δ at 66 °C is out of the range of measurement, one can only estimate its position from the shape of the curve on the low frequency side. In this way one is led to a relaxation frequency of $\approx(1\text{-}2) \times 10^7$ s^{-1}.

QUASIELASTIC NEUTRON SCATTERING

The scattering of slow neutrons by molecules containing protons is nearly entirely determined by the large incoherent cross section of the protons. The scattering intensity is proportional to the incoherent scattering law $S_{inc}(Q, \omega)$, where $\hbar Q$ and $\hbar\omega$ are the momentum and energy transfer during the scattering process. According to van Hove[19] $S_{inc}(\boldsymbol{Q}, \omega)$ can be written as the Fourier transform of the self-correlation function $G_s(\boldsymbol{r}, t)$ of the protons with respect to space and time:

$$S_{inc}(\boldsymbol{Q}, \omega) = \frac{1}{(2\pi)^3} \iiiint d\boldsymbol{r}\, dt\, e^{i(\boldsymbol{Qr} - \omega t)}\, G_s(\boldsymbol{r}, t). \tag{5a}$$

Sometimes it is more convenient to use instead of eqn (5*a*) the intermediate scattering law

$$I(\boldsymbol{Q}, t) = \frac{1}{(2\pi)^3} \iiint d\boldsymbol{r}\, e^{i\boldsymbol{Qr}}\, G_s(\boldsymbol{r}, t). \tag{5b}$$

As can be seen from eqn (5) it is possible to separate longitudinal and rotational components of motion if uniaxially oriented samples are used with the vector of momentum transfer oriented parallel or perpendicular to the chain axes.

Under the assumption of instantaneous uncorrelated jumps with negligible jump times the time evolution of $G_s(\boldsymbol{r}, t)$ can be calculated from rate equations and in the limiting case of continuous diffusion from the diffusion equation.

The resulting scattering law

$$S_{inc}(\boldsymbol{Q}, \omega) = I_0(\boldsymbol{Q})\delta(\omega) + \sum_{k=1}^{N-1} I_k(\boldsymbol{Q}) \frac{1}{\pi} \frac{1/\tau_k}{(1/\tau_k)^2 + \omega^2} \tag{6}$$

in general is a superposition of Lorentzians with line widths given by the eigenvalues of the rate equation system and weights which are connected with the eigenvectors of the system. N is the number of sites accessible for each proton. $I_0(\boldsymbol{Q})$, the so-called elastic incoherent structure factor, only contributes to eqn (6) if molecular motion is spatially restricted. It represents the Fourier transform of the equilibrium distribution reached for $t \to \infty$.

The measurements were performed at the Grenoble high flux reactor using the backscattering spectrometer IN10 and the multichopper time-of-flight machine IN5.[16,17] Spectra were obtained simultaneously at different Q-values for various orientations.

In modification C a distinct broadening was observed for both, $\boldsymbol{Q}$ parallel and $\boldsymbol{Q}$ perpendicular to the chain axis. This result can be taken as confirmation that molecular motion in modification C indeed is composed of both longitudinal and rotational components.

In the analysis of spectra two different jump mechanisms were considered which are compatible both with the X-ray and spectroscopic data as described above. The first one is a flip-flop process, where longitudinal and rotational motions are strongly correlated and described by one common jump rate v_f. The corresponding scattering law has the form of eqn (6); SAXS data suggest one should set $N = 5$ (see table 2).

Secondly, completely independent rotational and longitudinal motions are assumed. In this case the intermediate scattering law is given by the product of a longitudinal $[I^{trans}(\boldsymbol{Q}, t)]$ and rotational $[I^{rot}(\boldsymbol{Q}, t)]$ part. $I^{trans}(\boldsymbol{Q}, t)$ is equivalent to the intermediate scattering law of the flip-flop model if the jump vectors $\boldsymbol{r}_m$ are exchanged in a proper way. The rotational part is very simple[20] and has the form:

$$\begin{aligned} I^{rot}(\boldsymbol{Q}, t) &= I_0(\boldsymbol{Q}) + I_1(\boldsymbol{Q})e^{-2v_{rot}t} \\ &= \tfrac{1}{2}(1 + \cos \boldsymbol{Qr}_{rot}) + \tfrac{1}{2}(1 - \cos \boldsymbol{Qr}_{rot})e^{-2v_{rot}t} \end{aligned} \tag{7}$$

where v_{rot} is the rotational jump rate and $\boldsymbol{r}_{rot}$ the corresponding jump vector. The theoretical scattering laws were compared with the experimental results after an averaging process[14] accounting for the uniaxial orientational distribution in the samples.

TABLE 3.—JUMP RATES AND DIFFUSION COEFFICIENTS IN THE MODIFICATIONS C AND D OF n-$C_{33}H_{68}$, AS OBTAINED BY QUASIELASTIC NEUTRON SCATTERING

T/°C	v_{rot}/s^{-1}	v_{trans}/s^{-1}	D_{rot}/s^{-1}	D_{trans}/cm^2 s^{-1}
66	$\leqslant 2 \times 10^7$	$(2.1 \pm 0.3) \times 10^8$	—	—
67	$(5 \pm 2) \times 10^7$	$(2.6 \pm 0.4) \times 10^8$	—	—
70	—	—	$(6.3 \pm 0.1) \times 10^{10}$	$(1.2 \pm 0.2) \times 10^{-5}$

From this data analysis it became evident that there was no rigid coupling between the longitudinal and rotational components of motion. Systematically the jump rates v_f derived from the spectra with $\boldsymbol{Q}$ perpendicular to the chain axis were smaller than those for a parallel $\boldsymbol{Q}$-orientation. This observation definitely excludes pure flip-flop jumps. The jump rates obtained assuming independent 180° rotational jumps and longitudinal jumps over distances of 1.27 Å are listed in table 3.

In fig. 7 some spectra of a series, where $\boldsymbol{Q}$ is mainly parallel to the axis of orientation, are fitted with the corresponding scattering law.

The rotational jump rates v_{rot}, which lie just on the border of the instrumental resolution, are in good agreement with the results of the dielectric measurements, described above.

An explanation for the different jump rates follows from the energy maps calculated by McCullough.[21] They show that the hindrance potential for 180° rotational jumps is indeed larger and exhibits a stronger temperature dependence than that of the motion in chain direction.

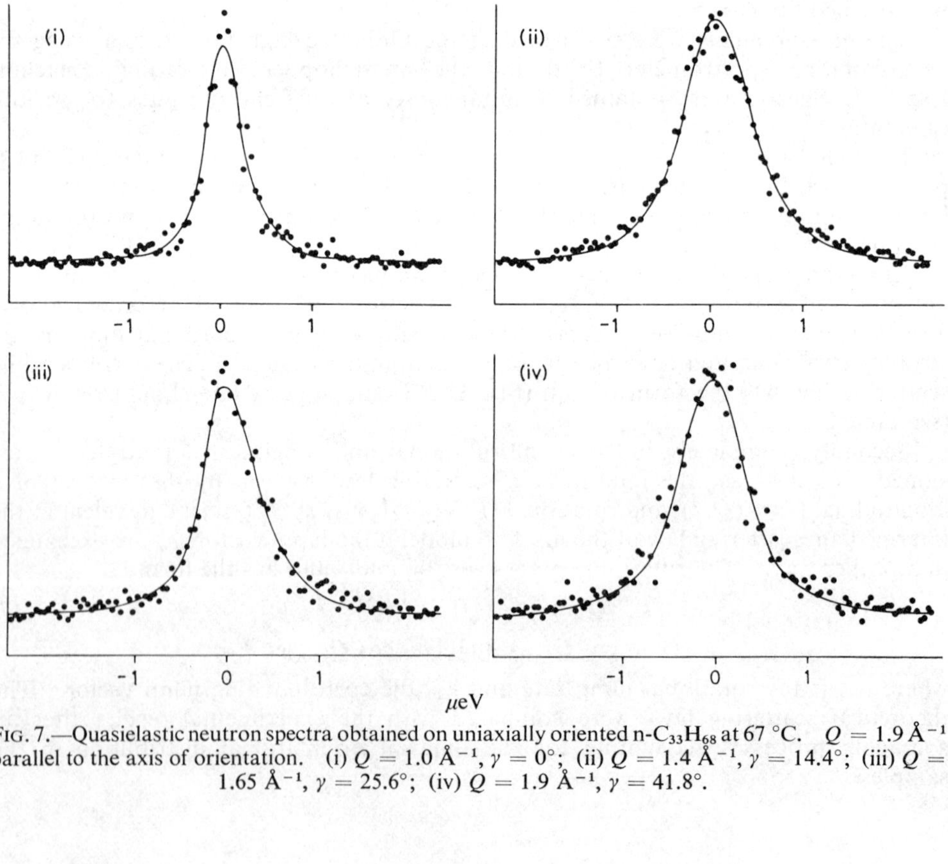

FIG. 7.—Quasielastic neutron spectra obtained on uniaxially oriented n-$C_{33}H_{68}$ at 67 °C. $Q = 1.9$ Å^{-1} parallel to the axis of orientation. (i) $Q = 1.0$ Å^{-1}, $\gamma = 0°$; (ii) $Q = 1.4$ Å^{-1}, $\gamma = 14.4°$; (iii) $Q = 1.65$ Å^{-1}, $\gamma = 25.6°$; (iv) $Q = 1.9$ Å^{-1}, $\gamma = 41.8°$.

In modification D the quasi-elastic broadening is considerably larger than in modification C. Here the elastic incoherent structure factor (see fig. 8) could be measured separately using the IN10, while the quasi-elastic spectra were obtained with the IN5. Analysis of the experimental data was based on a dynamic model where the motional process was assumed to be composed of a rotational diffusion of stretched

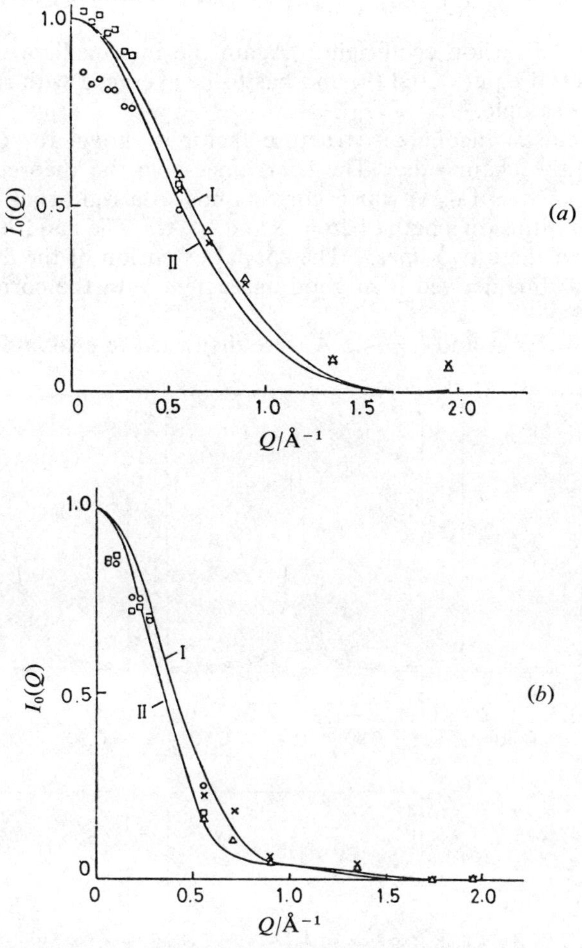

FIG. 8.—Elastic incoherent structure factor in modification D. (*a*) $\boldsymbol{Q}$ oriented nearly perpendicular, (*b*) $\boldsymbol{Q}$ oriented nearly parallel to the axis of orientation. ○, × experimental points before and □, △ after multiple scattering corrections. Solid lines: theoretical curves. Translational amplitude: I, $b = 4$ Å; II, $b = 5$ Å.

chains about the chain axis and of a diffusive motion in chain direction within a limited range $2b$, both occurring independent of each other.

The rotational part of the scattering law has the form[22]

$$I^{\text{rot}}(\boldsymbol{Q}, t) = J_0^2(\boldsymbol{Q}, \boldsymbol{a}) + 2 \sum_{n=1}^{\infty} J_n^2(\boldsymbol{Q}, \boldsymbol{a}) e^{-n^2 D_{\text{rot}} t} \tag{8}$$

where J_i are the Bessel functions of the ith order, D_{rot} is the rotational diffusion constant and a denotes the radius of the hydrogen rotation.

The translational part, calculated for a linear diffusion in a limited range of length $2b$ may be written as:[23]

$$I^{\text{trans}}(Q, t) = \frac{\sin^2 Qb}{(Qb)^2} + \sum_{n=1}^{\infty} I_n(Q) e^{-\left(\frac{n\pi^2}{2b}\right) D_{\text{trans}} t} \tag{9}$$

with

$$I_n(Q) = \frac{(4Qb)^2}{[(2Qb)^2 - (n\pi)^2]^2} [1 - \cos n\pi \cos(2Qb)];$$

D_{trans} is the linear diffusion coefficient. Again the intermediate scattering law becomes the product of eqn (8) and (9) and has to be averaged with respect to uniaxial orientation of the sample.

In fig. 8 the elastic incoherent structure factor is shown for $\boldsymbol{Q}$ parallel and $\boldsymbol{Q}$ perpendicular to the chain axis. The solid lines give the theoretical curves for a rotational radius of $a = 1.4$ Å, which corresponds to a rigid-rod-like rotation, and for a translational diffusion length of $2b = 8$ and 10 Å. The majority of experimental points lies between these two lines. The spatial extension of the diffusional process in chain direction thus derived is in good agreement with the corresponding SAXS result.

The values $a = 1.4$ Å and $b = 4.5$ Å were then used to evaluate the time-of-flight

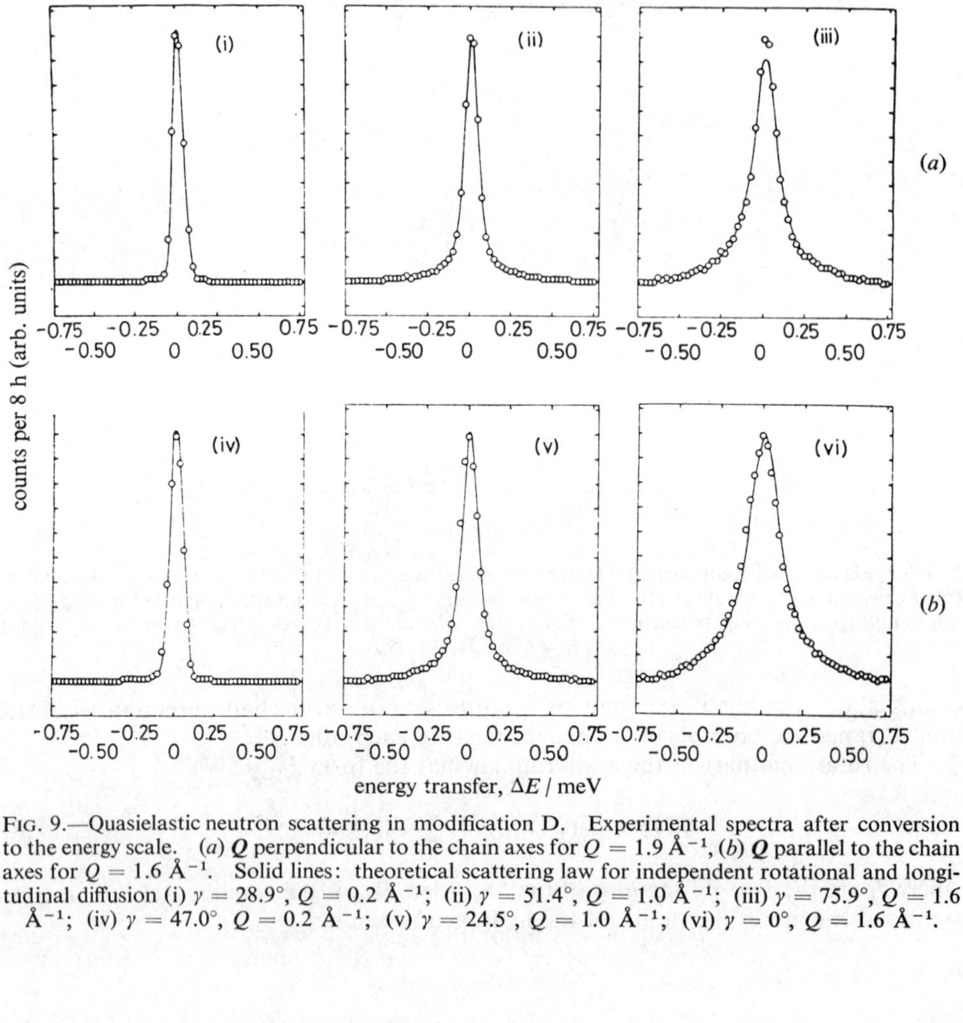

FIG. 9.—Quasielastic neutron scattering in modification D. Experimental spectra after conversion to the energy scale. (*a*) $\boldsymbol{Q}$ perpendicular to the chain axes for $Q = 1.9$ Å^{-1}, (*b*) $\boldsymbol{Q}$ parallel to the chain axes for $Q = 1.6$ Å^{-1}. Solid lines: theoretical scattering law for independent rotational and longitudinal diffusion (i) $\gamma = 28.9°$, $Q = 0.2$ Å^{-1}; (ii) $\gamma = 51.4°$, $Q = 1.0$ Å^{-1}; (iii) $\gamma = 75.9°$, $Q = 1.6$ Å^{-1}; (iv) $\gamma = 47.0°$, $Q = 0.2$ Å^{-1}; (v) $\gamma = 24.5°$, $Q = 1.0$ Å^{-1}; (vi) $\gamma = 0°$, $Q = 1.6$ Å^{-1}.

spectra. In fig. 9 some typical experimental spectra are shown together with corresponding theoretical curves. The diffusion coefficients used are given in table 3. In all cases, the theoretical scattering law fits the experimental data very well. This result shows that the molecular dynamics in modification D can be described predominantly as a diffusive process with independent rotational and longitudinal components. The creation of intrachain defects does not lead to a general increase of the radius of rotation of the protons, compared to rotating chains with an extended chain conformation. Otherwise the rotational part of the incoherent structure factor [see fig. 8(*a*)] would decrease faster with respect to Q.

The authors thank their colleagues Dipl.-Phys. K. Malzahn, Dr. W. Piesczek and Dipl.-Phys. T. Trzebiatowski for their important contributions to this work. They also thank Dr. A. Heidemann and Dr. R. Lechner for their advice during the neutron scattering experiments. Thanks are due in particular to Prof. Dr. E. W. Fischer and Prof. Dr. T. Springer for their continuous interest in this work and for many helpful and stimulating discussions.

Finally financial support by the Deutsche Forschungsgemeinschaft (Sonderforschungsbereich 41, Physik und Chemie der Makromoleküle, Mainz/Darmstadt) and by the Bundesministerium für Forschung und Technologie is gratefully acknowledged.

[1] *Physics and Chemistry of the Organic Solid State*, ed. D. Fox, M. M. Labes and A. Weissberger (Interscience, New York, London and Sydney, 1963), vol. I.
[2] W. Heitz, T. Wirth, R. Peters, G. Strobl and E. W. Fischer, *Makromol. Chem.*, 1972, **162**, 63.
[3] A. Müller, *Proc. Roy. Soc. A*, 1932, **138**, 514.
[4] W. Piesczek, G. Strobl and K. Malzahn, *Acta Cryst.*, 1974, **B30**, 1278.
[5] G. Strobl, B. Ewen, E. W. Fischer and W. Piesczek, *J. Chem. Phys.*, 1975, **61**, 5257.
[6] W. Pechhold, W. Dollhopf and A. Engel, *Acustica*, 1966, **17**, 61.
[7] D. H. Reneker, *J. Polymer Sci.*, 1962, **59**, 539.
[8] B. Ewen, E. W. Fischer, W. Piesczek and G. Strobl, *J. Chem. Phys.*, 1975, **61**, 5265.
[9] R. G. Snyder, *J. Chem. Phys.*, 1967, **47**, 1316.
[10] G. Zerbi, *Pure Appl. Chem.*, 1971, **26**, 499.
[11] R. F. Schaufele and T. Shimanouchi, *J. Chem. Phys.*, 1967, **47**, 3605.
[12] G. Strobl and W. Hagedorn, *J. Polymer Sci., Polymer Phys. Ed.*, 1978, **16**, 1181.
[13] H. G. Olf and A. Peterlin, *J. Polymer Sci. A-2*, 1970, **8**, 771.
[14] H. G. Olf and A. Peterlin, *J. Polymer Sci. A-2*, 1970, **8**, 753.
[15] P. E. McMahon, R. L. McCullough and A. A. Schlegel, *J. Appl. Phys.*, 1967, **38**, 4123.
[16] D. Richter and B. Ewen, *Proc. IAEA Symp. Neutron Inelastic Scattering*, Paper M-219/35, Vienna 1977.
[17] B. Ewen and D. Richter, *J. Chem. Phys.*, 1978, **69**, 2954.
[18] T. Trzebiatowski, *Diplomarbeit* (Mainz, 1976).
[19] L. van Hove, *Phys. Rev.*, 1954, **95**, 249.
[20] J. D. Barnes, *J. Chem. Phys.*, 1973, **58**, 5193.
[21] R. L. McCullough, *J. Macromol. Sci. Phys.*, 1974, **9**, 97.
[22] A. Dianoux, F. Volino and H. Hervet, *Mol. Phys.*, 1975, **30**, 1181.
[23] P. L. Hall and D. K. Ross, *Mol. Phys.*, 1978, **36**, 1549.

The Phase Transition in t-Butyl Cyanide [$(CH_3)_3CCN$] †

By Jonathon C. Frost and Alan J. Leadbetter

Chemistry Department, University of Exeter, Exeter EX4 4QD

and Robert M. Richardson ‡

Institut Laue-Langevin, Grenoble

Received 18*th January*, 1980

Extensive neutron scattering measurements have been carried out on t-butyl cyanide over a wide temperature range. Crystal structures and the potential for methyl-group rotation have been determined and will be reported elsewhere; this paper is concerned with the whole-molecule motions as determined by incoherent quasielastic and inelastic scattering measurements. One transition only has been observed at 233 K and the whole-molecule reorientational motions determined between 190 and 280 K (m.p. 292 K). At all temperatures the molecules undergo rapid reorientational motion about their C—C—N axes in a 3-fold potential which is strongly temperature dependent. The transition is associated with an order of magnitude decrease in the rotational correlation time and the onset of a slower, strongly cooperative motion involving translational as well as rotational displacements of the molecules. These probably involve relaxations between different local structures each having essentially the monoclinic structure of the low-temperature phase and related by rotations of $\pi/2$ resulting in the average tetragonal symmetry of the high-temperature phase. The neutron results lead to a self-consistent reinterpretation of previous n.m.r. data.

Tertiary butyl cyanide [$(CH_3)_3CCN$] is an especially interesting member of the $(CH_3)_3CX$ series which shows a rich solid-state polymorphism. The dielectric studies of Clemett and Davies[1] indicated that in the highest temperature crystal phase (I) the molecules may rotate about the CCN axis but the dipolar axis itself does not reorientate until the liquid phase is reached. Extensive calorimetric measurements[2] gave the following phase behaviour:

$$\text{(Crystal III)}\ T = 213\ \text{K};\ \Delta S = 1.1\ \text{J K}^{-1}\ \text{mol}^{-1}\ \text{(Crystal II)}\ T = 233\ \text{K};\ \Delta S = 7.8\ \text{J K}^{-1}\ \text{mol}^{-1}\ \text{(Crystal I)}\ T = 292\ \text{K};\ \Delta S = 31.8\ \text{J K}^{-1}\ \text{mol}^{-1}\ \text{(liquid)}.$$

The transition III/II was not well-defined but the total entropy increment III to I is near $\boldsymbol{R}\ln 3$ (9.1 J K^{-1} mol^{-1}) which was taken to imply that I is a uniaxial rotator phase. Whether or not this interpretation of ΔS is correct the rotation of the molecules about their CCN axes in Crystal I has been confirmed by n.m.r.[3] and very recent incoherent neutron quasielastic (INQES) measurements.[4]

† Experiments carried out at I.L.L., Grenoble and A.E.R.E., Harwell.
‡ Now at Rutherford Laboratory, Chilton, Didcot, Oxon OX11 OQX.

A number of uncertainties remain, however. First, an earlier n.m.r. study[5] was interpreted as showing only methyl group rotation below the m.p., while in the later work[3] the single step decrease in linewidth near 110 K was interpreted (more reasonably) as due to the onset of both methyl and whole-molecule uniaxial rotations. The additional observation of a single T_1 minimum near 160 K led to the conclusion that the methyl and molecular rotations are coupled and describable by a single correlation time and energy barrier (16.3 kJ mol^{-1}). The motion is established in phase III and almost no changes are observed at the transitions (except a slight

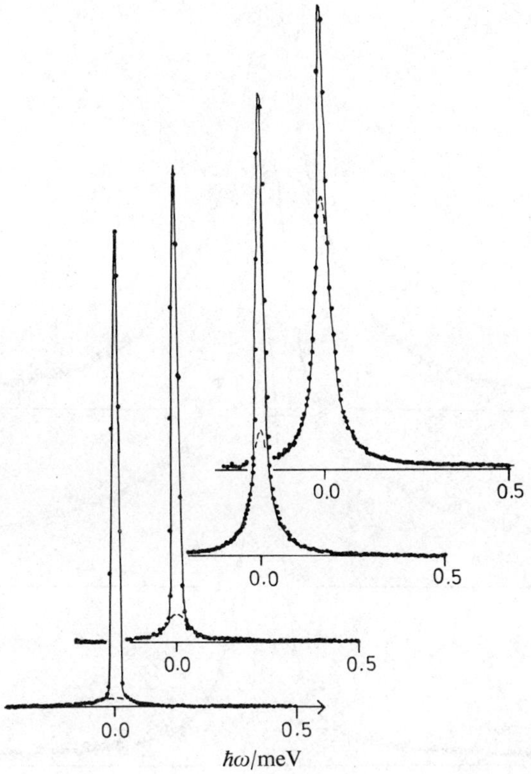

FIG. 1.—Scattering law, $S(Q, \omega)$ of t-butyl cyanide at 227 K in the Crystal II phase measured on IN5 with $\lambda_0 = 10.0$ Å. The spectra (which are not to the same vertical scale) have elastic Q values (from bottom to top) of 0.17, 0.38, 0.76 and 1.15 Å^{-1}. The solid curve is a fit of the 3-fold rotational jump model [eqn (6), (7) and (11)] to the experimental data. The dashed line separates the elastic from the quasielastic scattering.

hysteresis around III/II as found calorimetrically). This is a puzzling conclusion because the barriers to methyl rotation in t-butyl compounds have been thoroughly studied[6–8] and are largely intramolecular in origin whereas the whole-molecule reorientation must be controlled by intermolcular forces and should have a different temperature dependence. Secondly, the correlation time at the II/I transition from the n.m.r. work is $\approx 10^{-10}$ s whereas the INQES results gave $\approx 5 \times 10^{-12}$ s in I with a jump at the transition to values too great to be measured in II ($\gtrsim 2 \times 10^{-11}$ s). Questions remain, therefore, about the relation between the different experimental results, the nature and time scale of intra- and inter-molecular rotations in the various phases and the nature of the phase transition(s).

t-Butyl cyanide was chosen for an extensive set of measurements on a uniaxial rotator phase with the ultimate objective of understanding not only the single molecule motions but also the collective aspects of the reorientational motions and the nature of the transition. It was selected because both hydrogenous and deuterated versions may be obtained and the melting point being near room temperature should facilitate the preparation and handling of single crystals. Furthermore the strong cyano

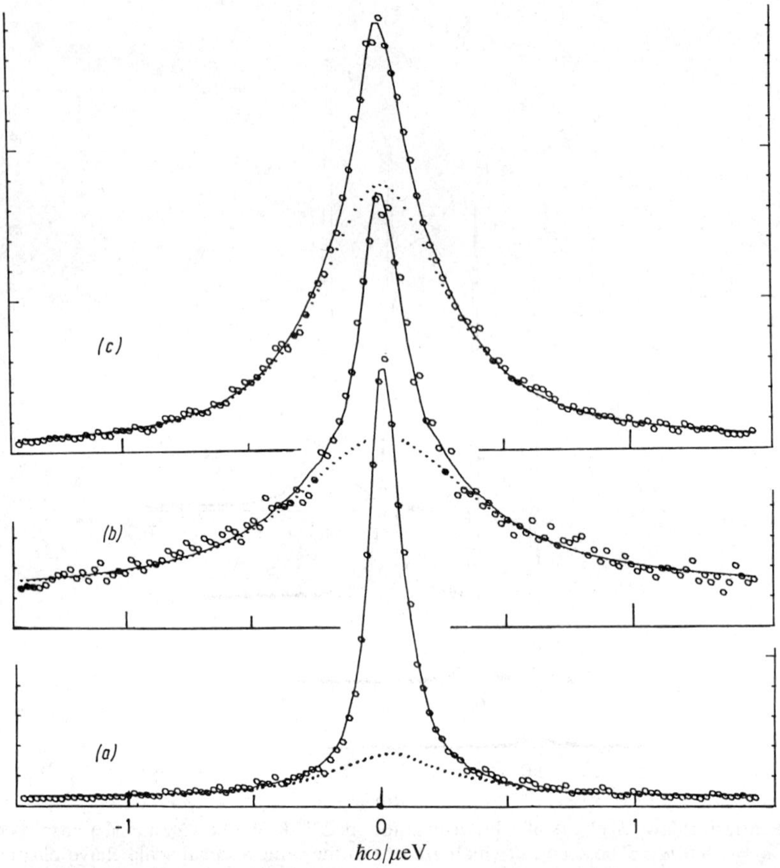

FIG. 2.—Scattering law, $S(Q, \omega)$ of t-butyl cyanide in the Crystal II phase measured on the back scattering spectrometer IN10. The solid curves are calculated for the 3-fold jump reorientation model [eqn (6), (7) and (11)] and the dotted curves separate elastic and quasi-elastic components: (*a*) $Q = 0.57$ Å^{-1}, $T = 208$ K; (*b*) $Q = 1.51$ Å^{-1}, $T = 208$ K; (*c*) $Q = 1.51$ Å^{-1}, $T = 190$ K.

dipole and the uniaxial rotations suggested that the dipoles are probably all aligned (along 0 and π) to give essentially a 2-d system of coupled rotors. So far we have completed low-resolution structural determinations of low- and high-temperature phases and INQES plus incoherent neutron inelastic scattering (INIS) on powder samples.[8] The structure from <5 to >227 K is monoclinic ($P2_1/m$) with two molecules per cell in a head to tail arrangement, all dipoles colinear and directed along ***c***. No structural or other changes have been found at the reported III/II transition so we hereafter designate the structure existing below 233 K as Crystal II. The structure of

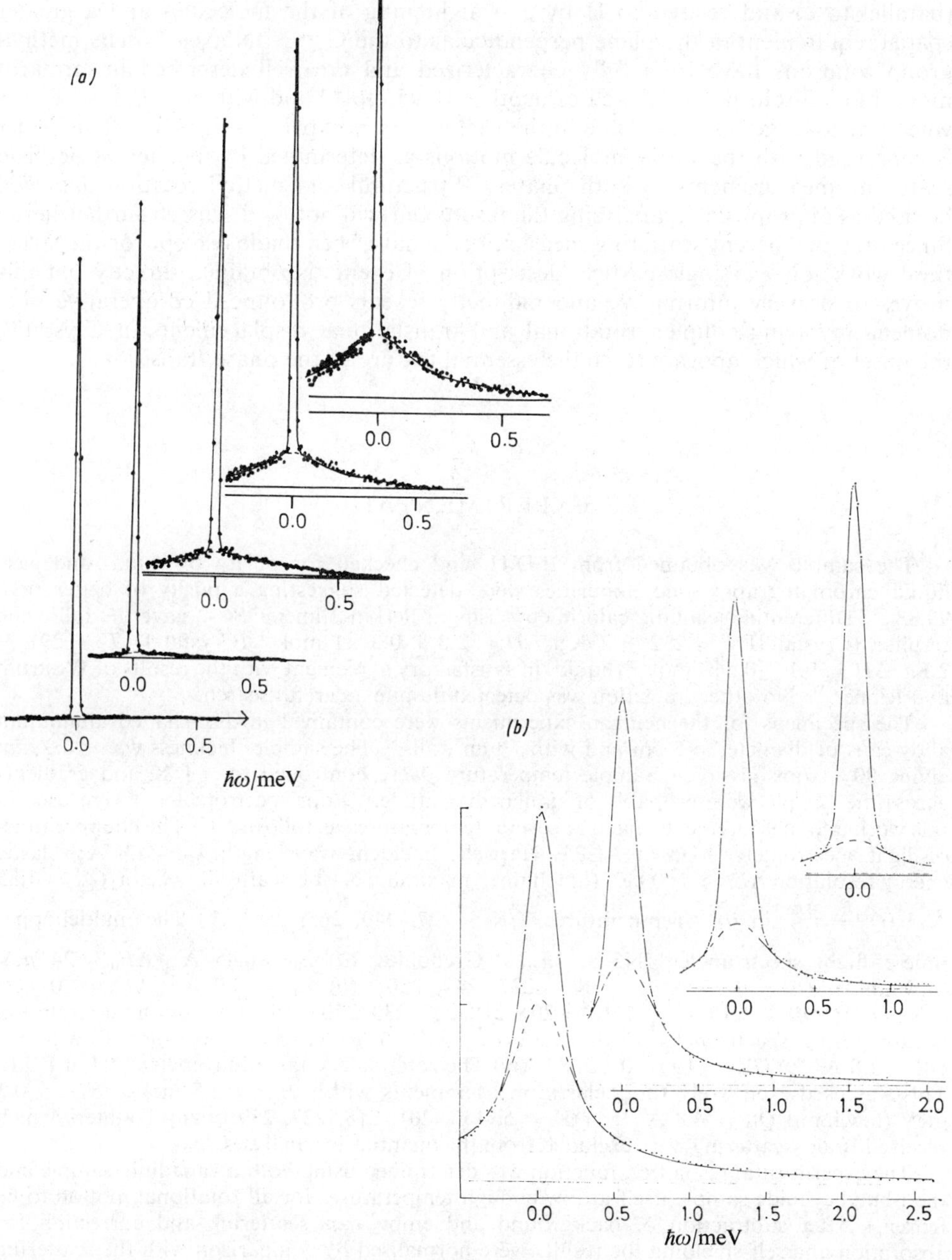

FIG. 3.—Scattering law, $S(Q, \omega)$ of t-butyl cyanide in the Crystal I phase. The solid curves are typical fits to the data using eqn (3), (5) and (11) with arbitrary elastic coefficients, the dashed curves separate elastic and quasielastic components and the horizontal straight lines show the flat inelastic background: (*a*) 270 K, measured using IN5 at $\lambda_0 = 10$ Å, Q values (bottom to top) of 0.17, 0.38, 0.76, 0.92 and 1.15 Å^{-1}; (*b*) 240 K measured using 4H5, Q values (bottom to top) of 1.7, 1.1, 0.75 and 0.53 Å^{-1}.

Crystal I is tetragonal (probably $P4/n$) with 2 molecules per cell, at the origin and body centre, again in a head to tail configuration with dipoles along the 4-fold axis (parallel to $\boldsymbol{c}$) and related to II by a straightening of the molecules and a greater spatial requirement in the plane perpendicular to the C—C—N axes.* The methyl-group rotations have been fully characterized and are well described in terms of motion in a 3-fold potential well of depth ~18 kJ mol^{-1} and with $\tau_0 \sim 0.3 \times 10^{-12}$ s where the average residence time in the well is $\tau = \tau_0 \exp(V_3 - E_0)/\boldsymbol{R}T$. This paper is concerned with the whole-molecule motions as determined by incoherent neutron scattering measurements in both phases. Structural and methyl rotation data will be used as appropriate in analysing the results but will not be discussed further here.[8] Since only incoherent scattering measurements have been made (except for the structural work) only a single-particle description of events is obtained directly but this proves to be very informative and indirectly reveals pronounced co-operative phenomena involving coupled rotational and translational displacements in Crystal I, the onset of which appears to be the essential feature of the phase transition.

EXPERIMENTAL

The sample was obtained from B.D.H. and checked for purity by n.m.r. and gas–liquid chromatography: no impurities were detected suggesting a purity of better than 99.8%. Differential scanning calorimetry using a Perkin–Elmer DSC-2 gave the following results: (Crystal II)$T = 232 \pm 2$ K; $\Delta H = 2.0 \pm 0.3$ kJ mol^{-1} (Crystal I) $T = 291 \pm 2$ K; $\Delta H = 9.1 \pm 0.5$ kJ mol^{-1} (liquid) in satisfactory agreement with the results of Westrum and Ribner.[2] No other transition was detected despite a careful search.

The specimens for the neutron experiments were contained in disc-shaped aluminium alloy cans of diameter ≈5 cm and with 1 mm walls. The sample thickness was ≈0.3 mm giving 90% transmission. Sample temperatures were controlled to ±1 K and gradients across the sample were probably of similar magnitude. Four spectrometers[9] were used in this work and measurements made at several temperatures as follows: (i) The chopper time-of-flight spectrometer 4H5 at AERE Harwell: incident wavelength $\lambda_0 = 4.8$ Å; elastic energy resolution $\Delta E_0 \approx 100$ μeV (h.w.h.m.); maximum elastic scattering vector $Q^{el}_m = 1.85$ Å^{-1} ($Q^{el} = \frac{4\pi}{\lambda_0} \sin \theta$). Temperatures T/K = 227, 240, 265, 284. (ii) The multichopper time-of-flight spectrometer IN5 at I.L.L., Grenoble: (a) $\lambda_0 = 6.45$ Å; $\Delta E_0 = 24$ μeV (h.w.h.m.); $Q^{el}_m = 1.75$ Å^{-1}; T/K = 227, 240, 250; (b) $\lambda_0 = 10.0$ Å; $\Delta E_0 = 10$ μeV (h.w.h.m.); $Q^{el}_m = 1.15$ Å^{-1}; T/K = 201, 218, 227, 239, 270. (iii) The very high resolution backscattering spectrometer IN10 at I.L.L.: $\lambda_0 = 6.28$ Å; $\Delta E_0 \approx 0.5$ μeV (h.w.h.m.); $Q^{el}_m = 1.8$ Å^{-1}; T/K = 190, 201, 208. (iv) The rotating crystal spectrometer IN4 at I.L.L. This was used exclusively for inelastic measurements with: $E_0 = 12.5$ meV; $\Delta E_0 = 0.3$ meV (h.w.h.m.) $Q^{el}_m = 4.8$ Å^{-1}; T/K = 5, 150, 201, 218, 227, 239. Any counters which received Bragg scattering were excluded from the quantitative analyses.

The experimental resolution function was determined using both a vanadium sample and the t-butyl cyanide sample itself at low enough temperatures for all rotational motion to be frozen. After subtraction of background and empty-can scattering and correction for absorption and self-shielding the results were normalised by comparison with the scattering from a vanadium sample and were converted, using standard procedures, to the symmetrised scattering law, $S(\boldsymbol{Q}, \omega)$. This involves only experimentally known quantities and $S(\boldsymbol{Q}, \omega)$ is a property only of the system, independent of the experiment; it is also the function most

* Crystal II (227 K)$a = 6.45$ Å, $b = 6.95$ Å, $c = 6.75$ Å, $\beta = 92.8°$; C—C—N axis parallel to $\boldsymbol{c}$. Crystal I (239 K) $a = b = 6.89$ Å, $c = 6.70$ Å; C—C—N axis parallel to $\boldsymbol{c}$.

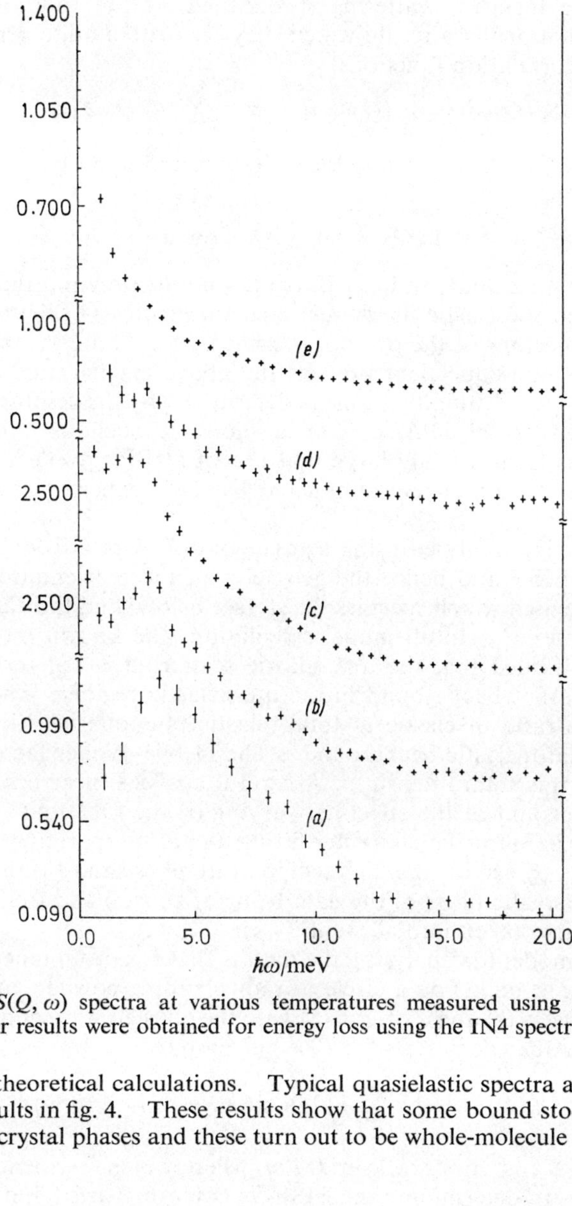

FIG. 4.—Inelastic $S(Q, \omega)$ spectra at various temperatures measured using IN5 in energy gain. Similar results were obtained for energy loss using the IN4 spectrometer.

simply related to theoretical calculations. Typical quasielastic spectra are shown in fig. 1–3 and inelastic results in fig. 4. These results show that some bound stochastic motions are occurring in both crystal phases and these turn out to be whole-molecule rotations about the dipole axis.

RESULTS AND DISCUSSION

(i) QUASIELASTIC DATA ANALYSIS

The incoherent scattering law from a proton in molecule undergoing reorientational motions may be written[10]

$$S_s(\boldsymbol{Q}, \omega) = e^{-2W}[S_s^R(\boldsymbol{Q}, \omega) + S_s^I(\boldsymbol{Q}, \omega)] \tag{1}$$

where e^{-2W} is a Debye–Waller factor

$$\{2W = \langle(\boldsymbol{Q} \cdot \boldsymbol{u})^2\rangle\}, \tag{2}$$

$S_s^I(\boldsymbol{Q}, \omega)$ is the inelastic scattering (broadened by the rotational motion) and $S_s^R(\boldsymbol{Q}, \omega)$ is the rotational scattering which may be written quite generally (assuming exponential time correlation functions):

$$S_s^R(\boldsymbol{Q}, \omega) = A_0(\boldsymbol{Q})\delta(\omega) + \pi^{-1} \sum_{l=1} A_l(\boldsymbol{Q})\mathscr{L}(\tau_l^{-1}) \tag{3}$$

where $\mathscr{L}(\tau_l^{-1}) = \tau_l^{-1}/(\tau_l^{-2} + \omega^2)$ and the $A_l(\boldsymbol{Q})$ depend upon the type of motion and $\sum_{l=0} A_l = 1$.

$$A_0(\boldsymbol{Q}) = |\int \exp i\boldsymbol{Q}.\boldsymbol{r}P(\boldsymbol{r})\, d\boldsymbol{r}\,|^2 \tag{4}$$

where $P(\boldsymbol{r})$ is the (time independent) distribution function for the proton. $A_0(\boldsymbol{Q})$ is commonly called the elastic incoherent structure factor (EISF) and gives a direct measure of the trajectory of the proton, averaged over " infinite " time, as the molecule rotates. For non-equivalent protons the above results must be averaged over the different protons. " Infinite " time is determined by the resolution of the experiment, so effectively $t_\infty \approx 5\hbar/\Delta E_0$. For a molecule such as t-butyl cyanide the number of protons is sufficiently large that except at Bragg reflections the observed scattering is generally completely dominated by the incoherent cross section of the protons.

The data were first analysed using eqn (1) and (3) with adjustable coefficients A_l to determine the EISF and hence the geometry of the reorientational motions. A model was then devised which specifies the A_l (see below) and the data fitted to obtain the correlation times τ_l. In all model calculations the known molecular geometry was used.[11,12] Provided that the quasielastic scattering is not too broad $S_s^I(\boldsymbol{Q}, \omega)$ may be taken as a flat background in the quasielastic region. The EISF was then determined as the ratio of elastic to total (elastic plus quasielastic) scattering after subtraction of a flat inelastic background. The Debye–Waller factor is given by the Q-dependence of this total intensity. A careful analysis of errors in the EISF was made by taking account of the effects of varying (i) the functional form (number of Lorentzians) in eqn (3) for the computer fitting and also separating elastic and quasielastic components by eye for the high temperature phase and (ii) the energy range of the fit. The very small effects of the data being at constant scattering angle and not constant Q were also taken into account.

A very useful model for analysing the data is that for instantaneous jump motion among m equidistant points on a circle of radius r for a powder sample,[13] for which the scattering law is of the form of eqn (3) with the quasielastic summation from $l = 1$ to $l = m - 1$, and the coefficients for $l \geqslant 0$ given by:

$$A_l(Q) = m^{-1} \sum_{j=1}^{m} j_0(2Qr \sin \pi j/m) \cos (2\pi lj/m). \tag{5}$$

$j_0(X) = \sin X/X$ and $\tau_l = (\tau/2) \sin^2 (\pi l/m)$ where τ is the residence time (τ^{-1} is the jump frequency). In determining the EISF, m could be varied and the coefficient of the elastic term allowed to be an arbitrary variable: $D_0(Q)$. The model scattering law was fitted to the data after convolution with the experimental resolution function.

Reorientational motion between three equidistant sites is important both for methyl and whole-molecule rotations and for this eqn (3) becomes, for a powder sample,

$$S_s^R(Q, \omega) = A_0(Q)\delta(\omega) + [1 - A_0(Q)]\mathscr{L}(\tau_1^{-1}) \tag{6}$$

with

$$A_0(Q) = [1 + 2j_0(Qr\sqrt{3})]/3 \text{ and } \tau_1 = 2\tau/3, \tag{7}$$

τ being the residence time.

(ii) EFFECT OF METHYL ROTATIONS

For compounds like t-butyl cyanide both methyl and whole molecule rotations may occur. Assuming that these are independent then the total scattering law is a convolution of the two separate components which for a powder sample may be written

$$S_{\text{tot}}(Q, \omega) = \overline{S_{\text{A}}(Q, \omega) \otimes S_{\text{B}}(Q, \omega)} \tag{8}$$

the bar denoting the powder average.

The methyl rotation (A) has been shown to be described by eqn (6) and taking the general form (3) for the whole-molecule motion (B) gives

$$S_{\text{tot}}(Q, \omega) = \overline{A_0 B_0}\delta(\omega) + \overline{B_0(1 - A_0)}\pi^{-1}\mathscr{L}(\tau_{\text{A}}^{-1}) \tag{9}$$
$$+ \overline{A_0 \sum_{l=1} B_l}\pi^{-1}\mathscr{L}(\tau_{\text{B}_l}^{-1}) + \overline{(1 - A_0) \sum_{l=1} B_l}\pi^{-1}\mathscr{L}(\tau_{\text{B}_l}^{-1} + \tau_{\text{A}}^{-1})$$

where the A and B are functions of Q.

Explicit expressions for the coefficients have been given by Schlaak[14] for the case where the whole-molecule motion is also a 3-fold uniaxial rotation. For the present purpose, however, we note the fortunate circumstance that the correlation times for methyl rotation in t-butyl cyanide are at least one order of magnitude slower than those for the uniaxial molecular rotation for $T \gtrsim 190$ K, which confirms their independence in this case. This has been established by investigating the apparent elastic peaks such as those of fig. 3 using an order of magnitude higher resolution. The methyl rotation is thus so slow that it is virtually unobservable using IN5 (and 4H5). For a truly unobservable methyl rotation the EISF would be given by

$$\text{EISF} = \overline{A_0 B_0} + \overline{(1 - A_0) B_0} = \bar{B}_0. \tag{10}$$

In fact, although it made little difference to the value of the EISF, especially in the high T phase, a significantly better fit to $S(Q, \omega)$ was obtained by taking account of the very small broadening contribution of the methyl rotation (just observable in the wings of the elastic component). This was simply achieved by writing the scattering law as:

$$S_{\text{tot}}(Q, \omega) = \bar{B}_0(Q)\{a(Q)\delta(\omega) + [1 - a(Q)]\pi^{-1}\mathscr{L}(\tau_{\text{A}}^{-1})\} + \sum_{l=1} \bar{B}_l(Q)\pi^{-1}\mathscr{L}(\tau_{\text{B}}^{-1}) \tag{11}$$

where the expression in brackets describes the experimental spectrum observed within the limited energy window of the very high resolution spectrometer IN10 so that $a(Q)$ and τ_{A} are experimentally determined quantities. This approach may be used with $B_0(Q)$ as a variable to determine the EISF, or to fit a given model [*e.g.*, eqn (6)] to $S(Q, \omega)$, and means that a self-consistent treatment is used to analyse all data at all resolutions.

(iii) MULTIPLE SCATTERING

The correction of experimental data for the effects of multiple scattering is very complicated and expensive of computer time. We have adopted the simpler procedure of incorporating the effects of multiple scattering into the model calculations. These calculations were based on the Monte Carlo programme of Johnson[15] and parameters such as the EISF or τ_l values from the resultant scattering laws can be used to assess the effects of multiple scattering on the experimental results.

CRYSTAL II

With the 4H5 spectrometer ($\Delta E_0 \approx 100$ μeV h.w.h.m.) no quasielastic broadening was observed, in agreement with previous low-resolution work on this phase.[4] However, with higher resolution a rotational motion is clearly seen and by also using the back-scattering spectrometer this was observed down to ~190 K. Analysis of the data as described above yields the EISF values shown in fig. 5. This shows that

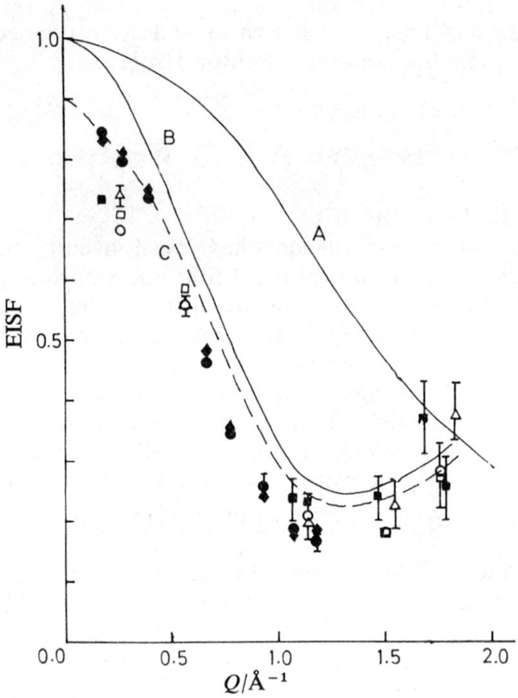

FIG. 5.—EISF results for the Crystal II phase. Experimental values from the IN5 data are shown as filled shapes and from IN10 as open shapes, with the following instrumental resolutions and sample temperatures.

	$\Delta E/\mu$eV	T/K
●	24	227
♦	10	227
■	10	218
○	0.5	208
△	0.5	201
□	0.5	190

The error bars (some of which have been omitted for clarity) have been estimated by extracting the EISF for the whole body motion with the value of the time constant for methyl reorientations [τ_A in eqn (11)] changed by $\pm 100\%$ of its most probable value. Line A shows the EISF that would be obtained if only methyl group reorientation were being observed. Line B shows the EISF expected from whole body reorientation about the CCN axis between 3 equivalent sites [eqn (6)] and line C represents the same model with the estimated multiple scattering correction.

throughout the low temperature phase the average proton trajectory remains to a good approximation the same. Furthermore the EISF is independent of resolution as well as temperature, which suggests immediately that $S(Q, \omega)$ is of a rather simple form. Model calculations show clearly that the motion observed is a rotation of the molecule about its CCN axis between three equivalent sites. This model fits the EISF data very well although the higher resolution and higher T results may be slightly low at the higher Q values indicating some additional motion, probably a precursor of that observed in Crystal I to be discussed below. However, any such effect is difficult to quantify uniquely in Crystal II and indeed the full $S(Q, \omega)$ is fitted very well at all Q values by the simple 3-fold reorientation model (see fig. 1 and 2). If the residence time is allowed to vary with Q in the fitting computations then except for the lowest Q values, where the EISF tends to unity, τ is generally constant to better than 5%. For comparison the EISF expected for rotation of the methyl groups alone is shown in fig. 5 which clearly demonstrates that this is not the motion being seen.

These results unequivocally demonstrate that the molecules of t-butyl cyanide are undergoing rapid reorientation in a 3-fold potential in the ordered low temperature phase. The correlation times τ_1 (eqn (7)) are plotted in fig. 6 where the correlation times for the methyl group reorientation are also shown for comparison. τ_1 varies linearly with T^{-1} and the line in the figure gives the relation

$$\tau_1/\text{s} = 1 \times 10^{-16} \exp 23.3 \text{ kJ mol}^{-1}/\boldsymbol{R}T.$$

The pre-exponential factor is totally unphysical since a value of the order of the free rotator correlation time $(I/\boldsymbol{k}T)^{1/2} \approx 10^{-12}$ s is expected.[16]

A more reasonable interpretation is possible using the inelastic data shown in fig. 4. By virtue of its high intensity, which also increases with T, the lowest energy

TABLE 1.—PARAMETERS FOR THE MOLECULAR ROTATION IN CRYSTAL II. COMPARISON OF INELASTIC AND QUASIELASTIC RESULTS *via* BARRIER HEIGHTS AND RESIDENCE TIMES.

T/K	$(E_1 - E_0)$/ meV	V_3/kJ mol^{-1}	V_3^{qe}/kJ mol^{-1}	τ_{exp}/ 10^{-11} s	τ_{calc}/ 10^{-11} s	$\tau_0(T)$/ 10^{-12} s
5	4.2 ± 0.2	10.0 ± 1.0				
150	3.7 ± 0.2	7.7 ± 1.0				
190		$(6.5) \pm 1.0$	7.2	36	34	7.5
201	3.2 ± 0.2	5.8 ± 1.5	6.5	23	13	7.2
208		$(5.6) \pm 1.5$	5.8	11.4	10.5	5.2
218	3.0 ± 0.3	5.1 ± 1.5	5.1	6.9	6.9	4.5
227	2.8 ± 0.3	4.5 ± 1.5	4.2	3.6	4.5	3.9
(239	2.8 ± 0.4	4.5 ± 1.5; Crystal I)				

$\tau(\text{calc})/\text{s} = 4.5 \times 10^{-12} \exp (V_3 - E_0)/\boldsymbol{R}T$
$\tau(\text{exp})/\text{s} = 4.5 \times 10^{-12} \exp (V_3^{qe} - E_0)/\boldsymbol{R}T = \tau_0(T) \exp(V_3 - E_0)/\boldsymbol{R}T$
and $\tau(\text{exp})$ is the experimentally determined residence time.

peak at 3-4 meV may be assigned to the libration of the molecule about its CCN axis. The peak is increasingly swamped by multiphonon scattering with increasing T but remains detectable into the high-temperature phase. Assuming a potential energy of the form

$$V(\theta) = V_3(1 - \cos 3\theta)/2$$

then in the harmonic approximation the depth of the potential well is given by $V_3 = 8\,\pi^2 I(E_1 - E_0)/9\boldsymbol{h}^2$.

The reorientation frequency is then given by the relation

$$\tau^{-1} = \tau_0^{-1} \exp - (V_3 - E_0)/\boldsymbol{R}T.$$

Using this very simple model the following calculations were made with the results given in table 1. Residence times were calculated using values of $(V_3 - E_0)$ derived from the inelastic results and $\tau_0 = 4.5 \times 10^{-12}$ s. [Note that dipole correlation times (τ_1) not residence times (τ) are shown in fig. 6 for ease of comparison with Crystal I results later.] The agreement between τ_{exp} and τ_{calc} is as close as could be expected. The difference is shown in two ways by giving (*a*) τ_0(T) values and (*b*) V_3^{ge} values required to give the solid line shown in the figure. It is obvious that a slight change of V_3, well within the uncertainty of that derived from the librational frequencies, or a slight temperature dependence of τ_0 gives perfect agreement with experiment for τ. Hence inelastic and quasielastic results are self-consistent and the apparent Arrhenius behaviour for crystal II shown in fig. 6 is an artefact caused by a barrier height to reorientation which is a strong function of T. It is in fact approximately linear in T for $190 < T/\mathrm{K} < 230$ and in precisely this region the unit cell dimensions become strongly temperature dependent.

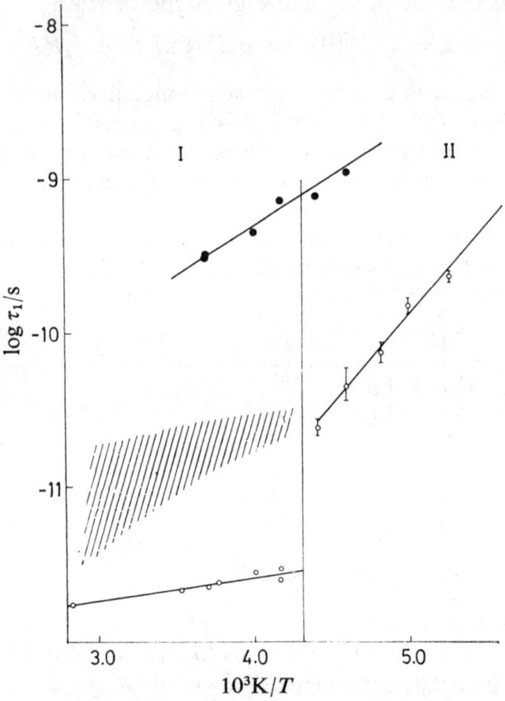

FIG. 6.—Correlation times τ_1 for t-butyl cyanide. Filled circles are for methyl-group reorientations. Open circles are for whole-molecule motions and show all the results obtained with three different spectrometers at 4 different experimental resolutions (see Experimental section). Vertical line shows the transition temperature. The hatched area shows the estimate of the correlation time for the cooperative translation/rotation relaxation discussed in the text. For Crystal I the line through the points is simply the best fit Arrhenius line. For Crystal II, see text and table 1.

CRYSTAL I

The EISF is determined very accurately from the IN5 data at both resolutions and also with reasonable accuracy from the lower resolution 4H5 data because of the width of the quasielastic components (200-300 μeV, h.w.h.m.). The results are shown in fig. 7 and 8 and for the two higher resolution experiments with the IN5 spectrometer the EISF is independent both of resolution and temperature showing that on a time-scale of $\approx 10^{-11}$ s the average proton trajectory is constant throughout the temperature range of Crystal I. However, with the lower-resolution instrument 4H5 the EISF is both considerably higher and also temperature dependent, but approaches the IN5 result with increasing temperature. This shows that the molecular motion is more complex than in Crystal II and this is confirmed by detailed model calculations.

The 4H5 data at the two lower temperatures (fig. 7) are in fair agreement with a model of jump reorientation in a 3-fold potential but in much better agreement with

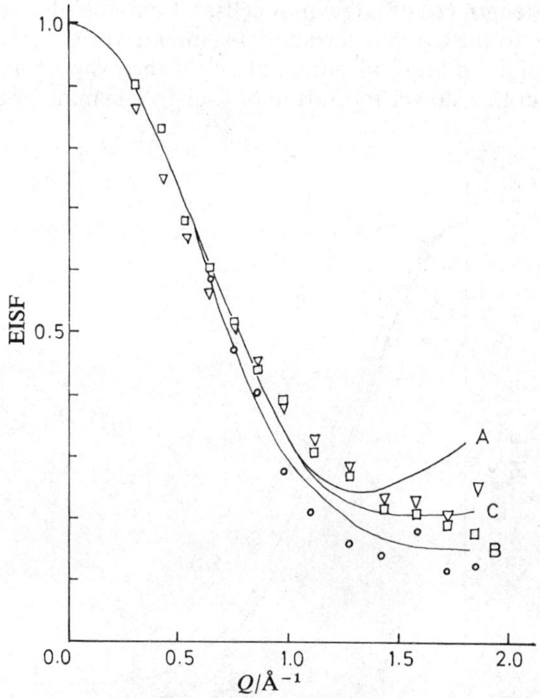

FIG. 7.—The EISF for the Crystal I phase of t-butyl cyanide determined from the 4H5 data. ○, 284; ▽, 265 and □, 240 K. The lines represent the EISF calculated for: A, 3-fold jump reorientation about the CCN axis [eqn (6)]; B, reorientation about the CCN axis between 6 or more equivalent sites [eqn (3) and (5)]; C, uniaxial rotational diffusion in a 3-fold cosine potential[17] with barrier height, $V = 2kT \simeq 4$ kJ mol^{-1}.

a model for diffusive motion in such a potential[17] (effectively a strong collision model) with $V_3 = 4$ kJ mol^{-1} which is in good agreement with the value deduced from the damped librational peak seen at the lowest temperatures in Crystal I, and with the Arrhenius activation energy deduced from the temperature dependence of the correlation times (see below). Furthermore, the simple jump model fits $S(Q, \omega)$ very well, with a residence time τ independent of Q to within 10%. Taken all together this evidence confirms that the faster stochastic motions in Crystal I are the molecular rotations in a 3-fold potential. This motion becomes increasingly diffusive in character (as opposed to librational plus jump reorientation) with increasing T because of the small barrier height: $V_3 \approx 2RT$.

At the highest temperature the 4H5 EISF is close to that for diffusion on a circle among m sites with $m \geqslant 6$, but the high resolution results are still lower at all temperatures. Thus it seems clear that there must be at least two components of motion present. The additional motion cannot be methyl-group rotation which is much too slow, and 3-dimensional molecular rotation is excluded by EISF calculations (fig. 8). In searching for the origin of the additional motion two facts must be noted: first the existence of steric hindrance to rotation even in the high-temperature phase and second the fact that the molecular C_3 axis lies parallel to the four-fold axis of the tetragonal unit cell of Crystal I. The latter can be taken into account by assuming rotation about the CCN axis among 12 sites on a circle but this cannot explain the results because motion *additional* to diffusion on a circle is required. These considerations lead to the suggestion that the *local* structure in Crystal I is close to that of Crystal II with a coherence length (L) of a few unit cells. Four-fold disorder of this structure in steps of $\pi/2$ leads to the average tetragonal symmetry of Crystal I. The molecules then rotate very rapidly in a 3-fold potential about their dipole axes as for Crystal II but in addition undergo a slower relaxation of four-fold symmetry among the different

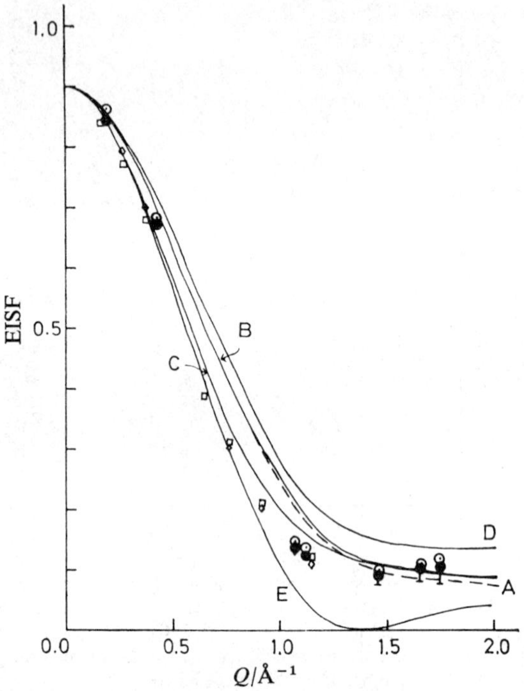

FIG. 8.—EISF for the Crystal I phase of t-butyl cyanide. Experimental results obtained using the IN5 spectrometer at the following temperatures and resolutions

	$\Delta E/\mu$eV	T/K
◇	10	270
□	10	239
○	24	240
●	24	250

The error bars represent the probable effect of contamination by coherent " elastic " diffuse scattering at $Q > 1.3$ Å^{-1}. A multiple scattering correction has been applied to all the model EISF calculations shown. The lines A, B and C represent the EISFs calculated for the models described in the text:

A (dashed) = 3-jump + 4-fold + $z = 1.0$ Å
B = 3-jump + 4-fold + $\langle u_x^2 \rangle^{\frac{1}{2}} = 0.30$ Å
C = 3-jump + 4-fold + 1.2 Å additional displacement parallel to a and b of fig. 9.

D is the EISF for a model of reorientation about the CCN axis between 6 or more sites [eqn (3) and (5)] and E is the EISF for isotropic tumbling of the molecule

local configurations. These may be generated by $\pi/2$ rotations of the monoclinic Crystal II structure about either $\boldsymbol{c}^*$ or $\boldsymbol{c}$: the former results in molecular rotation plus translational (centre of mass) displacement in the *ab* plane and the latter gives additional translational displacement of the molecules along their dipole axes. The actual magnitudes of the average molecular centre of mass displacements will depend on the coherence length of the local structure, and there must also be additional more-or-less random displacements at the boundaries between local structures of different orientation. The dynamic translational/rotational displacements have therefore been modelled as described below using the known crystal and molecular structures to determine a manifold of N sites for each proton, the molecules also being allowed to undergo 3-fold rotation about their CCN axes in all cases. The EISF was then calculated using the formula for a powder sample:

$$A_0(Q) = \sum_{i=1}^{N} \sum_{j=1}^{N} P_i P_j \sin Qr_{ij}/Qr_{ij}$$

where $r_{ij} = |\boldsymbol{r}_i - \boldsymbol{r}_j|$, $\boldsymbol{r}$ being the position of a proton site of occupation probability P, and in general we assumed $P = N^{-1}$. $A_0(Q)$ was averaged over non-equivalent protons.

The starting point was to rotate a unit cell of Crystal II about an axis through its centre of mass parallel to $\boldsymbol{c}^*$ in steps of $\pi/2$ and then displace the resultant structure

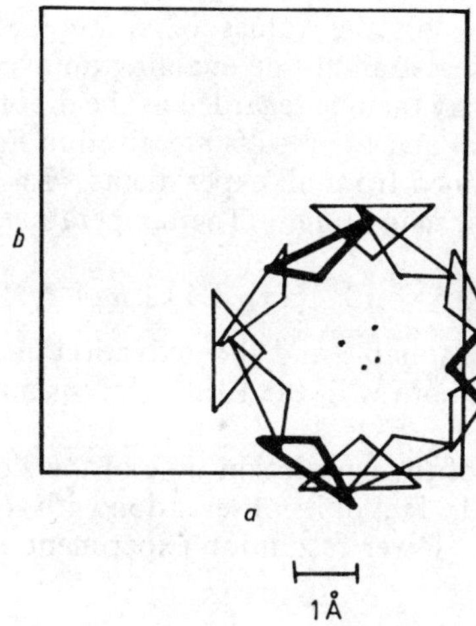

FIG. 9.—Model for translational/rotational disorder in Crystal I. The bold triangles represent the projections of the 3 methyl groups for one of the molecules in the unit cell onto the *ab* plane in solid II (or the *ab* plane of solid I). The light triangles are the methyl group positions generated by $\pi/2$ rotations about $\boldsymbol{c}^*$ and suitable translations to keep the centre of mass of the 2 molecules in the cell constant. This (arbitrary) procedure generates a pseudo-tetragonal local structure and gives both molecules small lateral displacements. The real displacement could be much larger depending on the coherence length of the Crystal II local structure in Crystal I (see text).

by appropriate combinations of $\boldsymbol{b}/2$ and/or $\boldsymbol{a}/2$ to give the resultant superposition of structures for each of the two molecules in the unit cell shown in fig. 9. This represents the minimum possible set of translational displacements for this model and it does not describe the EISF. A displacement z parallel to the dipole axis ($\boldsymbol{c}$ in Crystal I) which would result from $\pi/2$ rotations about $\boldsymbol{c}$ in Crystal II was therefore added as an adjustable parameter to give a total of 8 molecular positions and the EISF for $z =$ 1.0 Å is shown in fig. 8(A). An as alternative, an isotropic Gaussian displacement of the molecular centre of mass was added to the minimum set of four positions from the c^* rotation (and the 3-fold molecule rotation) and the EISF for $\langle u_x^2 \rangle^{1/2} =$ 0.3 Å is shown in fig. 8(B). Finally, on the grounds that displacements perpendicular to the direction of the opposed dipoles are perhaps the most likely, additional displacements x parallel to the unit cell axes a and b were added to the structure shown in fig. 9 and the resultant EISF for $x = 1.2$ Å is shown in fig. 8 (C). The fits in all cases, but especially for C, are reasonable when taking account of uncertainties (included in the error bars) at the highest Q arising from the possible presence of coherence contributions due to the strong coherent diffuse scattering in the disordered phase. An elastic coherence contribution of only a few percent would remove the discrepancies with the models at the highest Q. The parameters for models A, B and C imply a coherence length for the local structure in Crystal I of not more than ≈ 5 unit cells.

The geometrical nature of the molecular motions in the high temperature phase is thus established as rotation in a weak 3-fold well about the molecule C_3 axis, plus additional rotational/translational motions which are strongly cooperative, generate the correct high temperature symmetry and involve centre of mass displacements of $\lessdot 1$ Å. The dynamics of these two components of motion have also been determined as follows. The fastest and largest amplitude motion is undoubtedly the 3-fold rotation about the dipole axis which dominates the lower resolution data (see above). For the higher resolution data, using a scattering law of the form of eqn (3) and (5) then at low Q the term in τ_1 is the totally dominant quasielastic component. In fact using $m = 10$, the values of τ_1 were only weakly Q-dependent and for $Q \leqslant 1.5$ Å^{-1} fell on a straight line enabling an averaged low-Q value to be determined. These values may then be regarded as the dipole correlation time for the motion in the 3-fold potential and the results are given in fig. 6, where it is seen that self-consistent data are obtained from all experiments with experimental resolutions differing by up to an order of magnitude. The temperature dependence is described by

$$\tau_1 = 0.5 \times 10^{-12} \exp 3.4 \text{ kJ mol}^{-1}/RT.$$

The τ_0 value is totally reasonable and the activation energy implies $V_3 \approx 3.5$ kJ mol^{-1} which is in agreement both with the EISF analysis and the value derived from the librational frequency.

The order of magnitude of the correlation time of the slower co-operative relaxation may be estimated from the fact of its observation with the high resolution but not (at least at lower T) with the lower resolution experiment and this is also shown in fig. 6.

COMPARISON WITH OTHER EXPERIMENTS

The previous lower resolution INQES experiments[4] on Crystal I were interpreted in terms of a simple 3-fold jump reorientation model and gave τ values $\approx$40-50% greater than those reported here and an Arrhenius activation energy of ≈ 2 kJ mol^{-1}. In view of the much lower resolution of these experiments the agreement is as good as could be expected.

The n.m.r. measurements (at 16 MHz) showed a single step decrease in linewidth between 95 and 120 K and a single T_1 minimum (measurements made at 42 MHz) at 162 K from which it was concluded that methyl and whole-molecule uniaxial rotations are coupled and describable by a single correlation time τ_c and a single energy barrier (16.3 $\pm$ 0.8 kJ mol^{-1}). The INQES results show clearly that this is not correct but also provide an alternative explanation. Extrapolation to lower T of the neutron correlation times for methyl and whole-molecule reorientation (fig.

6) show that these become equal near 130 K at $\tau_1 \approx 10^{-7}$-10^{-8} s so that the correlation times indeed become equal at temperatures where their value is just that of the n.m.r. frequency; but this is fortuitous and the motions are not strongly coupled. The n.m.r. activation energy is in excellent agreement with the neutron value for *methyl* reorientation. The n.m.r. correlation time is about a factor of 10 less than the observed τ_1 values for methyl rotation but a factor of 2-3 is expected since the spin–lattice relaxation experiment gives τ_2 $[P_2(\cos\theta)]$. Hence the n.m.r. T_1 experiments are certainly dominated by the methyl reorientations. On the other hand an analysis by El Saffar *et al.*[3] of the linewidth transition gave values for a correlation time, for $T \leqslant 120$ K in agreement with the extrapolated INQES values for whole-molecule rotation and a corresponding activation energy (12 ± 1 kJ mol^{-1}) close to the 5 K value from the INIS data suggesting both that this analysis is sensitive to the whole-molecule motion and that the barrier height is not significantly temperature dependent for $T \leqslant 100$ K. Hence the onset of the line width transition is determined by the slower (at this T) molecular rotations but the spin–lattice relaxation is dominated by the methyl-group rotations.

THE PHASE TRANSITION

The t-butyl cyanide molecules undergo rotation about their 3-fold axes in a 3-fold potential at all temperatures. In the low temperature phase the residence time $\tau(>2 \times 10^{-11}$ s) is more than an order of magnitude longer than the flight time $\tau_f(8 \times 10^{-13}$ s) where this is defined by $\tau_f = (kT/I)^{-1/2}$ and also about an order of magnitude or more longer than the librational period $\tau_L = \nu_L^{-1}(\tau_L \approx 10^{-12}$ s). Hence the model of librational motion interspersed by relatively rare instantaneous jumps between wells is reasonable. The molecules may be envisaged as jumping to a new orientation when allowed by phonon-type thermal fluctuations.

In the high-temperature phase τ is only ≈ 2 or 3 times longer than τ_f and τ_L so that the simple reorientational model is no longer adequate, but more importantly neighbouring molecules must frequently be rotating simultaneously. Hence, strongly co-operative displacements of the centres of mass are required to allow this. Consideration of the molecule and site symmetry has led to a picture of relaxation among a set of local configurations comprising different $\pi/_2$ orientations of structure similar to that of the low-temperature phase. The correlation time for this motion is slower than that for the 3-fold molecular rotation.

The energy barrier for the 3-fold potential in the low temperature phase is strongly temperature dependent but does not collapse at the transition: in fact it changes rather little. Instead the decrease of the barrier, which is linked to a high thermal expansion, increases the rotation frequency until the concomitant growth of the co-operative translational displacements triggers the transition to the disordered phase. Thus it is the presence of the ≈ 1 Å co-operative displacements of the molecular centres of mass and much faster molecular rotation about the C—C—N axis which distinguishes the high temperature phase and not the *existence* of rotation of the molecules about their C—C—N axes.

[1] C. Clemett and M. Davies, *Trans. Faraday Soc.*, 1962, **58**, 1705.
[2] E. F. Westrum and A. Ribner, *J. Phys. Chem.*, 1967, **71**, 1216.
[3] Z. M. El Saffar, P. Schultz and E. F. Meyer, *J. Chem. Phys.*, 1972, **56**, 1477.
[4] S. Urban, J. Mayer, I. Natkaniec, J. Sciensinki and W. Nawrocik, *Acta Phys. Polon.*, 1978, **A53**, 379.
[5] S. L. Segel and A. Mansingh, *J. Chem. Phys.*, 1969, **51**, 4578.
[6] J. R. Durig, S. M. Craven and J. Bragin, *J. Chem. Phys.*, 1970, **53**, 38.

[7] C. I. Ratcliffe and T. C. Waddington, *J.C.S. Faraday II*, 1976, **72**, 1821.
[8] J. C. Frost, *Ph.D. Thesis* (University of Exeter, 1979).
[9] A. J. Dianoux, R. E. Ghosh, H. Hervet and R. E. Lechner, I.L.L. Internal Report 75 D16T (1975); W. S. Howells, I.L.L. Internal Report 75 H1 30T (1975); H. A. Baston, A.E.R.E. M5270 Harwell (1972);
[10] A. J. Leadbetter and R. E. Lechner, in *The Plastically Crystalline State*, ed. J. N. Sherwood (Wiley, Chichester, 1979, chap. 8).
[11] R. L. Livingston and C. N. R. Rao, *J. Amer. Chem. Soc.*, 1959, **81**, 3584.
[12] L. J. Nugent, D. E. Mann and D. R. Lide Jr, *J. Chem. Phys.*, 1962, **36**, 965.
[13] J. D. Barnes, *J. Chem. Phys.*, 1973, **58**, 5193.
[14] M. Schlaak, *Mol. Phys.*, 1977, **33**, 125.
[15] M. W. Johnson, A.E.R.E. Report 7682 (1974).
[16] C. Brot, *Chem. Phys. Letters*, 1969, **3**, 319.
[17] A. J. Dianoux and F. Volino, *Mol. Phys.*, 1977, **34**, 1263.

A Crystallographic Study of the Second-order Phase Transition in Bis(*p*-toluene sulphonate) Diacetylene Polymer Crystals

By Richard L. Williams, David Bloor and David N. Batchelder

Department of Physics,
Queen Mary College, Mile End Road, London E1 4NS

and Michael B. Hursthouse

Department of Chemistry,
Queen Mary College, Mile End Road, London E1 4NS

and William B. Daniels

Department of Physics,
University of Delaware, Newark, Delaware 19711, U.S.A.

Received 10*th January*, 1980

The second-order phase transition which occurs in bis(*p*-toluene sulphonate) diacetylene polymer crystals has been studied by X-ray diffraction. Measurements of the temperature dependence of the integrated intensity of Bragg reflections from several crystals provide evidence that the phase transition is very sensitive to the presence of defects. For one crystal several Friedel pair reflections have integrated intensities which are very asymmetric in the vicinity of the transition temperature. This suggests that the crystal was not centrosymmetric in this region and provides a possible explanation for the previous observation of pyroelectricity in *p*TS crystals.

The second-order phase transition which occurs in crystals of the polymer of the bis (*p*-toluene sulphonate) of 2,4-hexadiyne-1,6 diol (*p*TS) has been studied by a variety of experimental techniques. The crystal structures of both the high- [1] and the low-temperature [2,3] phases have been determined and the gradual onset of order in the low-temperature phase has been observed in X-ray studies.[4,5] Order parameter analysis has been applied to the splittings of the optical absorption and Raman lines observed in the low-temperature phase and to the intensities of " hard " vibrational modes observed by far-infrared spectrocopy.[6] The order parameter fitting suggested that at low temperatures the transition had a two-dimensional character while closer to the transition temperature it appeared to be three-dimensional. There were indications of a broad transition region which might be expected for a predominantly two-dimensional phase transition in which fluctuations dominate the behaviour of the system.

Fig. 1 shows a projection of the $P2_1/c$ unit cell of *p*TS onto the *ac* plane at 300 K in the high temperature phase.[6] The polymer chains have the chemical structure with R = $CH_2OSO_2C_6H_4CH_3$ and are oriented parallel

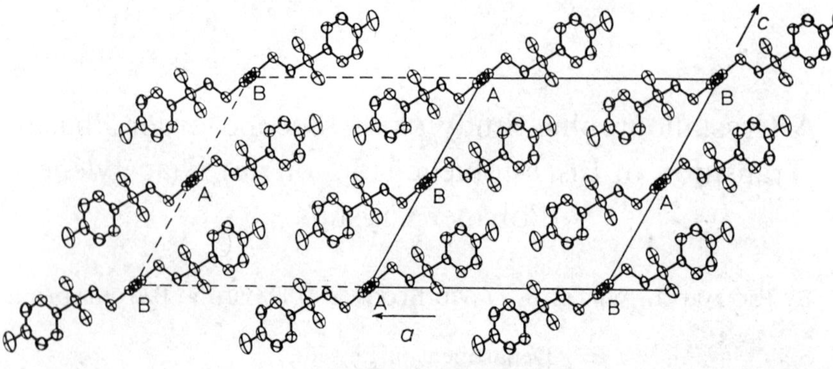

FIG. 1.—Projection of the $P2_1/c$ unit cell of *p*TS crystals at 300 K onto the *ac* crystallographic plane. The polymer backbone is parallel to the *b* axis which is perpendicular to the *ac* plane. The atoms, which are identified in fig. 2, are represented by thermal ellipsoids. The solid lines outline the unit cell in the high-temperature phase. In the low-temperature phase with $P2_1/n$ structure, the unit cell is doubled in the *a* direction (dotted lines) and there are two inequivalent polymer chains, A and B, in the unit cell.

to the *b* axis. The doubling of the unit cell in the *a* direction, which occurs in the low-temperature $P2_1/n$ structure, and the two types of polymer chain (those with type A and those with type B side-groups), are also indicated. Fig. 2 shows the anisotropic thermal motion of the non-hydrogen atoms in *p*TS.[6] The central figure is a plot of a backbone segment and one sidegroup at 300 K, and the plots labelled A and B are similar parts of the two distinct types of polymer chain at 120 K. Despite the fact

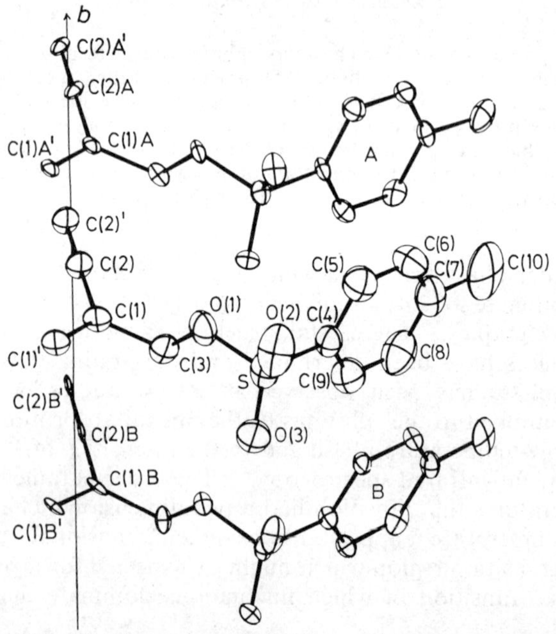

FIG. 2.—Comparison of a *p*TS side-group at 300 K (centre diagram) with that of the two structures (top and bottom diagrams labelled A and B) in which the side-groups are found in the low-temperature phase at 120 K. The anisotropic thermal motion is depicted by 50% probability ellipsoids for the non-hydrogen atoms. The atomic labels are: C, carbon; O, oxygen; S, sulphur. Atoms C(1)′, C(1), C(2), and C(2)′ are on the backbone with the vector *b* defining the polymer chain direction.

that the atoms in the side-groups labelled A and B occupy positions which are slightly displaced from the positions they would have occupied at room temperature, the thermal ellipsoids of side-groups A and B could be completely enveloped within the thermal ellipsoids of the room-temperature molecule.

A model for the phase transition has been suggested in which ordering in two dimensions predominates.[6] The side-groups along a single side of a polymer chain are considered to form an ordered stack which has either the A or B structure. In the low temperature phase these stacks are ordered in two dimensions (*ac* plane) while the high temperature phase would consist of a random array of A and B stacks. The third dimension would only be important for a stack in the process of changing from A to B or *vice versa*. This model would appear to satisfy in a qualitative way the requirements of both the order parameter analysis and the X-ray structural data.

The present X-ray study was undertaken in order to provide further evidence against which this model could be tested. In addition it was hoped to find a possible explanation for the observation of the pyroelectric effect in *p*TS crystals.[7] The structures of both the low- and high-temperature phases determined by X-ray analysis are centrosymmetric, while the pyroelectric effect can only be observed in crystals which lack a centre of symmetry.

EXPERIMENTAL

High-purity TS monomer crystals were grown by evaporation from acetone solution.[8] Small as-grown crystals of typical dimensions 0.2 × 0.8 × 0.2 mm were selected for study on a four-circle X-ray diffractometer. Larger crystals typically 2 × 6 × 4 mm in size were cleaved parallel to the (100) face to provide a clean surface for study by a rotating back-reflection X-ray camera. Both types of crystal were then thermally polymerized to 100% conversion at 333 K.[9]

INTENSITY MEASUREMENTS ON THE BACK-REFLECTION CAMERA

The large crystals were glued along one edge with low-temperature varnish to the aluminium cold finger of a continuous flow cryostat. The temperature was controlled to ±0.02 K and measured by an NBS calibrated platinum resistance thermometer located only 1 mm from the crystal. With the crystal fixed in the cryostat a rotating back-reflection camera[10,11] was used to scan the $(\overline{33}, 0, 3)$ Bragg reflection of the low-temperature phase. This difference reflection does not appear in the high-temperature phase. The normal to the $(\overline{33}, 0, 3)$ planes lies at about 12° to the (100) face and the Bragg angle was approximately 68.6° with Cu $K\alpha$ radiation. The pinhole X-ray collimator had an angular divergence of ±20′ and was 19 cm from the crystal. A scintillation counter with a 1.5 cm square NaI crystal was placed 20 cm from the crystal to measure the intensity of diffracted X-rays. Rotating the camera over a 2° range produced a profile of the diffracted $K\alpha_1$ and $K\alpha_2$ peaks on a chart recorder.[12] The integrated intensities at each temperature were then determined graphically.

INTENSITY MEASUREMENTS ON THE FOUR-CIRCLE DIFFRACTOMETER

The smaller crystals were studied on a Nonius CAD-4 X-ray diffractometer with computer control. Intensities were collected with Ni-filtered Cu $K\alpha$ radiation and an $\omega/2\theta$ scan mode. In each 96-step scan the outer 16 steps on each side constituted left (B_l) and right (B_r) backgrounds, and the central 64 steps the peak count (C). The integrated intensity (I_0) of a reflection was calculated from the equation $I_0 = |C - 2(B_l + B_r)|$. The temperature of the crystals was held constant during the measurements to ±0.5 K in a stream of cold nitrogen gas. The accuracy of temperature measurement by a thermocouple in the gas flow was estimated to be ±2 K. At each new temperature the lattice parameters were calculated by least-squares analysis of setting angles for 25 reflections which were automatically centred.

RESULTS

INTENSITY MEASUREMENTS ON THE BACK-REFLECTION CAMERA

Fig. 3 shows the integrated intensity of the ($\overline{33}$, 0, 3) Bragg reflection of the low-temperature phase, relative to the value at 80 K, as a function of temperature for two different *p*TS crystals. The intensities have been ratioed against those of the (12,0,6)

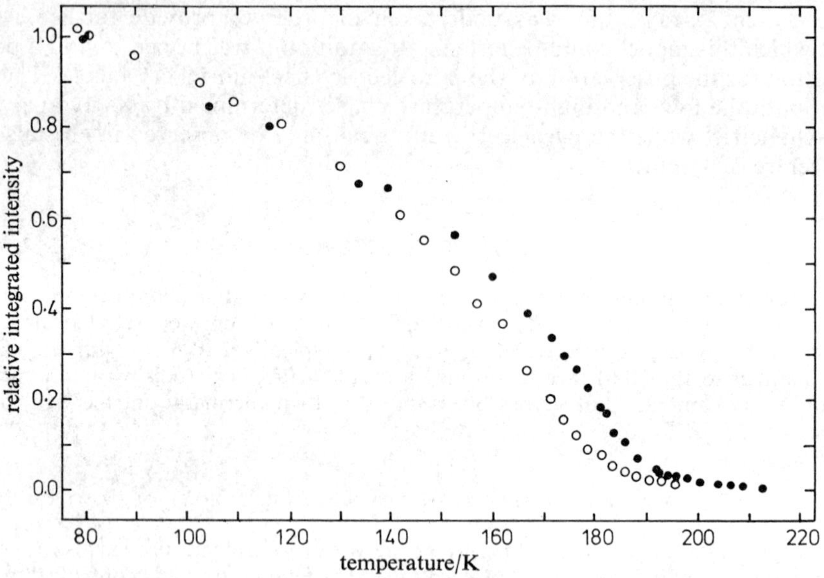

FIG. 3.—Relative integrated intensity of the ($\overline{33}$, 0, 3) Bragg reflection of the low-temperature phase as a function of temperature for two different crystals (open and closed circles). The data have been scaled to remove intensity changes associated with the Debye–Waller factor.

reflection, which appears in both the high-temperature and low-temperature phases, in order to remove the intensity changes associated with the Debye–Waller factor. The measurements were made with both increasing and decreasing temperature with no evidence of significant hysteresis. The estimated uncertainty in the intensity measurements increased continuously from 3% at 80 K to 20% at the highest temperatures.

The intensity data in fig. 3 suggest that the two *p*TS crystals have transition temperatures which differ by nearly 10 K. For both crystals the transition temperature is not well defined as the intensity has no sharp cut-off. There is even significant intensity in the ($\overline{33}$,0,3) reflection for one crystal as high as 210 K, 15 K above the transition temperature previously estimated from a photographic X-ray investigation.[5] For both crystals the $K\alpha_1$ and $K\alpha_2$ Bragg peaks were clearly resolved with no significant differences in peak width at 80 K between the two crystals. Thus it is not possible to say from these measurements alone which was the " better " crystal.

INTENSITY MEASUREMENTS ON THE FOUR-CIRCLE DIFFRACTOMETER

Fig. 4 illustrates the temperature dependences of the profiles of the (5,0,1) Bragg reflection of the low temperature phase which were recorded on the four-circle diffractometer for three different crystals. Photographic studies of the three crystals

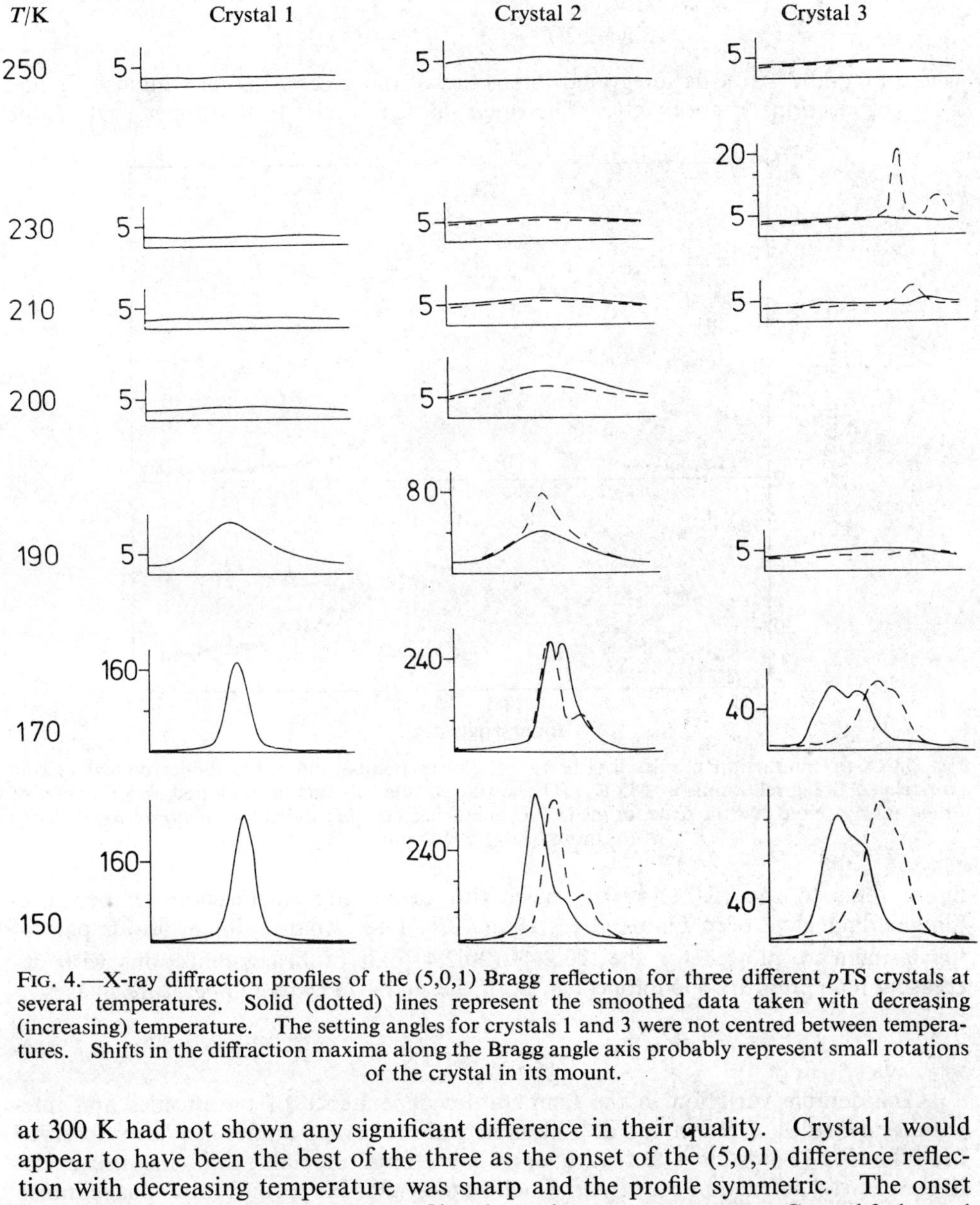

FIG. 4.—X-ray diffraction profiles of the (5,0,1) Bragg reflection for three different *p*TS crystals at several temperatures. Solid (dotted) lines represent the smoothed data taken with decreasing (increasing) temperature. The setting angles for crystals 1 and 3 were not centred between temperatures. Shifts in the diffraction maxima along the Bragg angle axis probably represent small rotations of the crystal in its mount.

at 300 K had not shown any significant difference in their quality. Crystal 1 would appear to have been the best of the three as the onset of the (5,0,1) difference reflection with decreasing temperature was sharp and the profile symmetric. The onset for crystal 2 is less sharp and the profiles showed some asymmetry. Crystal 3 showed a weak maximum in the (5,0,1) reflection at 235 K and the profiles are very asymmetric. There was some indication of hysteresis of the profiles for crystals 2 and 3 during temperature cycling while those of crystal 1 were reproducible. Similar results to fig. 4 were observed for another difference reflection, the (7,2,1), for the three crystals.

With crystal 2 on the diffractometer a study of the intensities of Friedel pairs was made. Fig. 5 is a comparison of typical profiles for the (5,0,1) and ($\bar{5}$,0,$\bar{1}$) at 185 K. It can be seen that the latter has much greater intensity with a narrower line-width. In fig. 6 data are plotted for the Bijvoet ratio Δ which is defined by the relation [13]

$$\Delta = 2(I^+ - I^-)/(I^+ + I^-)$$

where I^+ and I^- are the integrated intensities of the greater and less intense Friedel related reflections, respectively. The open (closed) circles in the upper part of the

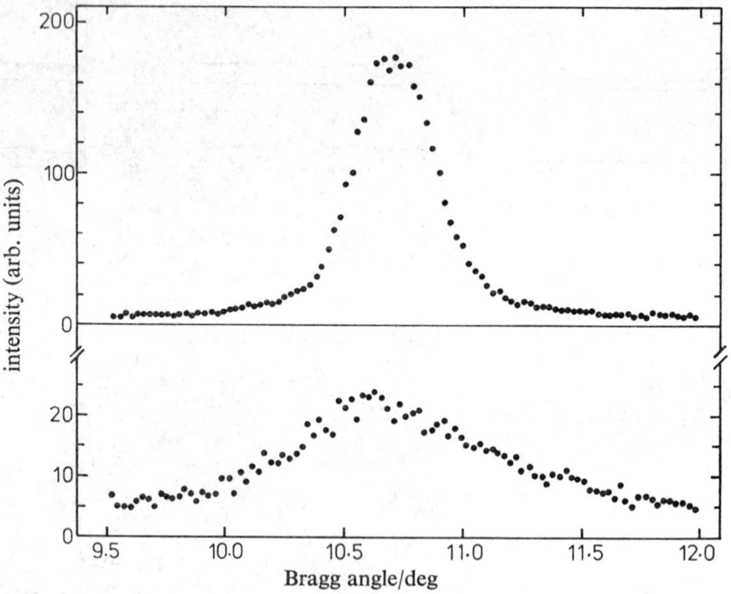

FIG. 5.—X-ray diffraction profiles for the (5,0,1) (lower points) and $\bar{5}$,0,$\bar{1}$) (upper points) Friedel pair related Bragg reflections at 185 K. The diffracted intensity has been plotted as a function of Bragg angle. Note that the scale for the (5,0,1) points has been expanded by a factor of four relative to those of the ($\bar{5}$,0,$\bar{1}$) points.

figure refer to the (5,0,1)–($\bar{5}$,0,$\bar{1}$) pair with decreasing (increasing) temperature- Similar data have been obtained for the (7,2,1)–($\bar{7}$,$\bar{2}$,$\bar{1}$) pair. In the lower part of fig. 6, data are plotted for the (20,2,$\bar{4}$)–($\overline{20}$,2,4) pair of Bragg reflections with decreasing temperature which appear in both the high- and low-temperature phases.

DISCUSSION

Considerable variation in the temperature dependence of the profiles and integrated intensities of reflections at both high and low Bragg angles has been observed for different *p*TS crystals. Both large crystals and small crystals exhibit inconsistencies in their X-ray diffraction properties which suggest that the second-order phase transition in *p*TS crystals is strongly affected by the presence of crystalline defects. Furthermore, these defects need only be present in relatively low concentrations as their effect on the X-ray diffraction properties of the crystals at 300 K, far from the transition temperature, was not readily apparent.

The sensitivity of the phase transition in *p*TS crystals to the presence of defects which is observed by X-ray diffraction should also be apparent in other properties. Thus far the various spectroscopic techniques which have been applied to the problem have not been sufficiently sensitive to show up any consistent differences among the variety of crystals studied. It is clear, however, that great caution must be used when applying order parameter analysis to properties of crystals which have not been characterized by X-ray diffraction.

The extraordinarily large asymmetry between the diffracted intensities of various Friedel pairs as illustrated in fig. 5 could be due to one of two reasons. The Renninger effect,[14] whereby the diffracted beam from one set of planes is the incident beam

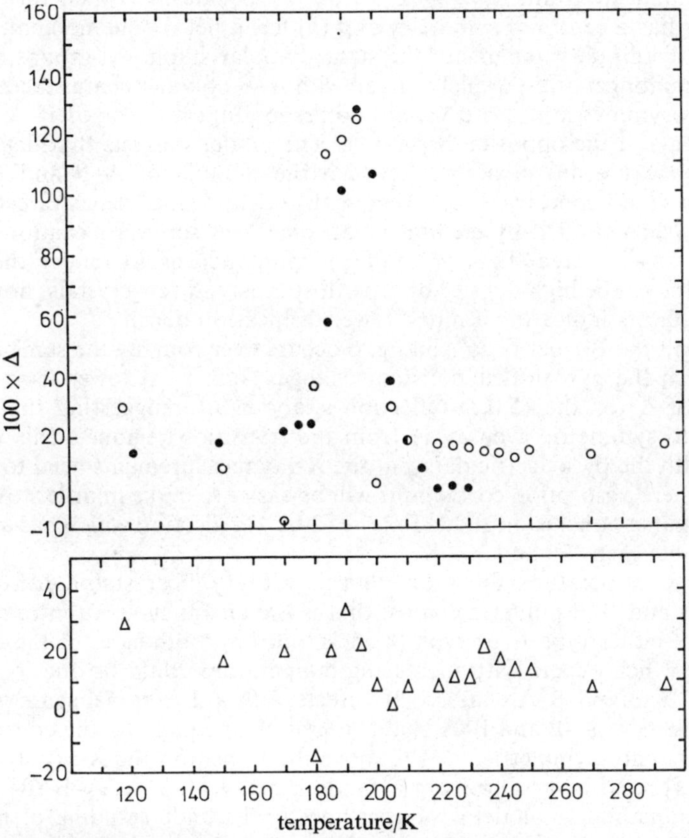

FIG. 6.—Temperature dependence of the Bijvoet ratio, $\Delta = 2(I^+ - I^-)/(I^+ + I^-)$, where I^+ and I^- are the integrated intensities of the greater and less intense Friedel related reflections, respectively. The open (closed) circles in the upper part of the figure refer to the $(\bar{5},0,\bar{1})$–(5,0,1) pair with decreasing (increasing) temperature. Above 220 K the reflections were so weak that the data points do not have too much significance. In the lower part of the figure the triangles refer to the $(20,2,\bar{4})$ $(\overline{20},\bar{2},4)$ pair of Bragg reflections which appear in both the high- and low-temperature phases.

for another, can be ruled out as a possible explanation since the scattering geometry was not appropriate. Thus the intensity asymmetry must have been caused by anomalous scattering[15] which occurs in non-centrosymmetric crystals when absorption by one or more atoms in the unit cell causes a phase shift in the X-rays scattered by those atoms. For most Bragg reflections in *p*TS crystals this effect should be very small as the 1.54 Å Cu $K\alpha$ radiation is very far from the 5.02 Å K absorption edge of sulphur, the heaviest atom in the unit cell. For difference reflections like the (5,0,1), however, the scattering amplitude becomes nearly zero in the vicinity of the phase transition and the effects of anomalous dispersion are greatly magnified. From currently available data it is impossible to tell whether this lack of a centre of symmetry is an intrinsic property of *p*TS crystals or associated with the presence of defects in crystal 2.

The pyroelectric effect observed in *p*TS crystals [7] must be closely associated with the loss of the centre of symmetry. A possible origin for the required temperature dependent dipole moment can be found in the A–B stack model of the phase transition described above. If a single polymer chain has the stack of side-groups on one side in the A configuration and the stack on the other side in B then this A–B chain: (*a*) no longer has a centre of symmetry and (*b*) has a net dipole moment. The latter occurs since the dipole moments of the strongly polar sulphonyl groups in the A and B stacks are no longer anti-parallel. Even with A–B polymer chains the crystal would still be centrosymmetric if there were a corresponding number of B–A chains with dipole moments in the opposite direction. The model suggests that defects must be present which cause an imbalance between the number of A–B and B–A chains. Such a defect could conceivably be a screw dislocation since it has directional sense. Monomer crystals studied by etching reveal emergent screw dislocations parallel to the polymer axis. Often these are sufficiently numerous to render the end-facets non-planar due to the high density of growth spirals. A few crystals, however, show specular end facets indicating a much lower dislocation density.[16]

The peak in the Bijvoet ratio Δ in fig. 6 occurs over roughly the same temperature range in which the pyroelectric constant changes sign.[7] At lower temperatures the mean value of Δ for the (5,0,1) reflection is about 0.13 suggesting that the crystal was non-centrosymmetric even away from the transition region. This would be in agreement with the pyroelectric data but the X-ray measurements need to be repeated on crystals where absorption corrections will be easier to make in order to confirm this result. The same conclusion holds for the $(20,2,\bar{4})$ reflection which has a mean value for Δ of 0.13 between 120 and 300 K.

At very low temperatures the X-ray data show that *p*TS crystals consist of a regular array of A–A and B–B polymer chains, that is the chains have a centre of symmetry with stacks of either type A or type B side-groups on both sides of the chain. The loss of order which occurs with increasing temperature could be due to the random formation of A–B and B–A chains. If defects with a directional sense were present then the numbers of A–B and B–A chains might be unequal and the crystal would no longer appear centrosymmetric. Well above the transition the X-ray structural data could be interpreted as arising from a random array of A–A, A–B, B–A, and B–B chains. This qualitative model is of assistance in the understanding of many of the effects which have been observed in pyroelectric, spectroscopic and X-ray diffraction investigations of the second-order phase transition in *p*TS crystals. Considerable further X-ray diffraction studies will be required to determine the role of defects in the phase transition and the origin of the unusually large asymmetry in the intensities of Bragg reflections from Friedel pairs.

This research was supported by grants from the S.R.C., the National Science Foundation and the University of Delaware. The authors are grateful to Mr. D. J. Ando for the preparation of the polymer specimens and to Mr. S. Mehta for assistance with data processing.

[1] D. Kobelt and E. F. Paulus, *Acta Cryst. B*, 1974, **30**, 232.
[2] V. Enklemann and G. Wegner, *Makromol. Chem.*, 1977, **178**, 635.
[3] V. Enklemann, *Acta Cryst. B*, 1977, **33**, 2842.
[4] B. Reimer, H. Bässler, and T. Debaerdemaker, *Chem. Phys. Letters*, 1976, **43**, 85.
[5] R. Clarke, personal communication.
[6] D. Bloor, D. A. Fisher, D. N. Batchelder, R. Kennedy, A. C. Cottle and W. F. Lewis, *Mol. Cryst. Liquid Cryst.*, 1979, **52**, 83.
[7] H. Kiess and R. Clarke, *Phys. Stat. Solidi a*, 1978, **49**, 133.
[8] G. C. Stevens, D. J. Ando, D. Bloor, and J. S. Ghotra, *Polymer*, 1976, **17**, 623.
[9] G. C. Stevens and D. Bloor, *J. Polymer Sci., Polymer Phys. Ed.*, 1975, **13**, 2411.
[10] D. N. Batchelder, *J. Polymer Sci., Polymer Phys. Ed.*, 1976, **14**, 1235.
[11] D. N. Batchelder and R. O. Simmons, *J. Appl. Phys.*, 1965, **36**, 2864.
[12] O. G. Peterson, D. N. Batchelder and R. O. Simmons, *Phil. Mag.*, 1965, **12**, 1193.
[13] S. Parthasarathy, *Acta Cryst.*, 1967, **22**, 98.
[14] M. Renninger, *Z. Phys.*, 1937, **106**, 141.
[15] A. F. Peerdeman, *Anomalous Scattering*, ed. S. Ramaseshan and S. C. Abrahams (International Union of Crystallography, Munksgaard International Publishers Ltd, Copenhagen, 1975), p. 3.
[16] D. Bloor, *J. Mat. Sci.*, 1979, **14**, 248.

^{13}C Nuclear Magnetic Resonance Study of Phase Transitions in a Lipid Bilayer Embedded in a Crystalline Matrix: $(C_{10}H_{21}NH_3)_2CdCl_4$ and $(C_{10}H_{21}NH_3)CuCl_4$

By Robert Blinc, Matjaž Koželj, Venceslav Rutar, Ivan Zupanicić and Boštjan Žecš

J. Stefan Institute,
E. Kardelj University of Ljubljana, Ljubljana, Yugoslavia

and Honza Arend and Raymond Kind
Laboratory of Solid State Physics, Swiss Federal Institute of Technology,

Hönggerberg, CH-8093 Zürich, Switzerland

and Gerhard Chapuis

Institute of Crystallography, BSP, University of Lausanne, 1015 Lausanne, Switzerland

Received 10*th December*, 1979

The hydrocarbon parts of $(C_{10}H_{21}NH_3)_2CdCl_4$ and $(C_{10}H_{21}NH_3)_2CuCl_4$ represent smectic lipid bilayers exhibiting two structural phase transitions analogous to those in biomembranes. A ^{13}C nuclear magnetic resonance study showed that in the Cd compound the low-temperature transition is connected with a disordering of the polar heads whereas the high-temperature transition corresponds to a partial melting of the alkyl chains. In the Cu compound the sequence of the two successive transitions is reversed. A Landau theory describing the two transitions in terms of order parameters used in the theory of smectic liquid crystals is presented.

The occurrence of phase transitions in cell membranes,[1,2] which are one of the principal organizational structures of living matter, has recently attracted a great deal of attention. It has been convincingly demonstrated[3,4] that the transitions in living cell membranes are due exclusively to the lipid component of the bilayer membrane. Most phase transition studies have therefore been performed on " model lipid membranes " which are similar in their structural properties to biomembranes, but can be much better characterized physically and exhibit sharper phase transitions. Lipid bilayers of dipalmitoyl phosphatidyl choline (DPPC) in particular clearly exhibit two first order-phase transitions,[5] the so-called main transition at $T_{c2} = 42$ °C connected with a melting of the chains and the pretransition at $T_{c1} = 33$ °C. Below T_{c1} the lipid chains are well ordered and tilted with respect to the bilayer normal. X-ray scattering,[6] calorimetric,[7] n.m.r.[8–10] and e.p.r.[11] studies have provided much information on the microscopic nature of these transitions, but several important questions are still unanswered. One of them is the nature of the pretransition at T_{c1}. Another is the detailed state of motion of the hydrocarbon chains in the high ($T > T_{c2}$) and the intermediate temperature phase ($T_{c1} < T < T_{c2}$) and the origin of the difference[12] in the values of the nematic order parameter $S_n = \frac{1}{2}\langle 3\cos^2\theta_n - 1\rangle$ where θ_n is the angular deviation of the nth C--C bond from its orientation in a hydrocarbon chain oriented parallel to a preferred direction, as measured by n.m.r.[9,10] and e.p.r.[11]

Here we report on a ^{13}C n.m.r. study of the phase transitions in the pseudo two-dimensional layer structure perovskite [13] $(C_{10}H_{21}NH_3)_2CdCl_4$ (henceforth designated C10Cd) and $(C_{10}H_{21}NH_3)_2CuCl_4$ (designated C10Cu) which we believe represent the first known examples of a smectic lipid bilayer embedded in a crystalline matrix. The two first-order phase transitions in C10Cd at $T_{c1} = 35$ °C and $T_{c2} = 39$ °C are closely analogous to the ones [5] in DPPC whereas the nature of the two successive transitions in C10Cu seems to be reversed ($T_{c1} = 36$ °C, $T_{c2} = 42$ °C). The results obtained yield a detailed microscopic picture of the two phase transitions in bilayer membranes and allow for the construction of a Landau theory of the " melting " of membranes in terms of order parameters used in the theory of liquid crystals.

Perovskite-type layer structures of the general formula $(C_nH_{2n+1}NH_3)_2MCl_4$ (short notation C*n*M) show a large variety of structural and magnetic phase transitions.[13] The inorganic layers contain corner-sharing Cl_6 octahedra with divalent metal ions in their centres ($M = Cu^{2+}$, Mn^{2+}, Cd^{2+}, . . .) thus forming a rigid two-dimensional crystalline matrix. The NH_3 groups of the alkyl ammonium chains are attached to the layers by weak N—H · · · Cl hydrogen bonds and, depending on the length of the chain, different packings are realized.[13] For each packing scheme there are several different equivalent orientations of the keys (*i.e.* NH_3 groups) in the octahedral cavities.

The structural phase transitions observed in these systems can be divided into two classes: (*a*) order–disorder transitions of the rigid alkyl ammonium chains, (*b*) conformational transitions of the chains. For short-chain systems ($n \leqslant 3$) the hindering potential is mainly determined by the key (≡polar head): key hole interaction, whereas for long chains ($n > 3$) the interaction among neighbouring chains becomes more important.

The inorganic layers behave like a stable but elastic matrix and affect the phase transitions only indirectly through the coupling between the MCl_6 octahedra and the NH_3 " heads " of the chains.

The projection of the structure of C10Cd on the *bc* plane at room temperature ($T < T_{c1}$) is shown in fig. 1(*a*). The structure consists of $CdCl_4$ layers sandwiched between well ordered alkylammonium chains which are tilted by 40° with respect to the normal to the layer. The ammonium end of each chain is linked to the layer by N—H · · · Cl hydrogen bonds and each chain is coordinated to six others. In C10Cd the chains in subsequent bilayers form a zig-zag arrangement parallel to the crystal *c*-axis [fig. 1(*a*)], whereas all the chains are parallel to a single direction in C10Cu and C10Mn [fig. 1(*b*)]. In C10Cd the layers consist of nearly ideal, corner-sharing CdCl octahedra, whereas they consist of bipyramids in C10Cu. The entropy change at T_{c1} [$\Delta S = (0.9 \pm 0.3)\boldsymbol{R}$ per mole chains] can be explained by an order–disorder transition of rigid chains between two equivalent sites. The entropy change at T_{c2}, on the other hand, corresponds to $\Delta S = 0.8\ \boldsymbol{R}$ per R—C—C—R bond and can be only explained by a " melting " of the chains, or equivalently, by rapid chain isomerization *via* kink diffusion.

In the intermediate phase of C10Cd the polar NH_3 groups are flipping between two possible orientations,[13] $p = \pm 1$. The alkyl chains also flip by 90° around their chain axes so that neighbouring chains move in opposite directions like a 2-dimensional array of connected gears.

The onset of the disordering of the polar heads in the low-temperature phase of C10Cd can be quantitatively followed [14] by measuring the temperature dependence of the ^{14}N quadrupole coupling constant e^2qQ/h and asymmetry parameter η. Since the motion between the two sites with e.f.g. tensors **T**(1) and **T**(2) is fast on the n.q.r.

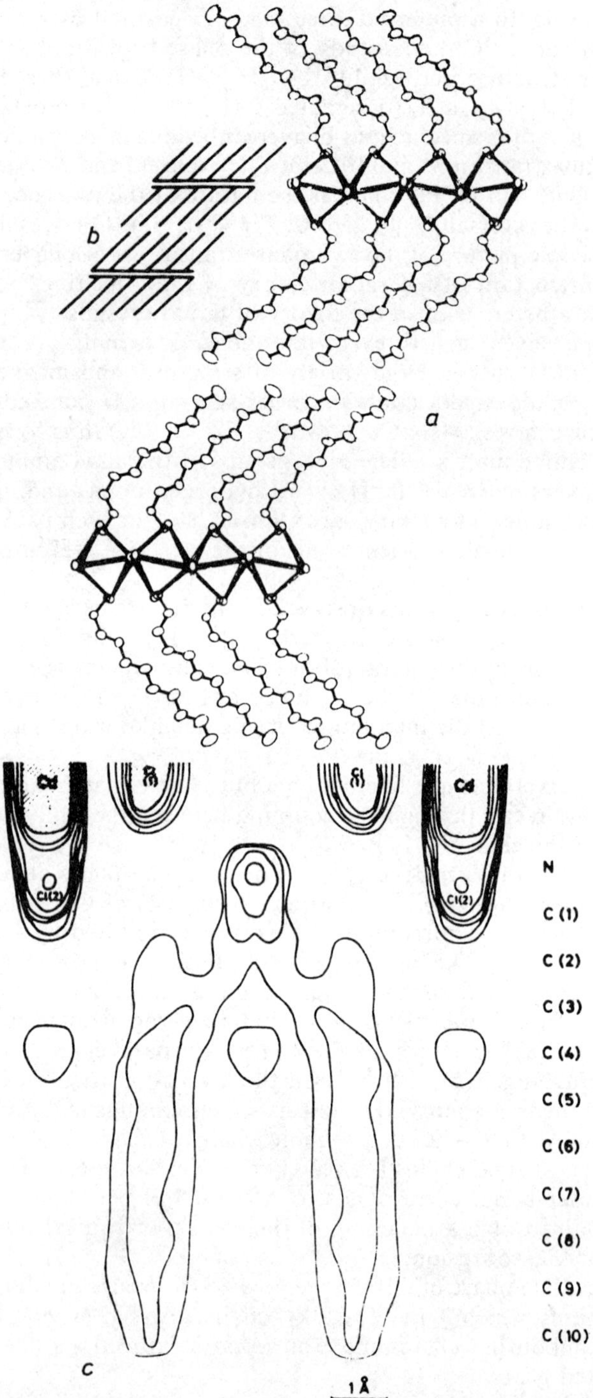

FIG. 1.—Projection of the low-temperature structure of $(C_{10}H_{27}NH_3)_2CdCl_4$ (*a*) and $(C_{10}H_{21}NH_3)_2$-$CuCl_4$ (*b*) on the *bc* plane. The projection of the electron density map of ClOCd on the *ac* plane in the high-temperature phase is shown in fig. 1(*c*). Due to 90° flips there is a dynamical disorder between the two displayed positions of the chain which are occupied with a 50% probability.

time-scale, the ^{14}N n.q.r. spectrum is determined by the time averaged value of the e.f.g. tensor:

$$\langle \mathbf{T}(t) \rangle = \tfrac{1}{2}(1+p)\mathbf{T}(1) + \tfrac{1}{2}(1-p)\mathbf{T}(2)\,. \tag{1}$$

Knowing the e.f.g. tensors **T**(1) and **T**(2) from the completely ordered state, the temperature dependence of $p = \langle p(t) \rangle$ can be determined from $\langle \mathbf{T}(t) \rangle$.

The Fourier map in the high-temperature phase of C10Cd [fig. 1(*c*)] shows in addition to the $CdCl_4$ octahedra well defined N and C(1) positions whereas the remaining part of the chain is parallel to the *c*-axis and splits symmetrically on both sides of the mirror plane with a statistical weight 0.5 for each chain due to the 90° flipping. The average positions of C(2) and C(3) are weakly defined, whereas the terminal part of the chain [C(4) to C(10)] forms a continuous distribution parallel to the *c*-axis. There is no evidence for a cone-like motion around the layer normal.

The ^{13}C spectra of the low-temperature and the intermediate-temperature phases of C10Cd are nearly identical but there is a sharp change in the ^{13}C spectrum on going into the high-temperature phase. The angular dependence of the $-CH_2$ group spectra in the low-temperature phase is characteristic of two sets of rigid $-(CH_2)_n-$ chains which are tilted by $\pm 40°$ with respect to the layer normal. The chain methylene groups display the full chemical shift anisotropy similar to that observed in n-eicosan[15] or polyethylene.[1b] In the intermediate-temperature phase the chains are still tilted by $\pm 40°$ and almost rigid (fig. 2). The bulk of the data can be described by an axially symmetric ^{13}C chemical shift tensor: $\delta_\perp = \tfrac{1}{2}(\delta_{11} + \delta_{22}) = 43 \pm 3$ p.p.m., $\delta_\parallel = \delta_{33} = 15 + 5$ p.p.m. with respect to TMS. The symmetry axis $\delta_\parallel = \delta_{33}$ is parallel to the chain direction. The axial symmetry of the ^{13}C tensor is due to the 90° flips around the chain axis which result in a rapid exchange of δ_{11} and δ_{22} but do not

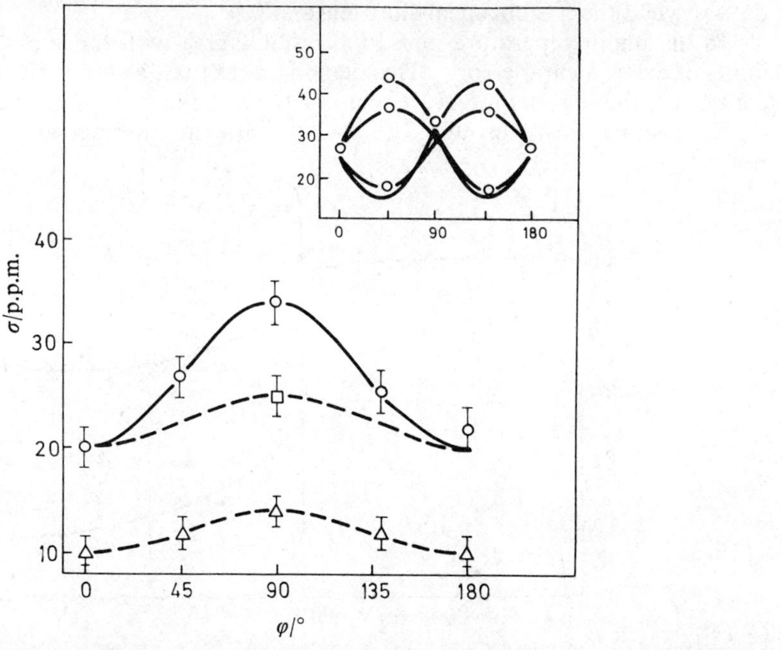

FIG. 2.—Angular dependence of the ^{13}C spectra in the high temperature (42 °C) and intermediate temperature (36 °C, insert) phases of C10Cd. φ stands for the angle between the *c* axis and the direction of the external magnetic field. $\nu(^{13}C) = 67.9$ MHz, $a \perp H$.

affect δ_{33}. The nematic order parameter S with respect to the preferred chain direction, $\theta = \pm 40°$, is here about one, $S \approx 0.95$-0.99. There is another set of ^{13}C lines belonging to the less rigid terminal methylene groups which is also axially symmetric and can be described by $S \approx 0.75$.

The angular dependence of the ^{13}C spectra in the high-temperature phase is completely different (fig. 2) from that in the low- and intermediate-temperature phases. It shows that all chains are equivalent and normal to the layers. Three different sets of ^{13}C lines can be resolved. The most intense line belongs to the bulk of the —CH_2 groups. It can be described by $\delta_{\parallel} = 20 \pm 2$ p.p.m. and $\delta_{\perp} = \frac{1}{2}(\delta_{11} + \delta_{22}) = 34$ p.p.m. The anisotropy of the chemical shift tensor is only partially removed by the molecular motion (chain isomerization *via* fast kink diffusion) and the nematic order is still significant:

$$S = \frac{(\delta_{\parallel} - \delta_{\perp})_{\text{obs}}}{(\delta_{\parallel} - \delta_{\perp})_{\text{rigid}}} = 0.55 + 0.10. \tag{2}$$

Another weak line belongs to the terminal methylene group, —CH_2—CH_3, and can be described by $\delta_{\parallel\alpha} = 20$ p.p.m., $\delta_{\perp\alpha} = 25 \pm 1$ p.p.m. The third line is due to the methyl group. Here $\delta_{\parallel CH_3} = 10 \pm 1$ p.p.m., $\delta_{\perp CH_3} = 14 \pm 1$ p.p.m. With the help of the rigid lattice ^{13}C shift tensors of n-eicosane [15] (n-$C_{20}H_{42}$) we find that $S(\alpha\text{-}CH_2) \approx 0.25 \pm 0.05$ and $S(\text{—}CH_3) \sim 0.20 \approx 0.05$. The above results and the absence of a significant flexibility gradient along the main part of the chain is at variance with the more " liquid like " picture of the bilayer interior suggested by e.p.r. spin label studies,[11] but agrees with the deuteron resonance data.[9]

The interpretation of the high-resolution ^{13}C spectra of C10Cu is complicated by the fact that the paramagnetic contribution of the Cu^{2+} ions is superimposed on the ^{13}C chemical-shift anisotropy. The contribution of the Cu^{2+} ions to the magnetic field at a given ^{13}C position has been evaluated and subtracted from the experimental data to yield the ^{13}C chemical-shift tensors.

In the high-temperature phase the results agree with those of C10Cd within the limits of experimental error. This demonstrates that, as far as the chain dynamics is concerned, the high-temperature phase is the same in both C10Cd and C10Cu.

On cooling down through $T_{c1} = 42$ °C into the intermediate temperature phase

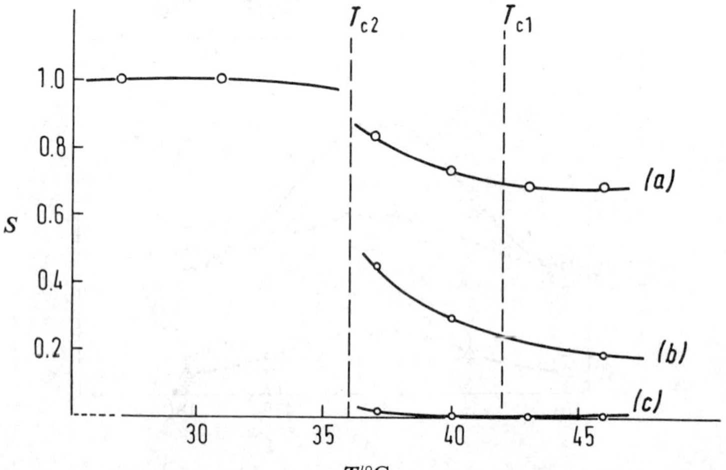

Fig. 3.—Temperature dependence of the nematic order parameter S of the bulk —CH_2 groups (*a*), the terminal —CH_2 group (*b*) and the —CH_3 group (*c*) in C10Cu.

the ^{13}C spectra change only very little. The angular dependence is the same as in the high temperature phase at $T_{c2} > 46$ °C $> T_{c1}$. This result is significantly different from the case of C10Cd where the chains are well ordered and tilted in the intermediate temperature phase.

At $T_{c2} = 36$ °C the ^{13}C spectra of C10Cu abruptly change. The lines broaden and the angular dependence of the ^{13}C chemical shifts qualitatively agrees with the data obtained in the low temperature phase of C10Cd if we suppose that all chains are parallel to a single direction and tilted by 40° with respect to the *c*-axis.

The temperature dependence of the nematic order parameter S of C10Cu is plotted as a function of temperature in fig. 3 for the bulk —CH_2 groups, the terminal methylene group and the methyl group (fig. 3).

The hydrocarbon parts of C10Cd and C10Cu thus represents a smectic liquid crystal with structure similar to the interior of the bilayer lipid membranes.[17-19] Let us therefore try to describe the two phase transitions in C10Cd and C10Cu by a Landau type expansion of the non-equilibrium free energy F in terms of order parameters θ, $S - S_c$ and p, used in the theory of smectic liquid crystals:[17]

$$F = \tfrac{1}{2}A\theta^2 + \tfrac{1}{4}B\theta^4 + \alpha(T - T_{c2})(S - S_c) - \tfrac{1}{2}c(S - S_c)^2 + \tfrac{1}{4}d(S - S_c)^4 + \\ + \tfrac{1}{2}ap^2 + \tfrac{1}{4}bp^4 - \tfrac{1}{2}A'\theta^2(S - S_c) - \tfrac{1}{2}a'p^2(S - S_c). \quad (3)$$

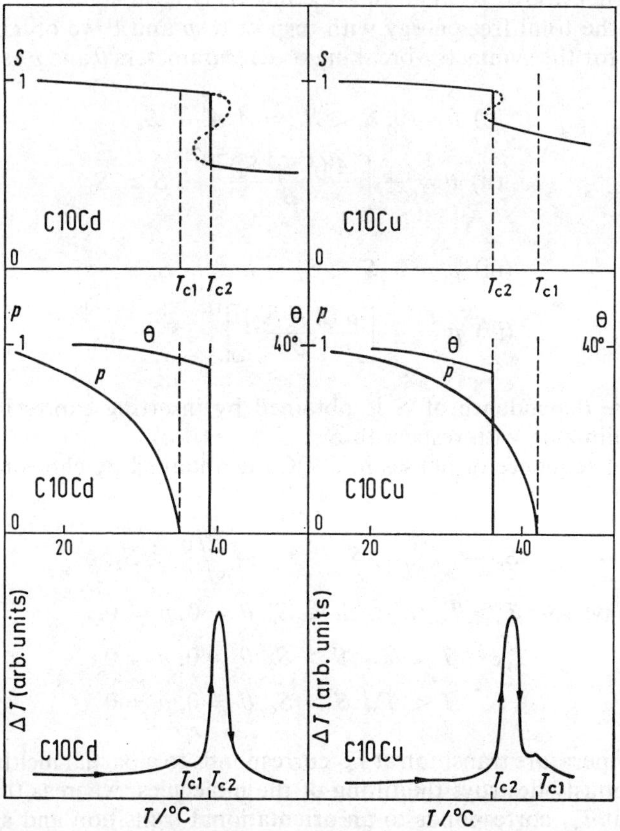

FIG. 4.—Schematic temperature dependences of the order parameters S, p and θ for C10Cd and C10Cu. The d.t.a. curves for these two systems are shown for comparison.

Here S stands for the average nematic order parameter and A, B, α, c, d, b, A' and a' are all positive constants whereas $a = \beta(T—T_c)$. The smectic-C order parameter θ gives the average tilt of the molecules with respect to the layer normal whereas p is the orientational order parameter for the 90° flipping of the chains and the terminal NH_3 groups between the two equilibrium orientations corresponding to $p = \pm 1$. S_c is the critical value of the average nematic order parameter S, *i.e.*, the arithmetic average of S in the melted and rigid phase coexisting at the first-order transition temperature T_{c2} (fig. 4). In deriving expression (3) we assumed that there are two important driving mechanisms which induce the transitions: the melting of the chains and the disordering of the keys. If $T_{c2} \gg T_{c1}$, as seems to be the case in C10Cd, $a \approx$ const. > 0 and the melting of the chains is the important driving mechanism which induces the transitions in S as well as in θ and p. In contrast to the isotropic–nematic transition, here S is not a symmetry-breaking order parameter and is not zero, both above and below T_{c2}. Because of that the linear term in the expansion of F in powers of $(S - S_c)$ is not identically equal to zero as in the case of nematic liquid crystals [17] but has the same structure as the free energy at liquid–gas phase transitions. The first coupling term $-\frac{1}{2}A'\theta^2(S - S_c)$ accounts for the fact that rigid chains show a stronger tendency for tilting because of the attractive van der Waals interaction between the chains. The second coupling term $-\frac{1}{2}a'p^2(S - S_c)$ describes the fact that " melted " molecules are axially symmetric and the potential hindering the rotation around the long axis is smaller above T_{c1} than in the completely rigid state.

Minimizing the total free energy with respect to p and θ we obtain the following stable solutions for the symmetry-breaking order parameters θ and p as functions of S:

$$\text{(i)}\ \theta = 0,\ S < S_1 = A/A' + S_c \tag{4}$$

$$\text{(ii)}\ \theta = \pm\left[\frac{A'(S - S_1)}{B}\right]^{1/2},\ S > S_1 \tag{5}$$

and

$$\text{(iii)}\ p = 0,\ S < S_2 = a/a' + S_c \tag{6}$$

$$\text{(iv)}\ p = \pm\left[\frac{a'(S - S_2)}{b}\right]^{1/2},\ S > S_2. \tag{7}$$

The temperature dependence of S is obtained by inserting expressions (4)-(7) into eqn (3) and minimizing with respect to S.

The observed sequence of phases in C10Cd is obtained by choosing $S_2 > S_1$, *i.e.*, $a/a' > 0$ so that

$$S_c - \sqrt{\frac{c}{d}} < S_1 < S_c + \sqrt{\frac{c}{d}} < S_2. \tag{8}$$

In this case we have for: $T > T_{c2}: S < S_1 < S_c,\ \theta = 0,\ p = 0,$ (9*a*)

$T_{c1} < T < T_{c2}: S > S_c,\ \theta \neq 0,\ p = 0$ (9*b*)

and for $T < T_{c1}: S > S_c,\ \theta \neq 0,\ p \neq 0.$ (9*c*)

The high-temperature transition at T_{c2} corresponds to a partial melting of the chains which simultaneously destroys the tilting of the molecules, whereas the low-temperature transition at T_{c1} corresponds to an orientational transition and a disordering of the polar " heads " in analogy to DPPC and biomembranes.

If, however, $a/a' < 0$, so that $S_2 < S_1$ and

$$S_2 < S_c - \sqrt{\frac{c}{d}} < S_1 < S_c + \sqrt{\frac{c}{d}} \tag{10}$$

we get the reversed order of the two phase changes. With increasing temperature we get first the chain melting, and at still higher temperatures the disordering of the keys as observed in C10Cu (fig. 4). Here we have:

$$T < T_{c1}:\ S < S_2 < S_c,\ \theta = 0,\ p = 0 \tag{11a}$$

$$T_{c2} < T < T_{c1}:\ S < S_c,\ \theta = 0,\ p \neq 0 \tag{11b}$$

$$T < T_{c2}:\ S > S_c,\ \theta \neq 0,\ p \neq 0\ . \tag{11c}$$

The reversed order of successive phase transitions in C10Cd and C10Cu is thus the result of a difference in the key–key-hole interaction.

Until now we have analysed only one bilayer. To get the observed ordering of the whole crystal one has to introduce an interaction between the two neighbouring hydrocarbon bilayers on the opposite sides of the same MCl_4 layer. The additional term in the free energy can be written as

$$F' = -K\theta_1\theta_2 \tag{12}$$

where θ_1 and θ_2 are the smectic-C order parameters on the opposite sides of the MCl_4 layer. For $K > 0$ one obtains $\theta_1 = \theta_2$ as observed in C10Cu and for $K < 0$, $\theta_1 = -\theta_2$ as observed in C10Cd (fig. 1).

[1] D. Chapman, *Quart. Rev. Biophys.*, 1975, **8**, 185.
[2] J. F. Nagle, *J. Membrane Biol.*, 1976, **27**, 233; *J. Chem. Phys.*, 1973, **58**, 252; *Proc. Roy. Soc. A*, 1974, **337**, 569.
[3] D. L. Melchior and J. M. Stein, *Biochim. Biophys. Acta*, 1977, **466**, 148.
[4] S. Marčelja, *Biochim. Biophys. Acta*, 1974, **367**, 165.
[5] E. Sackmann, *Ber. Bunsenges. phys. Chem.*, 1978, **82**, 891 and references therein.
[6] A. Tardieu, V. Luzzati and F. C. Reman, *J. Mol. Biol.*, 1973, **75**, 711.
[7] M. C. Philips, R. M. Williams and D. Chapman, *Chem. phys. Lipids*, 1969, **3**, 234.
[8] J. Charvolin, P. Manneville and B. Deloche, *Chem. Phys. Letters*, 1973, **23**, 345.
[9] A. Seelig and J. Seelig, *Biochemistry*, 1974, **13**, 4839.
[10] J. Urbina and J. S. Waugh, *Proc. Nat. Acad. Sci.*, 1974, **71**, 5062.
[11] B. J. Gaffney and H. M. McConnell, *J. Magnetic Resonance*, 1974, **16**, 1.
[12] N. O. Petersen and S. I. Chan, *Biochemistry*, 1977, **16**, 2657.
[13] R. Blinc, B. Žekš and R. Kind, *Phys. Rev. B*, 1978, **17**, 3409 and references therein.
[14] J. Seliger, R. Blinc, H. Arend and R. Kind, *Z. Phys.*, 1976, **B25**, 189.
[15] D. L. Van der Hart, *J. Chem. Phys.*, 1976, **64**, 830.
[16] S. J. Opella and J. S. Waugh, *J. Chem. Phys.*, 1977, **66**, 4919.
[17] P. G. de Gennes, *The Physics of Liquid Crystals* (Clarendon Press, Oxford, 1974).
[18] R. Kind, S. Pleško, H. Arend, R. Blinc, B. Žekš, J. Seliger, B. Ložar, J. Slak, A. Levstik, C. Filipič, V. Žagar, G. Lahajnar, F. Milia and G. Chapuis, *J. Chem. Phys.*, 1979, **71**, 2118.
[19] R. Blinc, M. I. Burger, V. Rutar, B. Žekš, R. Kind, H. Arend and G. Chapius, *Phys. Rev. Letters*, 1979, **43**, 1679.

Intermolecular Forces and Orientational Phase Transitions in Molecular Crystals

By Alfred Hüller

Institut für Festkörperforschung,
Jülich, West Germany

Received 11*th December*, 1979

The conditions under which one expects orientational phase transitions are discussed in the introduction. Then we consider the interactions between molecules and construct a Hamiltonian which is compared with the spin Hamiltonians that are in common use in the field of critical phenomena. A definition of the order parameter is given with special attention to the ordering of higher multipoles and with a discussion of its experimental determination. Finally the opportunities of obtaining information about critical exponents are considered, but the prospect is rather unpromising.

---

1. INTRODUCTION

In this paper we consider crystals which consist of molecules, polyatomic ions or side groups. These units will be denoted by the common name " molecules " and the crystals they form by the name " molecular crystals ". Methane (CH_4) is an example of a proper molecular crystal, whereas in the ammonium halides (*e.g.*, $NH_4^+Cl^-$) the " molecule " is an ammonium ion, and in dimethylacetylene the " molecule " is a methyl group. If the molecules are tightly bound units, one may distinguish between the high-frequency internal modes and the low-frequency external modes.[1] For our purposes the internal modes may be neglected, whereby the number of degrees of freedom per molecule is reduced to three translational and one, two or three rotational degrees of freedom: one rotation angle for a side group like CH_3, two for a linear molecule like N_2 and three for a three-dimensional molecule like CH_4.

Depending on the question as to whether the translational and rotational degrees of freedom are ordered or disordered, we distinguish four main classes of condensed molecular phases: (1) isotropic liquids with translational and rotational disorder, (2) liquid crystals which are rotationally ordered but translationally disordered, (3) orientationally disordered crystals (formerly also called plastic crystals) with translational order and (4) proper crystals (brittle crystals) where both translational and rotational degrees of freedom are ordered.

We are concerned with the orientationally disordered phases (3) and transitions towards complete order. Order may be achieved in several stages. Then we observe several orientational phase transitions with a stepwise reduction of the orientational disorder.

Orientationally disordered phases are characteristic of crystals consisting of globular molecules.[2] The globular shape is evident for cage-like molecules such as adamantane or the carbonanes. Small and highly symmetric molecules such as methane may also be considered to be spherical. Electron density contours in the outer parts of the electron shell (where the overlap between neighbours occurs) are very closely spherical. The weak non-isotropic interactions mainly originate in the relatively weak multipole–multipole interactions. The same argument holds true for hydrogen, but in addition to that the quantum-mechanical uncertainty principle prevents a sharp localization of the molecular axis, thereby amplifying the globular appearance of H_2. For globular molecules the angular average of the intermolecular forces is much stronger than the variation of these forces with respect to rotations of the molecules at a fixed distance. The radical part of the potential is responsible for the translational order, *i.e.*, for the formation of a regular lattice of the molecular centres of mass. The angle-dependent part of the potential is responsible for the orientational ordering of the molecules. Consequently, for globular molecules orientational order sets in at lower temperatures than translational order. One observes the phenomenon of orientational disorder in crystals (ODIC).

Unfortunately, in most of the phase transitions between crystal phases of different orientational order, the centre of mass structure also changes. In hydrogen, for example, the orientationally disordered phase is h.c.p. whereas the ordered phase is f.c.c. The simultaneous transformation of the centre-of-mass structure distorts the picture of a rotational phase transition. Therefore, the few examples where the symmetry of the centre-of-mass structure is not changed at the phase transition are of crucial importance. Our discussion will be based on the I/II phase transition in solid methane and on the II/IV transition in the ammonium halides (see fig. 1). In both cases the centre of mass structure is cubic above and below the phase transition.

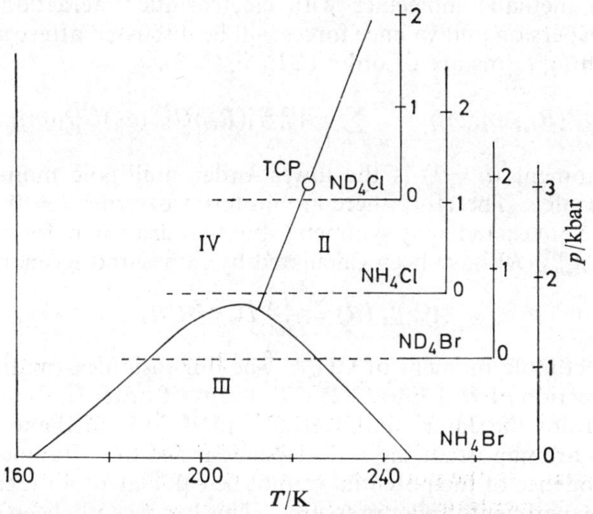

FIG. 1.—Generalized Stevenson plot for the p–T phase diagram in the ammonium halides which has been introduced by Press *et al.*[14] The roman numbers denote the three phases which differ in the orientational order of the ammonium ions: II is disordered, IV is ordered in a parallel way (ferro), and in III there is antiparallel order (anti-ferro). TCP denotes the tricritical point. For pressures higher than the tricritical pressure the II/IV phase transition is continuous. All other lines denote first-order transitions.

2. INTERMOLECULAR FORCES

The angle-dependent intermolecular forces are of the same nature as the central forces between atoms. They arise from the interactions between the permanent electrostatic monopole and multipole moments (electrostatic forces), the interactions between mutually induced multipoles (dispersion forces) and the overlap between the electron shells (valence forces). With the exception of the interaction between two hydrogen molecules,[3] the intermolecular forces cannot be calculated from first principles. Therefore, with the exception mentioned, all the potential functions found in the literature, as *e.g.*, atom–atom potentials[4] and the Kihara core model,[5] are phenomenological expressions containing a number of adjustable parameters. Here we only discuss the expansion of the intermolecular potential $V(\boldsymbol{R}_{ij}, \omega_i, \omega_j)$ between two molecules i and j in a twofold multipole series:[6,7]

$$V(\boldsymbol{R}_{ij}, \omega_i, \omega_j) = \sum_{l,m,n,l',m',n'} A^{l,l'}_{mn,m'n'}(\boldsymbol{R}_{ij}) D^{(l)}_{mn}(\omega_i) D^{(l')}_{m'n'}(\omega_j); \tag{2.1}$$

$\boldsymbol{R}_{ij}$ denotes the vector distance between the centres of mass of the two molecules, ω_i collectively denotes the set of Euler angles (ξ_i, ζ_i, η_i) which define the orientation of molecule i in space. The functions $D^{(l)}_{mn}(\omega)$ are a complete set of orthonormal functions in the Euler angle space, *e.g.*, the Wigner functions. Then $0 \leq l \leq \infty$ and $-l \leq m, n \leq l$ where l stands for the total angular momentum, m and n are projections of l. The coefficients $A^{l,l'}_{mn,m'n'}(\boldsymbol{R}_{ij})$ are phenomenological functions of $\boldsymbol{R}_{ij}$ which comprise the effect of all three types of interactions. For purely electrostatic forces the distance dependence is $|R_{ij}|^{-l-l'-1}$. In many cases it is advantageous to replace $D^{(l)}_{mn}(\omega)$ by a complete and orthonormal set of symmetry-adapted functions,[7] *e.g.*, the cubic rotator functions $U^{(l)}_{\mu\nu}(\omega)$ where l is still an angular momentum, but μ and ν do not have a physical interpretation. High molecular symmetry and a corresponding choice of the proper basis set $U^{(l)}_{\mu\nu}(\omega)$ greatly reduces the number of non-vanishing coefficients in expansion (2.1). Let us consider expansion (2.1) for two neutral methane molecules with electrostatic interactions only. Modifications due to dispersion and valence forces will be discussed afterwards. The lowest order non-vanishing terms are of order $(l, l') = (3, 3)$.

$$V(\boldsymbol{R}_{ij}, \omega_i, \omega_j) = \sum_{\mu,\mu'=1}^{7} A^{(3,3)}_{\mu 1,\mu' 1}(\boldsymbol{R}_{ij}) U^{(3)}_{\mu 1}(\omega_i) U^{(3)}_{\mu' 1}(\omega_j). \tag{2.2}$$

The octopole moment $(l = 3)$ is the lowest-order multipole moment of a neutral tetrahedral molecule. Therefore there are no terms of order $l = 0$, 1 and 2. Terms of order $l = 0$ are not excluded by symmetry, but they are absent for neutral molecules. The functions $A^{(3,3)}_{\mu 1,\mu' 1}(\boldsymbol{R})$ have been calculated by James and Keenan:

$$A^{(3,3)}_{\mu 1,\mu' 1}(\boldsymbol{R}) = \frac{I_3^2}{R^7} D_{\mu,\mu'}(\boldsymbol{R}/R) \tag{2.3}$$

where I_3 is the octopole moment of CH_4. The dimensionless matrix $D_{\mu,\mu'}$ only depends on the direction of $\boldsymbol{R}$. Eqn (2.2) with terms of order $(l, l') = (3,4), (4, 4)$ *etc.*, neglected constitutes the James and Keenan[7] model for methane. If valence and dispersion forces are important the $1/R^7$ dependence of the octopole term is changed. The new R dependence of the potential cannot be calculated but it can be determined from tunnelling experiments under pressure. This has recently been demonstrated for $(NH_4^+)_2SnCl_6^{2-}$.[8] An experiment on CH_4 is under way.[9] More important than the different R-dependence of the coefficients $A^{(l,l')}_{\mu\nu,\mu'\nu'}$ is the fact that now all symmetry-allowed coefficients are non-vanishing, as *e.g.*, the octopole–monopole interaction

$$V^{(3,0)}_{ij} = \sum_{\mu=1}^{7} A^{(3,0)}_{1\mu,11}(\boldsymbol{R}_{ij}) U^{(3)}_{1\mu}(\omega_i) U^{(0)}_{11}(\omega_j). \tag{2.4}$$

As $U^{(0)}_{11}(\omega_j) = 1$ is a constant $V^{(3,0)}$ does not depend on the orientation of molecule j. Summing eqn (2.4) over all neighbours j of molecule i one obtains a contribution to the crystal field experienced by molecule i:

$$V_c(\omega_i) = \langle \sum_j V^{(3,0)}_{ij}(\omega_i, \boldsymbol{R}_{ij}) \rangle, \tag{2.5}$$

where the angular brackets denote an average over the intermolecular distances. The crystal field persists into the disordered phases. It is distinguished from the molecular field $V_m(\omega_i)$ which is obtained from a summation of the interactions which depend on the orientation of both partners. The octopole–octopole term then reads:

$$V_m(\omega_i) = \langle \sum_j V_{ij}^{(3,3)}(\omega_i, \omega_j, R_{ij}) \rangle \tag{2.6}$$

where the angular brackets denote an average over molecular distances and over ω_j the orientations of the neighbours of molecule i. In disordered phases the angular average vanishes; therefore the molecular field is zero. Our discussion of V_c and V_m has been based on the molecular-field approximation which neglects correlations of the intermolecular distances and of the orientations of the molecule j with the orientation of molecule i.

The crystal field in methane seems to be rather weak.[10] This can be understood in terms of the nearly spherical shape of the outer parts of the electron shell and the charge neutrality of the molecules. This is different in $NH_4^+Cl^-$. The eight nearest-neighbour Cl^- ions of a NH_4^+ tetrahedron create a strong crystal field with pronounced minima. The strong crystal field leaves the molecule a choice between two different orientations (more precisely between two groups of 12 equivalent orientations each). Around the two equilibrium orientations, given by Euler angles $\omega_1 = (0,0,0)$ and $\omega_2 = (0,\pi/2,0)$ the molecule performs small librations with 8° amplitude.[11] The ammonium–ammonium interaction also contains an $(l, l') = (3, 0)$-term which contributes to the crystal field. The (3, 3)-term (2.2) is the same as in CH_4. The electrostatic approximation should be excellent as the NH_4^+ ions do not overlap. For the two equilibrium orientations allowed by the crystal field six of the seven functions $U_{\mu 1}^{(3)}(\omega)$ are zero. $U_{11}^{(3)}(\omega)$ is equal to 1 for ω_1 and equal to -1 for ω_2. For the librational ground state we obtain $\langle U_{11}^{(3)}(\omega)\rangle = s\xi$ with $\xi = 0.893$ for NH_4^+ and $\xi = 0.908$ for ND_4^+.[11] $s = \pm 1$ for wavefunctions centred around ω_1 and ω_2. The octopole–octopole interaction [eqn (2.2)] in the librational ground state thus reduces to:

$$V(\boldsymbol{R}_{ij}, s_i, s_j) = \frac{I_3^2}{R^7} \xi^2 D_{1,1}(\boldsymbol{R}_{ij}/R) s_i s_j. \tag{2.7}$$

For the nearest-neighbour distance $\boldsymbol{R}_{ij} = a(1, 0, 0)$ one finds[7] $D_{1,1} = -6$ and

$$V(s_i, s_j) = -6\xi^2 \frac{I_3^2}{a^7} s_i s_j \tag{2.8}$$

which shows that the orientational ordering in $NH_4^+Cl^-$ can be described in terms of a pseudo-spin Ising model with a nearest-neighbour ferromagnetic exchange constant $J = 6\xi^2 I_3^2/R^7$. This is, however, not the whole story: apart from direct interactions to more distant neighbours there are indirect interactions mediated by the halide ions. There are two contributions to the effect, namely the spin–phonon coupling[12] and the coupling *via* the electronic polarizabilities[13] of the halide ions. The indirect interactions favour the antiferromagnetic ordering which is found in $NH_4^+Br^-$ and $NH_4^+I^-$ for zero and moderate pressures. All these interactions contain a factor ξ^2. Therefore phase transitions in the deuterated ammonium halides should generally occur at temperatures which are higher by a factor $(0.908/0.893)^2 = 1.034$ than in protonated samples. This is indeed true for disordered to ferro-ordered transition with transition temperatures of 242 and 249 K for NH_4Cl and ND_4Cl, respectively. At the disordered to anti-ferro-ordered transition the isotope effect is reversed. In deuterated samples the transition is 30 K lower (!) than in protonated samples.[14,15] It is the opinion of the author that this reversed isotope effect is not understood on a microscopic basis, but that there are two approaches towards an understanding. The first one starts from the generalized Stevenson plot[14] for the ammonium halides (see fig. 1). One finds that the replacement of Br^- in $NH_4^+Br^-$ by Cl^- corresponds to the application of an external pressure of 1.8 kbar. Deuteration corresponds to a pressure of 1.3 kbar both in $NH_4^+Br^-$ and in $NH_4^+Cl^-$. A shift of the zero-pressure line on deuteration not only explains the different sign of the isotope effect for the transition into the ferro-ordered and antiferro-ordered phases, but also its size which depends on the slope of the transition line in the p–T diagram. The problem with this explanation is that we do not understand on a microscopic basis why deuteration should correspond to the application of a pressure of 1.3 kbar. The lattice constant change on deuteration is $\Delta a/a \approx 0.0005$, an effect which can be achieved by an externally applied pressure of 0.25 kbar. The second explanation of the isotope effect relies on the difference in the hydrogen bond with protons and deuterons.[16] The reasons for a reduction of the hydrogen bonding on deuteration are dynamic in nature. The theory of hydrogen bonding is, however, not developed to the point where a quantitative comparison is possible.

In molecular crystals the interactions typically fall off like $1/R^n$ with $n = 5, 7$ *etc.* This is in sharp contrast to magnetic systems where the exchange interaction falls off much more rapidly with distance. In many orientational phase transitions, interactions with distant neighbours are important, and if these interactions are competing as in the ammonium halides they lead to a complicated phase diagram and thus may influence the critical behaviour. We have given a detailed discussion of the octopole–octopole interaction which is responsible for the ordering in CH_4 and $NH_4^+Cl^-$, two important examples for rotational phase transitions. In the case of dumb-bell molecules as, *e.g.*, H_2 or N_2 the interaction is a little less complicated than eqn (2.2). It retains, however, its basic quality—the tensorial character. The orientation of a dumb-bell is defined by two angles, the polar coordinates θ and φ. Therefore, the interaction is expanded into a double series of functions (the spherical harmonics) defined in that space. The quadrupole moment has only five components, so the 7×7 matrix $D_{\mu\mu'}(\boldsymbol{R}/R)$ is replaced by a 5×5 matrix. Apart from these fine details the tensorial character of the interaction implies that $D_{\mu\mu'}$ depends on the direction of the intermolecular distance. In this respect the interaction is radically different from the isotropic exchange interaction in a Heisenberg ferromagnet. The interaction not only depends on the relative orientation of the two molecules, but on the orientation of the molecules with respect to $\boldsymbol{R}_{ij}$. Therefore the phase transition in molecular crystals cannot be mapped on a Heisenberg model. Contrary to the Heisenberg case the transverse excitations (librations) in the ordered phase always have a gap. This is true even for the case of vanishing crystal field which is almost ideally realized in solid hydrogen.

On the other hand, for strong crystal fields ($NH_4^+Cl^-$, adamantane, K^+CN^-, *etc.*) an Ising model represents the phase-transition properties almost ideally.

3. THE ORDER PARAMETER

ITS DEFINITION

In the Ising limit (strong crystal field) the definition of the order parameter η is obvious. For a ferromagnetically ordered phase we define

$$\eta = \frac{1}{N}\sum_i \langle s_i \rangle \tag{3.1}$$

where $\langle s_i \rangle$ is the expectation value of the ith pseudo-spin. In cases with weak crystal fields where we cannot construct a pseudo-spin Ising model such a definition is impossible.

We first claim that a purely rotational phase transition (*i.e.*, without rearrangement of the centre-of-mass structure) can only take place when the molecular symmetry is lower than the symmetry of the centre-of-mass structure. This statement will be discussed afterwards. Methane molecules with tetrahedral symmetry on a cubic (f.c.c.) lattice are an example. The hamiltonian [kinetic energy plus the interaction given in eqn (2.2)] then has the full symmetry of the lattice; in our example this symmetry is cubic. In the high-temperature phase the symmetry of the hamiltonian is not broken by the orientational order of the molecules. If we expand $f(\omega)$ the orientational distribution function (also called a probability distribution function p.d.f.) for a molecule into a series of symmetry-adapted functions (adapted to the symmetry of the lattice) only those functions are allowed which are simultaneously invariant under the symmetry operations of the molecule and of the lattice. For the CH_4 example such an expansion reads:[7,17]

$$f(\omega) = \sum_{l=0}^{\infty} (2l+1) \sum_{\mu,\mu'=1}^{2l+1} A_{\mu\mu'}^{(l)} U_{\mu\mu'}^{(l)}(\omega) \tag{3.2}$$

the cubic rotator functions $U_{\mu\mu'}^{(l)}(\omega)$ have been introduced in Section 2. There are $(2l+1)$ functions of order l but only very few are allowed functions. In our example the non-vanishing coefficients are: $A_{11}^{(0)}$, $A_{11}^{(4)}$, $A_{11}^{(6)}$, *etc.* For vanishing crystal field $f(\omega)$ is a constant, then $A_{11}^{(4)}$, $A_{11}^{(6)}$, . . . are all zero. The functions $U_{11}^{(4)}(\omega)$, $U_{11}^{(6)}(\omega)$, . . . modulate the orientational probability density $f(\omega)$, their coefficients $A_{11}^{(4)}$, $A_{11}^{(6)}$, . . . depend on the strength of the crystal field and for finite crystal fields they are non-zero at all temperatures.

When the cubic symmetry of the Hamiltonian is broken, the symmetry at a methane site is spontaneously reduced from cubic to tetrahedral. Then the functions $U_{11}^{(3)}(\omega)$, $U_{11}^{(7)}(\omega)$, . . . become symmetry-allowed. Their coefficients $A_{11}^{(3)}$, $A_{11}^{(7)}$, . . . are zero above the phase-transition temperature and non-zero below. Normally the lowest order coefficient $A_{11}^{(3)}$ is identified with the order parameter η. The new definition of η is valid both for weak and strong crystal fields. For strong crystal fields it is identical with the former definition by eqn (3.1). When the molecular symmetry and the site symmetry are identical, spontaneous symmetry breaking is not possible.

EXPERIMENTAL DETERMINATION OF THE ORDER PARAMETER

In a hypothetical example of polar molecules that undergo a phase transition into a phase with parallel orientation of the dipoles, one would observe a macroscopic electric moment which is proportional to the order parameter. If higher-order multipoles such as quadrupoles, octopoles, . . . order, no macroscopic moment is observed. The same is true for antiparallel dipole order. Therefore, apart from the hypothetical case of parallel dipole orientation, there is no means of observing the order parameter macroscopically. Neutron and X-ray diffraction is the only possibility for its determination.

A diffraction experiment determines $S(Q) = \int d\omega S(Q, \omega)$ which is the Fourier transform of the equal-time density–density correlation function $\langle \rho(r_1, t)\rho(r_2, t)\rangle$. More precisely stated, $\rho(r_1, t)$ is the density of the scattering power, *i.e.*, the scattering length density if one considers the scattering of neutrons by nuclei, the electronic spin density in the case of magnetic neutron scattering and the electron density in the case of X-ray scattering.

We divide the density into its contributions from the N molecules in the crystal:

$$\rho(\boldsymbol{r}_1, t) = \sum_{i=1}^{N} \rho_i(\boldsymbol{r}_1, t). \tag{3.3}$$

Then the correlation function contains terms of the type $\langle \rho_i(\boldsymbol{r}_1, t)\rho_j(\boldsymbol{r}_2, t)\rangle$ which we separate

$$\langle \rho_i(\boldsymbol{r}_1, t)\rho_j(\boldsymbol{r}_2, t)\rangle = \langle \rho_i(\boldsymbol{r}_1)\rangle\langle \rho_j(\boldsymbol{r}_2)\rangle + \langle [\rho_i(\boldsymbol{r}_1, t) - \rho_i(\boldsymbol{r})][\rho_j(\boldsymbol{r}_2, t) - \rho_j(\boldsymbol{r})]\rangle. \tag{3.4}$$

The first term comes from the time-independent single-molecule distribution (it leads to Bragg diffraction, the second term contains the correlations); it is responsible for the diffuse intensity between the Bragg points. The Bragg intensity is connected with the probability distribution of scattering centres in a coordinate system fixed in the crystal. To relate $\langle \rho_i(\boldsymbol{r})\rangle$ with the expansion (3.2) which defines the order parameter, we introduce a second (primed) coordinate system fixed in the molecule which rotates with it. If the molecular structure is known, $b(\boldsymbol{r}')$, the scattering length density in the primed coordinate system is also known, and can be expanded in terms of symmetry-adapted surface harmonics:

$$b(\boldsymbol{r}') = \sum_{l'm'} b_{l'm'}(r')K_{l'm'}(\theta', \varphi') \tag{3.5}$$

with known coefficients $b_{l'm'}(r')$. In our examples the proper set of functions are the cubic harmonics. $K_{31} \sim x'y'z'$ and $K_{41} \sim x'^4 + y'^4 + z'^4 - 3/5$ are the lowest-order functions with tetrahedral and cubic symmetry, respectively. Using the definition of the cubic rotator functions

$$K_{l'm'}(\theta', \varphi') = \sum_{m'} K_{l'm''}(\theta, \varphi)U^{(l')}_{m''m'}(\omega) \tag{3.6}$$

one obtains

$$b(\boldsymbol{r}) = \sum_{l'm'm''} b_{l'm'}(r)K_{l'm''}(\theta, \varphi)U^{(l')}_{m''m'}(\omega) \tag{3.7}$$

the scattering-length density in the crystal frame for a given orientation ω of the molecule. $b(\boldsymbol{r})$ is averaged with the orientational p.d.f. from eqn (3.2) to yield:

$$\begin{aligned}\langle \rho(\boldsymbol{r})\rangle &= \int d\omega f(\omega)b(\boldsymbol{r}) \\ &= \sum_{lm} a_{lm}(r)K_{lm}(\theta, \varphi)\end{aligned} \tag{3.8}$$

where

$$a_{lm}(r) = \sum_{m'} A^{(l)}_{mm'}b_{lm'}(r) \tag{3.9}$$

follows from the orthogonality relations of the cubic rotator functions. The coefficients $a_{lm}(r)$ can be measured in a diffraction experiment; they determine the order parameter. If $A^{(3)}_{11}$ is the order parameter then $a_{31}(r)$ is zero above T_c and finite below. Its temperature dependence, *i.e.*, the critical exponent β, can be determined from the intensity of the Bragg points (with h, k, l all odd) to which it contributes.

The diffuse scattering caused by the second term in eqn (3.4) can be used to obtain information about correlations, in particular about critical correlations in the vicinity of a phase transition. The intensity of the critical scattering is related to the magnitude of the fluctuations, *i.e.*, to the critical exponent γ. The width of the scattering in Q-space is inversely proportional to the correlation length ξ, *i.e.*, to the critical exponent ν.

Information about the critical slowing-down is obtained from the energy dependence of the critical scattering, that is to say from a determination of $S(\boldsymbol{Q}, \omega)$.[18]

4. CRITICAL PHENOMENA

Towards the end of the preceding Section, methods to determine critical exponents for orientational phase transitions in molecular crystals have been discussed. The presentation has perhaps been misleading. It may have left the impression that there is a wealth of examples of continuous phase transitions accompanied by critical phenomena. This is not the case. The phase transitions in molecular crystals are almost exclusively discontinuous phase transitions. The use of critical laws like $A_{11}^{(3)} \approx [(T_c - T)/T_c]/\beta$ is misleading,[19–21] since one obtains spurious values of the critical exponents (*e.g.*, $\beta = 1/6$) which have nothing to do with critical fluctuations or low dimensionality. The arguments why a critical exponent β should not be defined at a discontinuous phase transition have been given in ref. (22) andwill not be repeated here. So far the only example where a continuous orientational phase transition has definitely been found is the II/III phase transition of $NH_4{}^+Cl^-$ and $ND_4{}^+Cl^-$ under pressure. Here measurements of critical exponents are very difficult. This is seen from the two determinations [23] of β (for p = const and for T = const) which differ by 30%.

The phase transitions in molecular crystals are not interesting from the point of view of critical phenomena. These systems do not belong to the domain where scaling laws or universality can be tested.

The study of phase transitions in molecular crystals is important for a better understanding of intermolecular forces. Very little is known about these forces, which are the basis for all properties of condensed molecular phases. Phase diagrams are a very sensitive test for the angle dependence of these forces.

For the description of the phase transitions themselves a molecular-field theory is mostly sufficient. A very elegant and extremely successful example of such a calculation is the James and Keenan [7] theory of solid CH_4. The reason for the applicability of molecular are-field theory to slightly discontinuous phase transitions is the fact that the incipient phase transitions are interrupted by the discontinuity before the fluctuations become divergent. The non-divergent fluctuations that occur may be treated within the Ornstein–Zernicke approximation.

[1] G. Venkataraman and V. C. Sahni, *Rev. Mod. Phys.*, 1970, **42**, 409.
[2] J. Timmermans, *J. Phys. Chem. Sol.*, 1961, **18**, 1.
[3] I. Silvera, *Rev. Mod. Phys.*, to be published.
[4] A. I. Kitaigorodosky, *Molecular Crystals and Molecules* (Academic Press, New York, 1973).
[5] T. B. MacRury and W. A. Steele, *J. Chem. Phys.*, 1976, **64**, 1288 and references therein.
[6] J. O. Hirschfelder, C. F. Curtiss, and R. B. Bird, *Molecular Theory of Gases and Liquids* (Wiley, New York, 1954).
[7] H. M. James and T. A. Keenan, *J. Chem. Phys.*, 1959, **31**, 12.
[8] W. Press and M. Prager, to be published.
[9] W. Press, personal communication.
[10] W. Press, *J. Chem. Phys.*, 1972, **56**, 2597.
[11] A. Hüller and J. W. Kane, *J. Chem. Phys.*, 1974, **61**, 3599.
[12] Y. Yamada, M. Mori and Y. Noda, *J. Phys. Soc. Japan*, 1972, **32**, 1565.
[13] A. Hüller, *Z. Phys.*, 1972, **254**, 456.
[14] W. Press, J. Eckert, D. E. Cox, C. Rotter and W. Kamitakahara, *Phys. Rev.*, 1976, **B14**, 1983.
[15] P. Brauer and I. R. Jahn, *Z. Naturforsch.*, 1978, **33a**, 1093.
[16] C. W. Garland, K. J. Lushington and R. C. Leung, *J. Chem. Phys.*, 1979, **71**, 3165.
[17] W. Press and A. Hüller, *Acta Cryst.*, 1973, **A29**, 252.
[18] W. Press, A. Hüller, H. Stiller, W. Stirling and R. Currat, *Phys. Rev. Letters*, 1974, **32**, 1354.
[19] B. Dorner, J. D. Axe, and G. Shirane, *Phys. Rev.*, 1972, **B6**, 1950.
[20] E. Banda, R. A. Craven, R. D. Parks, P. M. Horn and M. Blume, *Solid State Comm.*, 1974, **17**, 11.

[21] J. P. Bachheimer and G. Dolino, *Phys. Rev. B*, 1975, **11**, 3195.
[22] A. Hüller and W. Press, *Theoretical Aspects of Solid Rotator phases*, in *The Plastically Crystalline State* (Wiley, London, 1978).
[23] W. B. Yelon, D. E. Cox, P. J. Kortman and W. B. Daniels, *Phys, Rev. B*, 1974, **9**, 4843.
[24] A. Hüller and W. Press, *Phys. Rev. Letters*, 1972, **29**, 266.

Static and Dynamic Aspects of Order and Disorder in CBr_4

By Marcel More and René Fouret

Laboratoire de Dynamique des Cristaux Moléculaires (E.R.A. 465),
U.E.R. de Physique Fondamentale, Université de Lille I,
59655 Villeneuve D'Ascq Cedex, France

Received 3rd *December*, 1979

In the first part of this paper is presented a review of experimental work carried out on CBr_4: X-ray and neutron diffraction, neutron diffuse scattering and neutron coherent inelastic scattering. Comparison is made with CD_4, and it is deduced that translational and rotational motions are coupled in CBr_4. Then in the second part, some consideration is given to the development of a model of translation–rotation coupling using the theoretical work of Yasuda on CD_4.

1. INTRODUCTION

Below its melting point (365 K), carbon tetrabromide (CBr_4) has an orientationally disordered cubic phase (phase I). A first-order transition occurs near 320 K and leads to an ordered monoclinic phase (phase II). Owing to the similarity of their molecular symmetry (T_d), it is interesting to compare CBr_4 with methane, which has been extensively studied: structures,[1–4] critical scattering,[5,6] orientational order parameter,[7] phonon and librational excitations,[8,9] theoretically[10–13] *etc.* The fact that the transition does not appear at the same temperature for CBr_4 and CD_4 is significant. The phenomena are not identical in the two cases mainly because of differences in molecular interaction.

The solid–solid transition which occurs at 320 K is of first order and although phase II keeps some " memory " of phase I, the orientational order remains unexplained. Pretransitional effects have been seen in methane: critical slowing down for example; they have not been clearly recognized in CBr_4. The potential seems to contain components which vary in a different manner with temperature, leading to a few competing order parameters.

It thus appears interesting to study in more detail the high-temperature phase itself before trying to understand the transition mechanism. Comparison with methane will be made as often as possible.

The main experimental results are summarised in Section 2. Phases II and I, obtained by X-ray and neutron diffraction respectively, are briefly schematized in Section 2(*a*) while in 2(*b*), the short-range order is analysed in time and space using neutron scattering measurements. The pattern obtained by a neutron diffuse scattering experiment is compared with that calculated from a hard-core potential model. Time analysis indicates that correlations have a complicated dynamical origin. Section 2(*c*) is devoted to neutron coherent inelastic scattering. Examples are given of acoustical phonons measured in the [111] direction. They show that acoustic velocities disperse with frequency. The scattering cross-section has a " three-peak-like " soft-mode structure. This probably results from a translation–rotation coupling similar to the spin–lattice coupling of the compressible Ising model.[12,13]

Section 3 provides a basis for a possible explanation of experimental results. In particular a potential model based on a calculation by Yasuda[11] on CD_4 is outlined. This model is adapted to take account of translation–rotation coupling. Although calculations are long and sometimes tedious such a model can account for the complexity of the observed phenomena.

2. EXPERIMENTAL

2(*a*). STRUCTURES

The-structures of phases II and I have already been described.[16–18] We give only an outline.

Low-temperature phase II is monoclinic $C_{2/c}$ ($a = 21.43$, $b = 12.12$, $c = 21.02$ Å, $\beta = 110.88°$, $Z = 32$).

The structure was determined by X-ray diffraction. The 4 molecules in the assymmetric unit are in general positions. Vibrational motions have been analysed in terms of rigid-body tensors **T**, **L** and **S**. Tensor **S**, which takes account of correlations between translational and rotational motions has been shown to have no significant value. The final R factor has been found equal to $R = R_w = 0.07$.

Molecular centres of mass lay approximately on a face-centred cubic lattice, the molecules being tilted a little from the $\bar{4}2m$ symmetry.

The structure of phase I was obtained by neutron diffraction. This phase, extending from 320 to 365 K, is face-centred cubic ($a = 8.82$ Å, $Z = 4$). The refinement was made with a 6-orientation Frenkel model corresponding to a $\bar{4}2m$ local symmetry. It gives $R = 0.16$ with $\langle u_x^2 \rangle = 0.2$ Å^2 (isotropic) and an isotropic rotational Debye–Waller factor $\langle \omega_x^2 \rangle = 0.05$ rad^2. Introduction of anisotropic coefficients does not improve the fit. A new refinement has been made where the bromine coherent-scattering-length density[2] $a(\mathbf{r})$ is expanded as a set of symmetry-adapted functions. This gives a weighted reliability factor $R_w = 0.07$. The probability of orientation of the C—Br bond is a maximum in the [110] direction. During reorientation motions, molecules then have a maximum probability of being in a local orientation close to $\bar{4}2m$.

Nevertheless, Hüller and Press[19] have shown that the orientational distribution function $f(\omega)$ calculated from structure parameters is not always positive. Constraints should be imposed upon the expansion coefficients. The authors prove that $f(\omega)$ is automatically positive when coefficients are deduced from an appropriate physical potential. To a first approximation it would be an angular potential but, owing to the large value of the translational Debye–Waller factor $\langle u_x^2 \rangle = 0.2$ Å^2, translation–rotation coupling must be considered. Such a coupling has been introduced phenomenologically[20] and has given new coefficients for the expansion. The R factor is then reduced to $R = 0.037$. Nevertheless, we do not know if $f(\omega)$ is definitely positive, and related coefficients are not connected to a potential.

Apart from the fact that molecular centres of mass remain approximately on a f.c.c. lattice, the correspondence between the two phases is not evident. Particularly, orientational order in phase II remains unexplained. It could result from lattice distortion leading to a vanishing translation–rotation coupling, as suggested by the zero value of tensor **S**.

2(*b*). SHORT-RANGE ORDER

Measurements made on powder samples by Dolling *et al.*[18] in thc cubic phase have revealed in the spectrum a diffuse peak for $|Q| \simeq 2.1$ Å^{-1}. This peak is not explained by the structure and the authors assign it to correlations. A neutron diffuse-scattering experiment conducted by More *et al.*[21] shows that the scattering pattern has cigar-shaped contours near the (220) Bragg peak. The long axes of the cigars point in the [111] directions (fig. 1). In this paper,[21] we show that scattering cannot be explained by structure-factor fluctuations $(\Delta F)^2 = \langle F^2 \rangle - \langle F \rangle^2$. These fluctuations contribute to the scattering intensity in the same region of reciprocal space but the corresponding intensity would be lower and broader.

Coulon and Descamps[22] have given an explanation for the existence of this peak. They show that neighbouring molecules are correlated *via* a potential corresponding to steric hindrance between bromine atoms. This potential has a short-range action and correlates only nearest-neighbour molecules. The authors choose it to be in the form of a static hard-core potential. They use a 6-orientation Frenkel model for the structure. The use of a hard-core potential is justified: among the 6×6 possible molecular configurations only 32 give an averaged attractive energy of 2.4 kcal mol^{-1}. The last 4 give a repulsive energy of 10 kcal mol^{-1}. The results of their calculation are in good agreement with experiment. Calculated intensity contours are practically identical to those shown in fig. 1.

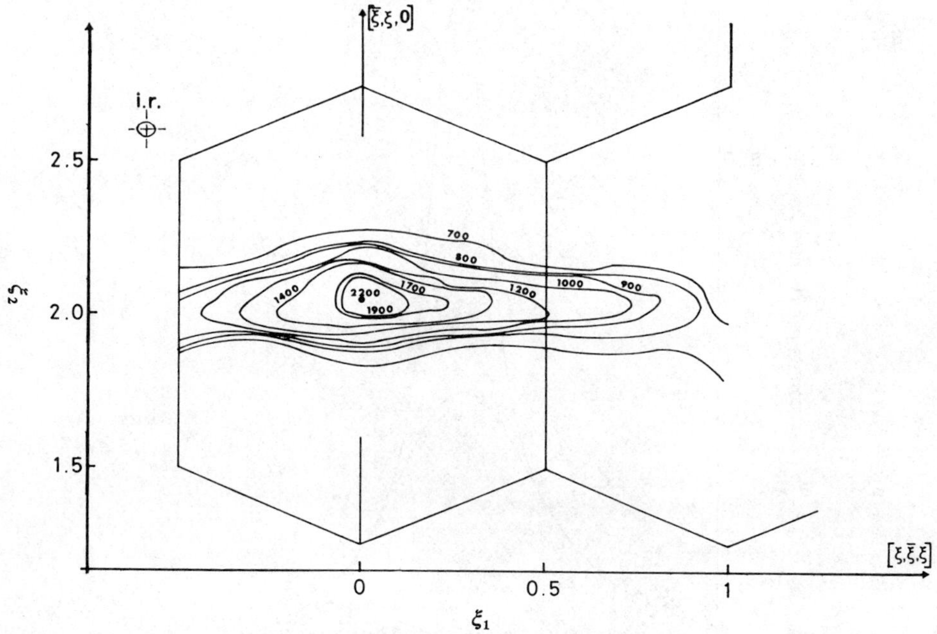

FIG. 1.—Diffuse scattering in the $(1, \bar{1}, \bar{2})$ plane. The scattering vector is $\boldsymbol{Q} = \xi_1(1, \bar{1}, 1) + \xi_2(1,1,0)$. (Sample 4 cm^3, background 400 counts, i.r. = instrumental resolution).

However, this model can describe nothing but static disorder with molecules in selected orientations taking account of their surroundings.

On the other hand one can imagine that a reorientating molecule translates and rotates in order to find a more stable equilibrium. Then correlations are dynamical. Because of dissipative processes, orientational fluctuations decrease in time as $e^{-\Gamma t}$ and give rise to quasi-elastic scattering.

The corresponding experiments have been made on spectrometer Hl at Saclay. Fig. 2 and 3 show energy scans for diffuse scattering at points $\boldsymbol{Q} = (2.15, 2.15, 0.)$ and $\boldsymbol{Q} = (2.5, 2.5, 0)$. We estimate the instrumental resolution width to be 0.054 THz at $\lambda_i = 4.05$ Å. Intensity comprises a purely elastic component arising from the quartz container of the crystal and one (or several) quasielastic component(s). This result is fitted with a Gaussian function for instrumental resolution and a Lorentzian centred at zero energy for the quasi-elastic peak. We then obtain:

$$\Gamma_1 = 0.03 \pm 0.005 \text{ THz (fig. 2)}$$
$$\Gamma_2 = 0.21 \pm 0.03 \text{ THz (fig. 3).}$$

Γ_1 is the inverse of the relaxation time of correlations responsible for intensity shown in fig. 1. In making this experiment (fig. 1) our instrumental resolution width was much larger than Γ_1 and the result was an integration over energy of correlation functions, that is to say the static susceptibility. The origin of the (Γ_2) Lorentzian is not understood at the present.

In the case of methane (CD_4, phase I), quasielastic diffuse scattering[5] additionally reveals anisotropic correlations. As with CBr_4, correlations are stronger in [111] planes than out of these planes. When the temperature is decreased toward the transition temperature T_c, orientational fluctuations slow down,[6] the intensity increases, becomes critical and finally we obtain a Bragg peak corresponding to a new order in phase II. Such critical scattering has not been observed in CBr_4. We have only noticed a significant hysteresis of 5 K when

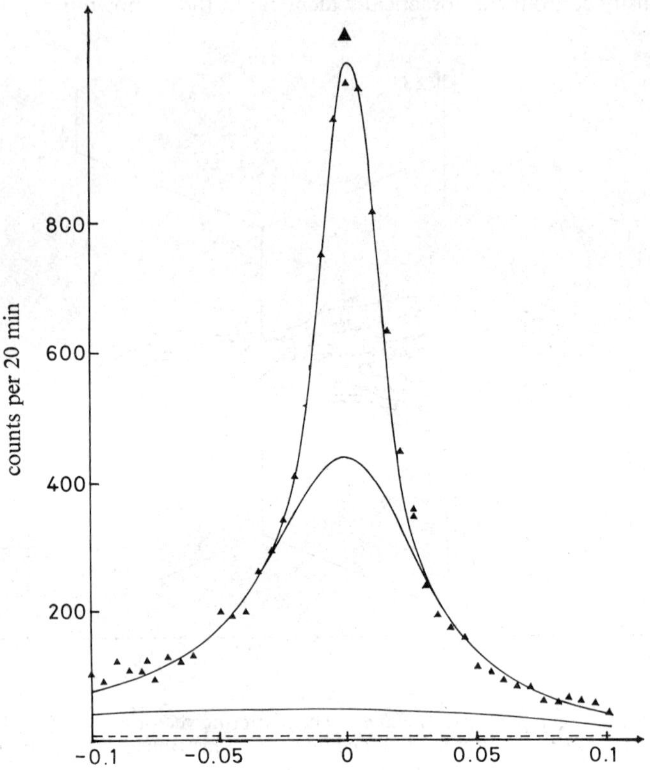

FIG. 2.—Constant Q scan (at $T = 325$ K) at the point $Q = (2.15, 2.15, 0)$. The elastic peak is mainly due to the quartz container. The Lorentzian has half width at half maximum = 0.03 THz. $\lambda_i = 5$ Å.

the temperature is lowered toward the transition. At this temperature the crystal breaks. However, when temperature is increased, Γ_1 increases and the quasi-elastic intensity decreases (for example: $\Gamma_1 = 0.05$ THz and intensity is divided by two at $T = 358$ K). While the transition is first-order, a second-order transition could take place but at $T_0 \ll T_c$.

2(c). PHONONS

Neutron coherent inelastic scattering experiments have been made at Saclay by M. More and B. Hennion. Acoustic phonons were measured in high-symmetry directions and with different polarizations. A few neutron groups are shown in fig. 4 and 5. They correspond to the longitudinal branch [111].

Intensities decrease rapidly and intrinsic widths increase when $\boldsymbol{q}$ increases (fig. 4). Fig. 5 is a constant Q-scan made at point $Q = (1.085, 1.085$ and $1.085)$ with an incoming wavelength $\lambda_i = 5$ Å. It shows a three-peak structure with maxima at $\pm\omega$ and 0. At zero energy, an elastic peak due to the quartz container is superimposed. The remaining intensity is clearly broader than the instrumental width (0.025 THz in this experiment). In a soft-mode study[23] as well as in a theoretical paper worked out for KCN by Michel and Naudts,[24] we find an expression for the scattering law:

$$S(Q, \omega) = \frac{\eta + \beta^2 f(\omega)}{[\omega^2 - \Omega^2 - \omega^2\beta^2 f(\omega)/\lambda]^2 + \omega^2[\eta + \beta^2 f(\omega)]^2} \quad (2.1)$$

where η represents phonon–phonon damping, $^2\beta = \omega_0^2 - \Omega^2$, Ω is the phonon frequency with interactions, ω_0 is the phonon frequency without interactions, λ^{-1} is the relaxation time of interactions and $f(\omega) = \lambda/(\omega^2 + \lambda^2)$.

Using this empirical expression, we have been able to take into account the observed results and then obtain for CBr_4 $\eta = 0$, $\lambda \simeq 0.04$ THz, $\Omega/\omega_0 = 0.6$, $\omega_0 = 0.24$ THz. A parametrization of this function for constant-energy scans has allowed us to fit the neutron groups of fig. 4. The full lines in fig. 4 and 5 represent this. λ is of the same order as Γ_1 previously determined to take account of dynamical correlations. The scattering law [eqn (2.1)] describes soft-mode behaviour. In CBr_4, longitudinal acoustic modes behave like this. In fact acoustic velocities are softening with frequency as presented in fig. 6.

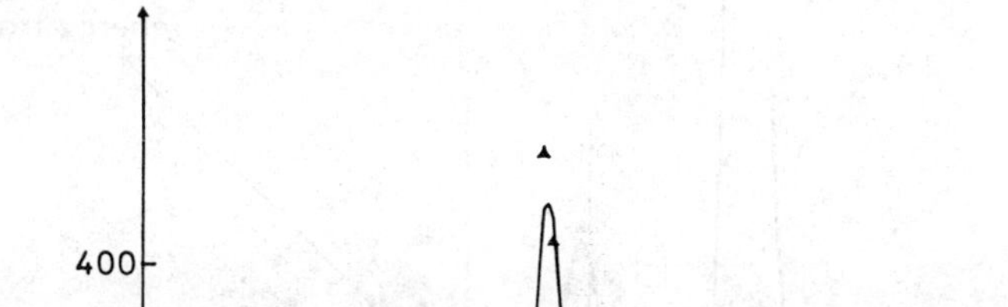

FIG. 3.—Constant Q scan (at $T = 325$ K) at the point $Q = (2.5, 2.5, 0)$. The Lorentzian has h.w.h.m. $= 0.21$ THz. $\lambda_i = 4.05$ Å.

The anisotropy which has already been seen in diffuse scattering is now characterized by the fact that velocities of longitudinal modes disperse more than transverse ones. The softening of velocities often appears in plastic crystals, for instance in KCN,[25] succinonitrile[26] or adamantane.[27]

Attempts to measure librational phonons have always failed in CBr_4. Spectra obtained at several points in reciprocal space (such as zone centres or zone boundaries) are all of the same kind. They show a more or less intense " background " centred at $\omega = 0$. A deconvolution with the experimental resolution function plus a Lorentzian give an intrinsic half width at half maximum of ≈1 THz. There seem to exist in CBr_4 nothing but diffusive modes characteristic of fluctuations exponentially decreasing with time.

In methane,[8] acoustic modes broaden out when the frequency increases, but the velocities do not seem to disperse. Librations have not been measured either. In the two cases,

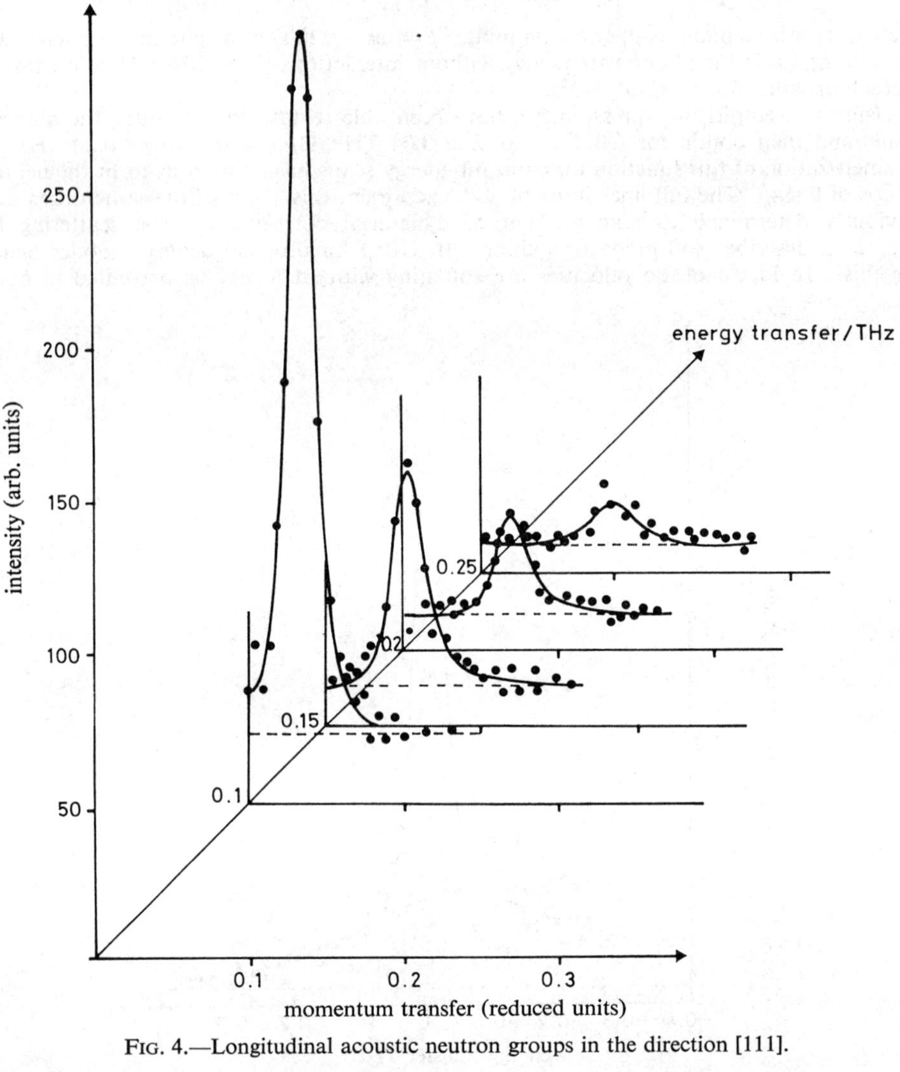

FIG. 4.—Longitudinal acoustic neutron groups in the direction [111].

molecules are reorientating with too high a frequency and neutrons cannot " see " their oscillations around temporary equilibrium positions. Libration frequencies would be of the order of 1 THz while correlation relaxation frequencies are close to 0.03 THz in CBr_4 and 0.09 THz in CD_4.[6]

3. THE MODEL

In order to explain these experimental features we develop in this Section a potential which takes account of translation–rotation coupling.

Potential energy can be calculated from atom–atom potentials between unbonded atoms or molecules. Using a 6-12 Lennard-Jones potential, Yasuda[11] has shown that the interaction potential between two tetrahedral molecules (1 and 2) such as CBr_4 or CD_4 is given by:

$$V_{12} = U(R) + V(R, \theta, \varphi; \omega_1) + V(R, \pi - \theta, \varphi + ; \pi\omega_2) + W(R, \theta, \varphi; \omega_1, \omega_2) \quad (3.1)$$

where (R, θ, φ) are the centre-of-mass polar coordinates of molecule 2 with respect to

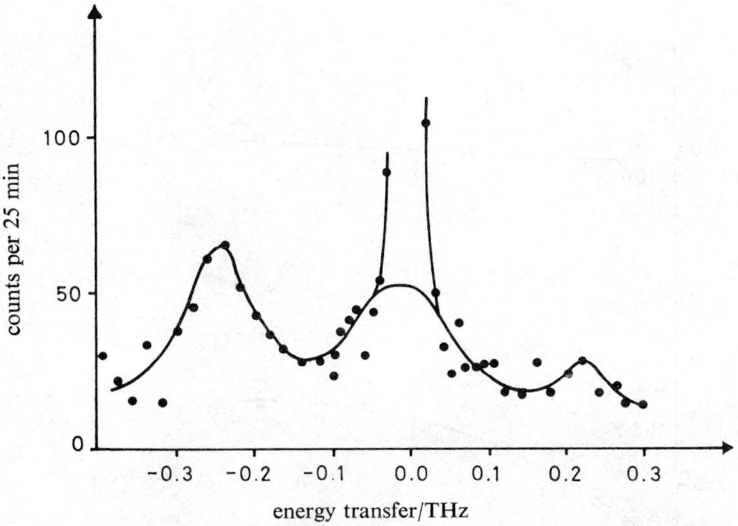

FIG. 5.—Constant Q scan (at $T = 325$ K) at the point $Q = (1.085, 1.085, 1.085)$. Instrumental width: 0.025 THz.

the centre of mass of molecule 1, $\omega_1(\omega_2)$ are three Euler angles setting the orientation of molecule 1(2) in a standard frame.

We define our standard orientation for the molecule like Yasuda.[11] $-U(R)$ is the isotropic part of the intermolecular potential. To obtain this term, we need an expansion in series of powers of r/R $[\sum_{p,q} A(r/R)^{p+q}]$ where $r = |r_{C\text{-}Br}|$ or $|r_{C\text{-}D}|$. In CD_4, we have $r/R = 0.26$ and convergence for the series is correct when $p + q = 24$. In CBr_4, we have $r/R = 0.31$ and series has to be taken up to $p + q = 36$. These expansions have been calculated numerically on a computer.

$$V = (R, \theta, \varphi; \omega) = \sum_{l \geqslant 3} v(R, l) \sum_{m} Y^{*}_{lm}(\theta, \varphi)\, T_{lm}(\omega) \quad (3.2)$$

where the coefficients $v(R, l)$ given in ref. (11) are numerically calculated. $Y_{lm}(\theta, \varphi)$ are the spherical harmonics, and $T_{lm}(\omega)$ are defined by:

$$T_{lm}(\omega) = \alpha_l^{-1} \sum_{k=1}^{4} Y_{lm}(\theta_k, \varphi_k) = \sum_{M} A_{l,M}\, \mathscr{D}^{l}_{M,m}(\omega). \quad (3.3)$$

(θ_k, φ_k) are polar angles for a position of a molecule at a given moment with respect to the original position (standard frame). α_l, $A_{l,m}$ are given in ref. (11).

Tetrahedral symmetry of the molecule implies that $A_{l,m}$ are different from zero only for $l = 0, 3, 4, 6, 7, 8, \ldots$ $\mathscr{D}^{(l)}_{M,m}(\omega)$ are the usual Wigner rotation matrices.

$$W(R, \theta, \varphi; \omega_1, \omega_2) = \sum_{\substack{l,l' \geqslant 3 \\ l'' \geqslant 0}} w(R; ll'l'') \sum_{m,m'} C(ll'l''; mm')\, Y^{*}_{l'',mm'}(\theta, \varphi) \times T_{lm}(\omega_1) T_{lm}(\omega_2) \quad (3.4)$$

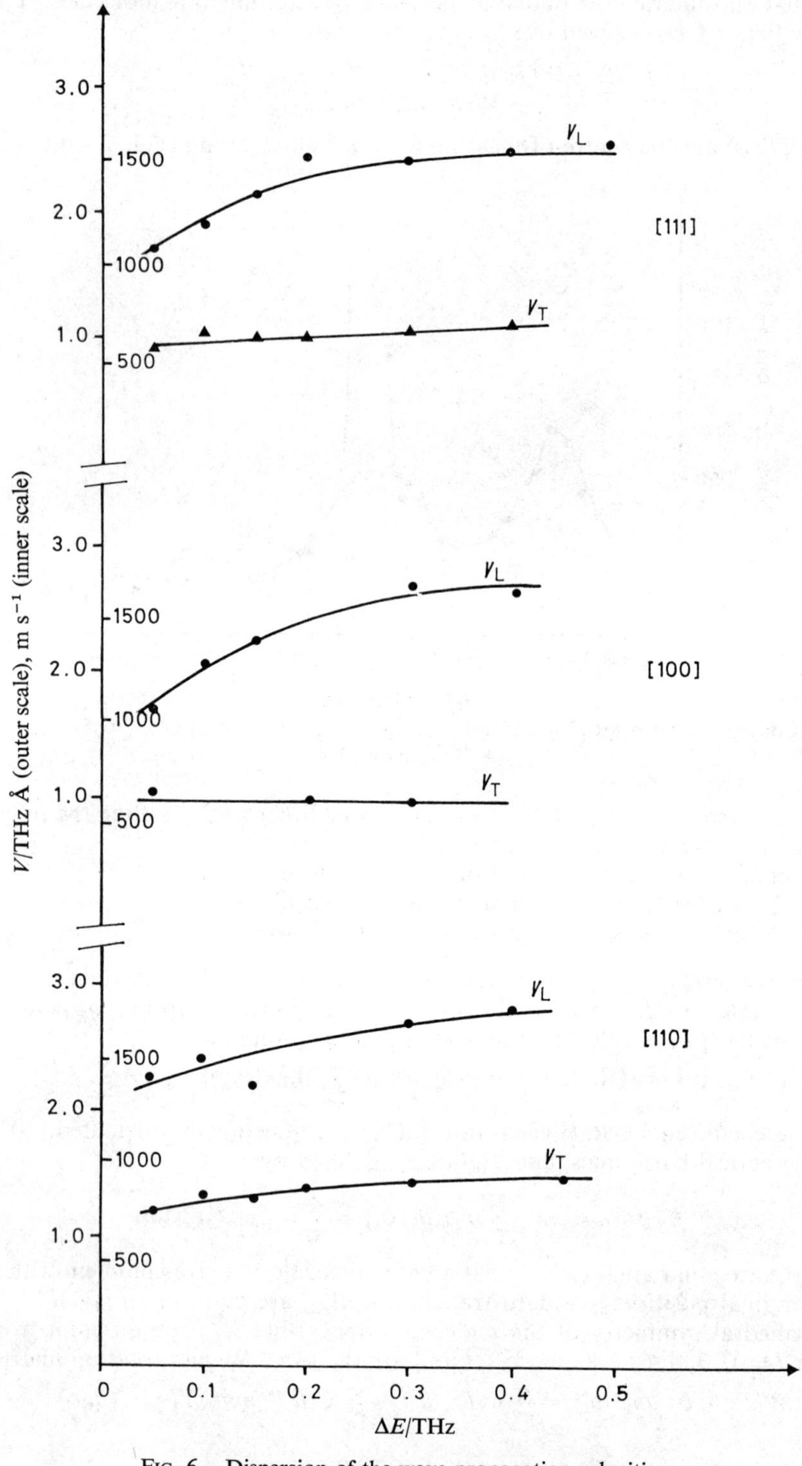

FIG. 6.—Dispersion of the wave propagation velocities.

$w(R;\ ll'l'')$ are numerical coefficients.[11] $C(ll'l'';\ mm')$ are the Clebsch–Gordan coefficients in Rose's notation.[28]

This term will not be used here but we can make some remarks about it: (i) The number of relevant coefficients is much greater in CBr_4 than in CD_4 where only a few are preponderant: in particular we notice that James and Keenan[10] have only used the (3, 3, 6) term leading to the octopole–octopole interaction. (ii) Many of these coefficients have opposite signs and can then cancel each other. (iii) We assume that, in spite of some differences, this term will give results not very different from those obtained in methane.

Numerical values (table 1) have been obtained using only interactions between

TABLE 1.—NUMERICAL VALUES OF $v(R, l)$ IN EQN (3.2) AND $w(R, ll'l'')$ IN EQN (3.4) IN kcal mol^{-1}

$v(R, 4) = 1.22$
$v(R, 6) = 0.688$
$v(R, 8) = 0.08$

l	l'	l''	$w(R, ll'l'')$
3	3	6	−1.96
3	3	4	1.275
3	3	0	−1.23
3	4	7	1.13
3	4	5	−0.649
3	4	3	0.547
3	4	1	−0.720
3	6	9	0.762
3	6	7	−0.362
3	6	5	0.254
3	6	3	−0.268
4	4	8	0.665
4	4	6	−0.342
4	4	4	0.254
4	4	2	−0.253
4	4	0	0.448
4	6	10	0.476
4	6	8	−0.206
4	6	6	0.128
4	6	4	−0.104
4	6	2	0.124

bromine atoms (Br–Br). Indeed we can assume that C–Br and C–C interactions are screened by the size of Br atoms. We use the coefficients given by Scott and Sheraga[29] for a Br–Br Buckingham potential

$$V_{\mathrm{Br-Br}} = A \exp(-Cr) - B/r^6$$

with $C = 2.78$ Å, $A = 34\,600$ and $B = 5180$ if V is expressed in kcal mol^{-1}. Yasuda's calculation needs this potential to be changed into a 6-12 Lennard-Jones potential $V = a/r^{12} - b/r^6$ where $a = 4.374 \times 10^6$ and $b = 3924$ with V in kcal mol^{-1}.

The crystal potential will now be expanded in series of displacements $\boldsymbol{u}_i$ of the centre of mass of the molecule with respect to the averaged equilibrium positions. At zero order, with respect to the displacements, the potential can be written:

$$V = \tfrac{1}{2}\sum_{\substack{i,j\\ i\neq j}} U(R_{ij}) + \sum_i V_c(\omega_i) + \tfrac{1}{2}\sum_{\substack{i,j\\ i\neq j}} W_{ij}(\omega_i, \omega_j). \tag{3.5}$$

R_{ij} is the " equilibrium " distance between molecules i and j and the second term corresponds to the crystalline field acting on molecule i.

$$V_c(\omega_i) = \sum_{l\geqslant 4} B_l V_l(\omega_i). \tag{3.6}$$

Due to disorder, the averaged equilibrium positions are not the true equilibrium positions of the molecules and there are terms of greater order in the expansion. However, owing to cubic symmetry the linear term cancels for the isotropic potential. The second and third terms of eqn (3.1) give:

$$\delta V = -\sum_i (\boldsymbol{F}_1^i + \boldsymbol{F}_2^i)\,\boldsymbol{u}_i. \tag{3.7}$$

If the summation is limited to nearest-neighbour molecules (r) at a distance R from molecule i, we obtain:

$$\boldsymbol{F}_1^i = \sum_r \operatorname{grad} V(R, \theta_r, \varphi_r;\ \omega_i) \tag{3.8}$$

$$\boldsymbol{F}_2^i = \sum_r \operatorname{grad} V(R, \pi - \theta_r, \pi + \varphi_r;\ \omega_r). \tag{3.9}$$

We neglect the linear term arising from the 4th term of eqn (3.1). Let us use Rose's gradient formula[28] to calculate eqn (3.8) and (3.9).

$$\operatorname{grad}\{v(R, l)\, Y^{*}_{lm}(\theta_r, \varphi_r)\} = \sum_{l'} \hat{\alpha}_{l'l}\, v(R, l)\, \boldsymbol{T}^{*}_{ll'm}(\theta_r, \varphi_r). \tag{3.10}$$

Here we have:

$$\hat{\alpha}_{l'l} = -C(l1l';\ 00)\left\{\frac{\partial}{\partial R} + \frac{Z(l, l')}{R}\right\}$$

$$Z(l, l') = \begin{cases} -l \text{ for } l' = l+1 \\ l+1 \text{ for } l' = l-1 \\ 0 \text{ for } l' \neq l \pm 1 \end{cases} \tag{3.11}$$

and

$$\boldsymbol{T}^{*}_{ll'm}(\theta_r, \varphi_r) = \sum_{\mu} C(l'1l;\ m+\mu, -\mu)\, Y^{*}_{l'm+\mu}(\theta_r, \varphi_r)\,\xi_{-\mu} \tag{3.12}$$

with $[\xi_1 = -(e_1 + ie_2)/\sqrt{2}, \xi_{-1} = (e_1 - ie_2)/\sqrt{2}, \xi_0 = e_3]$.

Let us define:

$$\sum_{r=1}^{12} Y^{*}_{l'm+\mu}(\theta_r, \varphi_r) = \mathscr{Y}_{l',m+\mu} \tag{3.13}$$

where $\mathscr{Y}_{l',m+\mu}$ are constants tabulated in ref. (11) and differ from zero only if $l' = 4$, 6, 8 . . .

and

$$\mathscr{C}_{ll'm} = \sum_{r=1}^{12} \boldsymbol{T}^{*}_{ll'm}(\theta_r, \varphi_r). \tag{3.14}$$

Then we obtain:

$$\boldsymbol{F}_1^i(R, \omega) = \sum_{l\geqslant 3}\sum_{l'}\sum_{m} \hat{\alpha}_{l'l}\, v(R, l)\, \mathscr{C}_{ll'm} T_{lm}(\omega_i). \tag{3.15}$$

Molecular symmetry gives us $l = 0, 3, 4, 6, 7, 8, \ldots$: crystalline symmetry gives $l' = 4, 6, 8, \ldots$. Taking eqn (3.10) into account we have $l = l' \pm 1$. Then non-zero terms will correspond to:

$$l = 3, l' = 4$$

$$l = 7, l' = 6$$

$$l = 7, l' = 8 \quad \ldots$$

For example, for $l = 3$, $l' = 4$ we have:

$$\boldsymbol{F}_1^{i(3)}(R, \omega) = \frac{7}{2\sqrt{4\pi}} \hat{\alpha}_{43} v(R, 3)[\mathscr{U}_2^{(3)}(\omega_i)\boldsymbol{e}_1 - \mathscr{U}_3^{(3)}(\omega_i)\boldsymbol{e}_2 + \mathscr{U}_4^{(3)}(\omega_i)\boldsymbol{e}_3], \qquad (3.16)$$

where the $\mathscr{U}_\alpha^{(3)}(\omega_i)$ are the (O_h) symmetry-adapted functions belonging to the irreducible representation T_{1u} for $l = 3$. They have been tabulated by Altmann and Cracknell:[30]

$$\mathscr{U}_2^{(3)}(\omega) = \tfrac{1}{4}\{\sqrt{3}[T_{3-1}(\omega) - T_{3-1}(\omega)] - \sqrt{5}[T_{33}(\omega) - T_{3-3}(\omega)]\}$$

$$\mathscr{U}_3^{(3)}(\omega) = -\frac{\mathrm{i}}{4}\{\sqrt{3}[T_{31}(\omega) + T_{3-1}(\omega)] + \sqrt{5}[T_{33}(\omega) + T_{3-3}(\omega)]\}$$

$$\mathscr{U}_4^{(3)}(\omega) = T_{30}(\omega). \qquad (3.17)$$

This result can be extended to any value of l.

The second term $\boldsymbol{F}_2^i$ implies a summation over all the orientations (ω_r) of neighbouring molecules. We can drop out this summation by taking the Fourier transform:

$$T_{lm}(\omega_r) = \tfrac{1}{\mathrm{N}} \sum_{\boldsymbol{q}} T_{lm}(-\boldsymbol{q}) \mathrm{e}^{\mathrm{i}\boldsymbol{q}(\boldsymbol{x}_i + \boldsymbol{R})}. \qquad (3.18)$$

We define:

$$\mathscr{V}_{l'm'}(\boldsymbol{q}) = \sum_{r=1}^{12} \mathrm{e}^{\mathrm{i}\boldsymbol{q}\boldsymbol{R}} Y_{l'm'}(\theta_r, \varphi_r) = \sum_{r=1}^{6} [\mathrm{e}^{\mathrm{i}\boldsymbol{q}\boldsymbol{R}} + (-1)^{l'} \mathrm{e}^{-\mathrm{i}\boldsymbol{q}\boldsymbol{R}}] Y_{l'm'}(\theta_r, \varphi_r). \qquad (3.19)$$

Then:

$$\boldsymbol{F}_2^i(R, \boldsymbol{q}) = \sum_{\boldsymbol{q}} \sum_{l \geqslant 3} \sum_{l'} \frac{(-1)^l}{\sqrt{N}} \hat{\alpha}_{l'l} v(R, l) \sum_m T_{lm}(-\boldsymbol{q}) \mathscr{C}_{ll'm}(\boldsymbol{q}) \mathrm{e}^{\mathrm{i}\boldsymbol{q}\boldsymbol{x}_i} \qquad (3.20)$$

where $\mathscr{C}_{ll'm}(\boldsymbol{q})$ is obtained by a generalization of eqn (3.14)-(3.18). For $\boldsymbol{q} = 0$, we obtain $\boldsymbol{F}_2^i(R, 0) = \boldsymbol{F}_1^i(R, \omega)$. For $\boldsymbol{q} \neq 0$, expression (3.19) cannot be reduced like eqn (3.13) and terms of the expansion will be: $\{(l = 3, l' = 2$ and $l' = 4), (l = 4, l' = 3$ and $l' = 5), \ldots\}$. $\boldsymbol{F}_2^i(R, \boldsymbol{q})$ can be expanded on sets of symmetry-adapted functions in the same manner as $\boldsymbol{F}_1^i(R, \omega)$.

Some coefficients $\hat{\alpha}_{l'l} v(R, l)$ have been calculated (table 2). A few of them have large values and for further calculations we will only take these main terms into account.

Let us call $U_\beta(\boldsymbol{q})$ a vector defined with Fourier components $\mathscr{U}_\tau^{(l)}(\boldsymbol{q})$ of symmetry adapted angular functions. l varies over all irreducible representations of the rotation group. τ numbers all the functions which are classified in irreducible representations of the O_h group. Geometrical factors and coupling constants can be put into a matrix $v_{\alpha\beta}(\boldsymbol{q})$. We denote by $u_\alpha(\boldsymbol{q})$ the Fourier components of displacements. Then, the translation–rotation potential can be written in a bilinear form:

$$V_{TR} = \sum_{q}\sum_{\alpha\beta} v_{\alpha\beta}(\boldsymbol{q})U_\beta(\boldsymbol{q})u_\alpha(\boldsymbol{q}) \tag{3.21}$$

$$= \sum_{q}\sum_{\alpha} N_\alpha(\boldsymbol{q})u_\alpha(\boldsymbol{q})$$

with

$$N_\alpha(\boldsymbol{q}) = \sum_{\beta} v_{\alpha\beta}(\boldsymbol{q})U_\beta(\boldsymbol{q}). \tag{3.22}$$

The hamiltonian can be written as

$$H = K + V_T + V_C + V_{TR}$$

where K is the kinetic energy of translation, V_T the translational potential energy, V_C the crystalline field and V_{TR} the translation–rotation potential. We have neglected the rotational kinetic energy, assuming that reorientations are slow enough. We have also neglected rotation–rotation coupling $W(\omega_1, \omega_2)$. One can see that V_C can easily be included in eqn (3.22). Then we obtain

$$H = \sum_{q}\{\tfrac{1}{2}p_\alpha^+(\boldsymbol{q})p_\alpha(\boldsymbol{q}) + \tfrac{1}{2}M_{\alpha\beta}(\boldsymbol{q})u_\alpha^+(\boldsymbol{q})u_\beta(\boldsymbol{q}) + N_\beta(\boldsymbol{q})u_\beta(\boldsymbol{q})\}. \tag{3.24}$$

$P_\alpha(\boldsymbol{q})$ are the conjugate momenta of $u_\alpha(\boldsymbol{q})$ and $M_{\alpha\beta}(\boldsymbol{q})$ is the usual dynamical matrix for translational motions.

TABLE 2.—NUMERICAL VALUES OF $\hat{\alpha}_{ll'}$, $v(R, l)$ IN EQN (3.15) IN kcal mol^{-1} Å^{-1}

$\hat{\alpha}_{23}v(R, 3) =$	-7.34
$\hat{\alpha}_{43}v(R, 3) =$	10.32
$\hat{\alpha}_{34}v(R, 4) =$	-3.64
$\hat{\alpha}_{54}v(R, 4) =$	5.36
$\hat{\alpha}_{56}v(R, 6) =$	1.84
$\hat{\alpha}_{76}v(R, 6) =$	3.04
$\hat{\alpha}_{67}v(R, 7) =$	-1.06
$\hat{\alpha}_{87}v(R, 7) =$	1.83

Let us denote:

$$\boldsymbol{u}_\beta(\boldsymbol{q}) = \boldsymbol{u}_\beta^{el}(\boldsymbol{q}) + \boldsymbol{u}_\beta^{vib}(\boldsymbol{q}) \tag{3.25}$$

where $\boldsymbol{u}_\beta^{el}(\boldsymbol{q})$ is an elastic deformation supposed to vary slowly with time and $\boldsymbol{u}_\beta^{vib}(\boldsymbol{q})$ the usual vibration coordinates.

From eqn (3.24) we derive the condition for instantaneous equilibrium which is:

$$\boldsymbol{u}_\alpha^{el}(\boldsymbol{q}) = -(M^{-1})_{\alpha\beta}N_\beta^+(\boldsymbol{q}). \tag{3.26}$$

This relation connects elastic displacements with angular vector $N_\beta(\boldsymbol{q})$. Using (3.24)-(3.26) we get:

$$H = \sum_{q}\{\tfrac{1}{2}p_\alpha^+(\boldsymbol{q})p_\alpha(\boldsymbol{q}) + \tfrac{1}{2}M_{\alpha\beta}(\boldsymbol{q})u_\alpha^{vib+}(\boldsymbol{q})u_\beta^{vib}(\boldsymbol{q}) + \tfrac{1}{2}N_\alpha(\boldsymbol{q})u_\alpha^{1}(\boldsymbol{q})\}. \tag{3.27}$$

In this form, the hamiltonian is a sum of independent components. The first two terms contain translational kinetic and potential energies, the third one has only terms of rotational origin. This result and the fundamental transformation denoted by eqn (3.25) are a generalization of the compressible Ising model already presented in ref. (14) and (15).

Although some numerical results have already been obtained, calculations have not yet been completed. However, we think that the theory presented here can explain some experimental features in CBr_4. As the translation–rotation potential is bilinear in the rotation and translation coordinates, the dynamical theory written for KCN[24] can be applied. In particular, rapid variations with $\boldsymbol{q}$ of matrix $v_{\alpha\beta}(\boldsymbol{q})$ could explain the softening of wave-propagation velocities when $\boldsymbol{q} \to 0$. At $\boldsymbol{q} = 0$, the potential is made of terms of order $l = 3$ and 7. This could account for additional terms found by Press[20] for the structure. The scattering law can be expanded using the hamiltonian derived from the generalized compressible Ising model. So we obtain the translation–translation, translation–rotation and rotation–rotation parts of scattered intensity. Special account of rotation–rotation terms $W(R, \theta, \varphi;\ \omega_1, \omega_2)$ should certainly be taken if we wanted to explain changes occurring at transition point in CBr_4 as it has been done in CD_4.

We thank J. Lefebvre (I.L.L., Grenoble), B. Hennion (L.L.B., Saclay), B. M. Powell (Chalk River, Canada) and M. Bee, J. L. Sauvajol and J. C. Damien (Lille) for their help in experiments and discussions.

1 W. Press, *J. Chem. Phys.*, 1972, **56**, 2597.
2 W. Press and A. Hüller, *Acta Cryst. A*, 1973, **29**, 252.
3 W. Press, *Acta Cryst. A*, 1973, **29**, 257.
4 W. Press and A. Hüller, in *Anharmonic Lattices, Structural Transitions and Melting*, Nato Advanced Study Institute Series (Noordhoft–Leiden, 1974), p. 185.
5 A. Hüller and W. Press, *Phys. Rev. Letters*, 1972, **29**, 266.
6 W. Press, A. Hüller, H. Stiller, W. Stirling and R. Currat, *Phys. Rev. Letters*, 1974, **32**, 1354.
7 W. Press and A. Hüller, *Phys. Rev. Letters*, 1973, **30**, 1207.
8 W. G. Stirling, W. Press and H. Stiller, *J. Phys. C*, 1977, **10**, 3959.
9 A. Hüller, *Phys. Rev. B*, 1974, **10**, 4403.
10 H. M. James and T. A. Keenan, *J. Chem. Phys.*, 1959, **31**, 12.
11 H. Yasuda, *Progr. Theor. Phys.*, 1971, **45**, 1361.
12 T. Yamamoto, Y. Kataoka and K. Okada, *J. Chem. Phys.*, 1977, **66**, 2701.
13 K. Maki, Y. Kataoka and T. Yamamoto, *J. Chem. Phys.*, 1979, **70**, 655.
14 H. Wagner and J. Swift, *Z. Phys.*, 1970, **239**, 182.
15 K. H. Michel and J. Naudts, *J. Chem. Phys.*, 1977, **67**, 547.
16 M. More, F. Baert and J. Lefebvre, *Acta Cryst. A*, 1977, **33**, 3681.
17 M. More, J. Lefebvre and R. Fouret, *Acta Cryst. A*, 1977, **33**, 3862.
18 G. Dolling, B. M. Powell and V. F. Sears, *Mol. Phys.*, 1979, **37**, 1859.
19 A. Hüller and W. Press, *Acta Cryst.*, *A* 1979, **35**, 876.
20 W. Press and A. Hüller, *Acta Cryst.*, *A* 1979, **35**, 881.
21 M. More, J. Lefebvre, B. Hennion, B. M. Powell and C. M. E. Zeyen, *J. Phys. C*, 1980, in press.
22 G. Coulon and M. Descamps, *J. Phys. C*, to be published.
23 J. D. Axe, S. M. Shapiro, G. Shirane and T. Riste, in *Anharmonic Lattices, Structural Transitions and Melting*. Nato Advanced Study Institutes Series. (Noordhoft-Leiden, 1974), p. 23.
24 K. H. Michel and J. Naudts, *J. Chem. Phys.*, 1978, **68** 216.
25 J. M. Rowe, J. J. Rush, N. J. Chesser, K. H. Michel and J. Naudts, *Phys. Rev. Letters*, 1978, **40**, 455.
26 H. Fontaine, R. Fouret, L. Boyer and R. Vacher, *J. Physique*, 1972, **33**, 1115.
27 J. C. Damien and G. Deprez, *Solid State Comm.*, 1976, **20**, 161.
28 M. E. Rose, *Elementary Theory of Angular Momentum* (John Wiley, New York, 1957).
29 R. A. Scott and H. A. Sheraga, *J. Chem. Phys.*, 1965, **42**, 2209.
30 S. L. Altmann and A. P. Cracknell, *Rev. Mod. Phys.*, 1965, **37**, 19.

Orientational Relaxation in Molecular Solids with Translation Rotation Coupling

By Bart De Raedt* and Karl H. Michel†

Institut für Theoretische Physik,
Universität des Saarlandes, D66 Saarbrücken, Germany

Received 3rd *December*, 1979

Starting from a Hamiltonian with bilinear coupling of translations and rotations and with single-particle orientational potential, we derive a closed transport equation for the dynamic displacement–displacement correlation function. The coupling to the rotation leads to a softening of the phonons and to an instability in the lattice at the orientational phase transition. The memory kernel which is due to the orientational velocity correlations is calculated in a systematic and consistent way. This leads to additional resonances in the displacement–displacement correlation function.

---

1. INTRODUCTION

In molecular liquids, the coupling of translational and rotational degrees of freedom is a dynamic effect[1] which manifests itself in the occurrence of mixed transport coefficients. On the other hand, in molecular solids there occurs a static coupling between translations and rotations. This coupling is of particular interest near a structural phase transition where an ordering in molecular orientations and a change in lattice structure occurs simultaneously. Dynamic equations for coupled pseudo-spin-phonon models have been proposed for the description of the dynamics of the ammonium halides.[2] A set of coupled equations for the lattice displacements and the anisotropic molecular polarizability has been put forward by Courtens,[3] following the lines of irreversible hydrodynamics.

Recently[4] a coupled system of dynamic equations for translations and rotations has been derived from a microscopic Hamiltonian.[5] In this Hamiltonian, the rotational degrees of freedom are described in terms of angular momenta and symmetry-adapted functions. In principle, all static and dynamic correlations can be calculated in terms of the microscopic model. In practice, only the calculation of static correlations is relatively easy. Previously one was not able to calculate the frequency dependent orientation–orientation transport coefficients. In fact these transport coefficients were replaced by frequency independent constants, which had to be determined from experiment.[6] The resulting dynamic equations then have the same structure (frequency dependence) as those of ref. (2) and (3).

The main advantage of the microscopic approach[4] consists of the fact that it allows one to go beyond the assumption of a constant transport coefficient. Since we know the commutation rules of the angular momenta and the symmetry-adapted functions, we should be able to perform a systematic derivation of kinetic equations and to

* Present address: Institut für Festkörperforschung, KFA Jülich, D5170 Jülich, Germany.
† Present address: Department Natuurkunde, University of Antwerp, B2610 Wilrijk, Belgium.

calculate explicitly the non-secular currents which enter the orientational dissipation function. We have done a similar calculation recently in order to describe the orientational dynamics of molecular impurities in crystals.[7] If the concentration of impurities is sufficiently small ($<1\%$), it should be sufficient to consider a single molecule and to neglect the indirect interaction between the impurities which is mediated by the lattice. Therefore, we have considered in ref. (7) the motion of a single impurity in a Devonshire-type[8] potential. This model corresponds to CN^- impurities in alkali halides.

In the present calculation we consider a model which corresponds to KCN and therefore we take into account a bilinear coupling of the rotations to the translations in addition to the single-particle Devonshire potential. The underlying model is specified in Section 2 where we give also results for the static correlations. In Section 3 we use the method of ref. (9) to calculate the transport coefficients and we apply it to the present model. A discussion of the resonances of the dynamic displacement–displacement correlation function is given in Section 4.

2. THE MODEL

We start from the Hamiltonian[5]

$$H = H^{\mathrm{T}} + H^{\mathrm{R}} + H^{\mathrm{TR}}. \tag{2.1}$$

Here H^{T} describes the pure translational part

$$H^{\mathrm{T}} = \sum_{\boldsymbol{k}} \left(\frac{p_i^{\dagger}(\boldsymbol{k})p_i(\boldsymbol{k})}{2m} + \frac{1}{2} M_{ij} s_i^{\dagger}(\boldsymbol{k}) s_j(\boldsymbol{k}) \right), \; i = x, y, z, \tag{2.2}$$

where $s_i(\boldsymbol{k})$ is the Fourier-transformed centre-of-mass displacement of mass m (we will take $m = 1$) per unit cell, and where $p_i(\boldsymbol{k})$ is the conjugate momentum with

$$[s_i^{\dagger}(\boldsymbol{k}), p_j(\boldsymbol{q})] = \mathrm{i}\delta_{i,j}\delta_{\boldsymbol{k}\boldsymbol{q}}. \tag{2.3}$$

The matrix M denotes the harmonic coupling coefficients. The rotational part of the Hamiltonian reads

$$H^{\mathrm{R}} = \sum_{\boldsymbol{k}} \frac{L_i^{\dagger}(\boldsymbol{k})L_i(\boldsymbol{k})}{2I} + \sum_{\boldsymbol{n}} V^{\mathrm{R}}[\Omega(\boldsymbol{n})]. \tag{2.4}$$

Here $L_i(\boldsymbol{k})$ is the angular momentum of the (linear) molecule with moment of inertia I. The single particle potential V^{R} is of octahedral symmetry;[8] it admits an expression in terms of cubic harmonics

$$V^{\mathrm{R}}(\Omega) = a_0 + a_4 K_4(\Omega) + \ldots \tag{2.5}$$

Here $\Omega(\boldsymbol{n}) = (\theta(\boldsymbol{n}), \varphi(\boldsymbol{n}))$ stands for the angular coordinates of the molecule at lattice site $\boldsymbol{n}$.

The coupling between translations and rotations is of the form

$$H^{\mathrm{TR}} = \mathrm{i} \sum_{\boldsymbol{k}} v_{i\alpha}(\boldsymbol{k}) Y_{\alpha}^{\dagger}(\boldsymbol{k}) s_i(\boldsymbol{k}). \tag{2.6}$$

Here the rotational coordinates Y_α, $\alpha = 1\text{-}5$ are represented by the Fourier-transformed linear combinations of spherical harmonics $Y_1(\Omega) = Y_2^0(\Omega)$; $Y_2(\Omega) = Y_2^{2,\mathrm{C}}(\Omega)$; $Y_3(\Omega) = Y_2^{2,\mathrm{S}}(\Omega)$; $Y_4(\Omega) = Y_2^{1,\mathrm{S}}(\Omega)$; $Y_5(\Omega) = Y_2^{1,\mathrm{C}}(\Omega)$, where we have used the definitions[10]

$$Y_l^{m,\mathrm{C}} = (Y_l^m + Y_l^{-m})/\sqrt{2} \tag{2.7a}$$

$$Y_l^{m,\mathrm{S}} = -\mathrm{i}(Y_l^m - Y_l^{-m})/\sqrt{2}\ . \tag{2.7b}$$

The functions Y_α are of E_g symmetry for $\alpha = 1, 2$ and of T_{2g} symmetry for $\alpha = 3, 4, 5$. The coupling matrix v reads

$$v_l = 2a\begin{pmatrix} k_xA & -k_xA & k_yB & 0 & k_zB \\ k_yA & k_yA & k_xB & k_zB & 0 \\ -2k_zA & 0 & 0 & k_yB & k_xB \end{pmatrix}, \tag{2.8}$$

where the elements A and B are given in terms of the microscopic potential[5] and where $2a$ is the lattice constant.

Previously, this model was used for the calculation of static and dynamic correlation functions. We define a scalar product by the static susceptibility $\chi_{\mu\nu}(z = 0)$:

$$(A_\mu, A_\nu) = \chi_{\nu\mu}(0), \tag{2.9a}$$

where[11]

$$\chi_{\mu\nu}(z) = \mathrm{i}\int_0^\infty \mathrm{d}t\, \mathrm{e}^{\mathrm{i}zt}\langle[A_\mu^+(t), A_\nu(0)]\rangle. \tag{2.9b}$$

Then the inverse of the static displacement–displacement susceptibility is given by

$$D_{ij}(k) = (s(k), s(k))_{ij}^{-1}. \tag{2.10}$$

For many experimental purposes, it is convenient and sufficient to consider particular symmetry directions in $\boldsymbol{k}$ space, say $\boldsymbol{k} = (0, 0, k)$. Then in a cubic crystal (NaCl-type structure), the matrices **M** and **D** are diagonal with elements

$$\mathbf{M} = [M_{11}, M_{11}, M_{33}] \tag{2.11a}$$

$$\mathbf{D} = [D_{11}, D_{11}, D_{33}]. \tag{2.11b}$$

We have

$$M_{11} = V_zk^2c_{44}^0 \equiv \omega_1^2;\quad M_{33} = V_zk^2c_{11}^0 \equiv \omega_3^2, \tag{2.12}$$

where c^0 are the bare elastic constants and where $V_z = 2a^3$ is the volume per unit cell. In ref. (5) it was shown that

$$D_{11} = M_{11}[1 - \delta\chi_{33}^{\mathrm{R}}] \equiv \Omega_3^2, \tag{2.13a}$$

and

$$D_{33} = M_{33}[1 - \gamma\chi_{11}^{\mathrm{R}}] \equiv \Omega^2. \tag{2.13b}$$

Here we have defined $\delta = 8a^2B^2/V_zc_{44}^0$, $\gamma = 16a^2A^2/V_zc_{44}^0$. χ_{33}^{R} and χ_{11}^{R} stand for the single-particle susceptibilities of T_{2g} and E_g symmetry respectively, defined with the single-particle potential V^{R}:

$$\chi_{\alpha\alpha}^{\mathrm{R}} \equiv \{\mathrm{Tr}[\exp(-V^{\mathrm{R}}/T)\,Y_\alpha(\Omega)\,Y_\alpha(\Omega)]/\mathrm{Tr}\exp(-V^{\mathrm{R}}/T)\}/T. \tag{2.14}$$

In eqn (2.14), Tr stands for trace and denotes the integration over the angular co-ordinates while T denotes the temperature (in energy units). We recall that χ_{11}^{R} has E_g symmetry and χ_{33}^{R} has T_{2g} symmetry. Numerical evaluations of these quantities are found in ref. (7).

3. DYNAMIC RESPONSE THEORY

It is convenient to introduce Kubo's[11] relaxation function

$$\Phi_{\mu\nu}(z) \equiv (A_\mu, (z - \mathscr{L})^{-1}A_\nu) = [\chi_{\mu\nu}(z) - \chi_{\mu\nu}(0)]/z. \tag{3.1}$$

Here we used the definition of the scalar product (2.9*a*). The Liouville operator $\mathscr{L}$ is defined by

$$A(t) = \exp(i\mathscr{L}t)A, \text{ where } \mathscr{L}A = [H, A]. \tag{3.2}$$

Given a set of secular variables $\{A_\nu\}$, Mori[12] has derived the matrix equation

$$[\mathbf{1}z - \mathbf{\Omega} + \mathbf{\Sigma}(z)]\mathbf{\Phi}(z) = \bar{X}(0). \tag{3.3}$$

Here Ω is the matrix of restoring forces

$$\Omega_{\mu\nu} = (A_\mu, \mathscr{L}A_\rho)\chi_{\rho\nu}^{-1}(0) \tag{3.4}$$

and $\Sigma(z)$ stands for the frequency-dependent dissipation matrix (transport-coefficient matrix):

$$\Sigma_{\mu\nu}(z) = -(Q\mathscr{L}A_\mu, (z - Q\mathscr{L}Q)^{-1}Q\mathscr{L}A_\rho)\chi_{\rho\nu}^{-1}(0). \tag{3.5}$$

Here Q is the projector onto the space of non-secular variables,

$$Q = 1 - P, \; PA = \sum_{\nu,\mu} A_\nu \chi_{\nu\mu}^{-1}(0)(A_\mu, A). \tag{3.6}$$

Since Σ, eqn (3.5), is again a relaxation function, it satisfies in turn an equation of type (3.3). Continuing this process one obtains a continued-fraction representation[13] for $\Phi(z)$. In order to get closed expressions, one has to terminate this series by some approximation. In ref. (4) we chose as set of dynamic variables $\{A_\nu\} \equiv \{s_i(\mathbf{k})\ p_i(\mathbf{k}), \bar{Y}_\alpha(\mathbf{k})\}$ with $\bar{Y}_\alpha(\mathbf{k}) = Y_\alpha(\mathbf{k}) - D_{ij}(\mathbf{k})(s_j(\mathbf{k}), Y_\alpha(\mathbf{k}))s_i(\mathbf{k})$. The dissipation matrix $\sum_{\alpha\beta}^{\bar{Y}\bar{Y}}$ was approximated by a frequency-independent constant $\lambda_{\alpha\beta}$. Since here we want to go beyond this simple approximation, we have to use a systematic method for calculating frequency-dependent transport coefficients.[9]

We consider a set of orthogonalized dynamic variables $\{A_\nu\} = \{A_0, A_1, \ldots A_i, \ldots A_n\}$ where A_n is related to $A_0 = A_0^\dagger$ by

$$A_1 = \mathscr{L}A_0,\ A_2 = \mathscr{L}A_1 - (A_1, A_1)(A_0, A_0)^{-1} A_0, \ldots A_n = \mathscr{L}A_{n-1} - (A_{n-1}, A_{n-1})(A_{n-2}, A_{n-2})^{-1} A_{n-2}. \tag{3.7}$$

Then only the transport coefficient $\Sigma_{nn}(z)$ is different from zero because $Q\mathscr{L}A_i = 0$ for $i \neq n$. The solution of eqn (3.3) then yields the continued fraction representation

$$\Phi_{00}(z) = \cfrac{(A_0, A_0)}{z - \cfrac{\Delta_1^2}{z - \cfrac{\Delta_2^2}{z - \ddots \cfrac{-\Delta_n^2}{z + \Sigma_{nn}(z)}}}} \tag{3.8}$$

where

$$\Delta_\nu^2 = (A_\nu, A_\nu)(A_{\nu-1}, A_{\nu-1})^{-1}. \tag{3.9}$$

We use the method of ref. (9) to determine Σ_{nn} by means of sum rules. The result reads

$$\Sigma_{nn}(z) = -\frac{\Delta_{n+1}^2}{z + i(\Delta_n^2 + \Delta_{n+1}^2)^{\frac{1}{2}}}. \tag{3.10}$$

We apply this scheme to the model which has been specified in Section 2.

We start with $A_0^i = s_i(\boldsymbol{k})$ and determine A_1^i, A_2^i, A_3^i according to eqn (3.7). Making use of the identity

$$(\mathscr{L}A, B) = \langle[A^\dagger, B]\rangle, \tag{3.11}$$

of the commutator (2.3) as well as of the definition (3.2) of $\mathscr{L}$, we find

$$A_0^i(\boldsymbol{k}) = s_i(\boldsymbol{k});\;\; A_1^i(\boldsymbol{k}) = -ip_i(\boldsymbol{k}); \tag{3.12a}$$

$$A_2^i(\boldsymbol{k}) = (M_{ij} - D_{ij})s_j(\boldsymbol{k}) - iv_{i\alpha}(\boldsymbol{k})Y_\alpha(\boldsymbol{k}); \tag{3.12b}$$

$$A_3^i(\boldsymbol{k}) = -iv_{i\alpha}(\boldsymbol{k})\mathscr{L}Y_\alpha(\boldsymbol{k}). \tag{3.12c}$$

The relevant susceptibilities are given by eqn (2.10) and by

$$(A_1^i, A_1^j) = (p_i(\boldsymbol{k}), p_j(\boldsymbol{k})) = \delta_{ij}, \tag{3.13a}$$

$$(A_2^i, A_2^j) = (M_{ij}(\boldsymbol{k}) - D_{ij}(\boldsymbol{k})), \tag{3.13b}$$

$$(A_3^i, A_3^j) = \sum_\alpha v_{i\alpha}(\boldsymbol{k})v_{\alpha j}(\boldsymbol{k})(\mathscr{L}Y_\alpha, \mathscr{L}Y_\alpha)_{\boldsymbol{k}=0}. \tag{3.13c}$$

Using translational invariance of the lattice, we have

$$(\mathscr{L}Y_\alpha, \mathscr{L}Y_\alpha)_{\boldsymbol{k}=0} = \frac{1}{N}\sum_{\boldsymbol{n}}(\mathscr{L}Y_\alpha(\boldsymbol{n}), \mathscr{L}Y_\alpha(\boldsymbol{n})) = (\mathscr{L}Y_\alpha, \mathscr{L}Y_\alpha), \tag{3.14}$$

which is independent of the lattice site $\boldsymbol{n}$. Here N denotes the total number of unit cells and $Y_\alpha(\boldsymbol{n})$ stands for $Y_\alpha[\Omega(\boldsymbol{n})]$. The quantity $(\mathscr{L}Y_\alpha, \mathscr{L}Y_\alpha)$ is now calculated within the single-particle approximation with potential V^{R}:

$$T(\mathscr{L}Y_\alpha, \mathscr{L}Y_\alpha)^{\mathrm{R}} = \mathrm{Tr}[\exp(-V^{\mathrm{R}}/T)(\mathscr{L}Y_\alpha)^\dagger \mathscr{L}Y_\alpha]/\mathrm{Tr}(\exp - V^{\mathrm{R}}/T). \tag{3.15}$$

Taking again $\boldsymbol{k} = (0, 0, k)$, $\mathbf{M}$ and $\mathbf{D}$ become diagonal [see eqn (2.11*a*)-(2.13*b*)] and eqn (3.13*c*) reduces to

$$(A_3^1, A_3^1) = (A_3^2, A_3^2) = 8k^2a^2B^2(\mathscr{L}Y_3, \mathscr{L}Y_3)^{\mathrm{R}} \tag{3.16}$$

$$(A_3^3, A_3^3) = 16k^2a^2A^2(\mathscr{L}Y_1, \mathscr{L}Y_1)^{\mathrm{R}}. \tag{3.17}$$

Writing $\Delta_{n,i}^2$ for Δ_ν^2 defined with A_n^i according to eqn (3.9), we obtain for $i = 1, 2$:

$$\Delta_{1,i}^2 = D_{11} \equiv \Omega_{T_{2g}}^2 \tag{3.18a}$$

$$\Delta_{2,i}^2 = M_{11} - D_{11} \equiv \beta_{T_{2g}}'^2 \tag{3.18b}$$

$$\Delta_{3,i}^2 = \frac{(\mathscr{L}Y_3, \mathscr{L}Y_3)^{\mathrm{R}}}{(Y_3, Y_3)^{\mathrm{R}}} \equiv \langle\omega^2\rangle_{T_{2g}} \tag{3.18c}$$

and

$$\Delta_{1,3}^2 = D_{33} \equiv \Omega_{E_g}^2 \tag{3.19a}$$

$$\Delta_{2,3}^2 = M_{33} - D_{33} \equiv \beta_{E_g}'^2 \tag{3.19b}$$

$$\Delta_{3,3} = \frac{(\mathscr{L}Y_1, \mathscr{L}Y_1)^{\mathrm{R}}}{(Y_1, Y_1)^{\mathrm{R}}} \equiv \langle\omega^2\rangle_{E_g}. \tag{3.19c}$$

The quantities $\beta_r'^2$, $r = E_g$ or T_{2g} are completely determined by means of eqn (2.12), (2.13*a*) and (2.13*b*). One finds, respectively, $\beta_{E_g}'^2 = 16k^2a^2A^2(Y_1, Y_1)^{\mathrm{R}}$ and $\beta_{T_{2g}}'^2 = 8k^2a^2B^3(Y_3, T_3)^{\mathrm{R}}$.

Eqn (2.13*a*) and (3.18*a*)-(3.18*c*) show that displacements in x and y directions

($i = 1, 2$) couple to T_{2g} orientations ($\alpha = 3$) while eqn (2.13*b*) and (3.19*a*)-(3.19*c*) show that displacements in the *z*-direction ($i = 3$) couple to E_g orientations ($\alpha = 1$).

Having determined all relevant coefficients, we obtain from eqn (3.8)

$$\Phi_{ss}^{ii}(z) = \cfrac{\Omega_r^{-2}}{z - \cfrac{\Omega_r^2}{z - \cfrac{\beta_r'^2}{z + \Sigma_r(z)}}} \tag{3.20}$$

where

$$\Sigma_r(z) = \frac{-\langle\omega^2\rangle_r}{z + \mathrm{i}(\langle\omega^2\rangle_r + \beta_r'^2)^{\frac{1}{2}}}. \tag{3.21}$$

Here we understand $r = T_{2g}$ for $i = 1, 2$ and $r = E_g$ for $i = 3$.

4. DISCUSSION

Eqn (3.20) looks more familiar in the form

$$\left(z^2 - \Omega_r^2 - \frac{z\beta_r'^2}{z + \Sigma_r'(z)}\right) \Phi_{ss}^{ii}(z) = \left(z - \frac{\beta'^2}{z + \Sigma_r'(z)}\right) \Omega^{-2}. \tag{4.1}$$

Eqn (3.20) with the frequency dependent dissipation $\Sigma(z)$, eqn (3.21), is the central result of our work. We first make a comparison with the corresponding result of ref. (4). Eqn (6.7) of ref. (4) is directly rewritten as:

$$\left(z^2 - \Omega^2 - \frac{z\beta_r'^2}{z + \mathrm{i}\lambda_r}\right) \Phi_{ss}^{ii}(z) = \left(z - \frac{\beta_r'^2}{z + \mathrm{i}\lambda_r}\right) \tag{4.2}$$

where λ_r was the frequency independent orientational relaxation coefficient:

$$\mathrm{i}\lambda_{\alpha\alpha} = -(Q\mathscr{L}\bar{Y}_\alpha, (\mathrm{i}\varepsilon - Q\mathscr{L}Q)^{-1}Q\mathscr{L}\bar{Y}_\alpha)/(\bar{Y}_\alpha\bar{Y}_\alpha), \tag{4.3}$$

with $\varepsilon \to 0$. Here $\alpha = 3$ corresponds to $r = T_{2g}$ or $i = 1, 2$ in Φ_{ss}^{ii} and $\alpha = 1$ to E_g or $i = 3$. Putting $z = \mathrm{i}\varepsilon \to 0$ in eqn (3.21), we then obtain by identification of eqn (4.1) and (3.22)

$$\lambda_{\alpha\alpha} = \langle\omega^2\rangle_r/(\langle\omega^2\rangle_r + \beta_r'^2)^{\frac{1}{2}}. \tag{4.4}$$

The second moments $\langle\omega^2\rangle_r$ entering this equation have been calculated in ref. (7). We rewrite eqn (A.24) and (A.27) of ref. (7)

$$\langle\omega^2\rangle_{E_g} = \frac{12T}{I}\left(\frac{1 - 3\langle x^4\rangle}{9\langle x^4\rangle - 1}\right), \tag{4.5a}$$

$$\langle\omega^2\rangle_{T_{2g}} = \frac{12T}{I}\left(\frac{\langle x^4\rangle}{1 - 3\langle x^4\rangle}\right). \tag{4.5b}$$

Here we only took into account the cubic harmonic K_4 in the expansion (2.5). Note that in Cartesian coordinates, $K_4(x, y, z) \propto (x^4 + y^4 + z^4)$. The thermal averages $\langle x^4\rangle$ are then evaluated numerically as a function of T/a_4.

Eqn (4.1), with λ independent of frequency, has been discussed in ref. (4). This situation describes relaxational behaviour of orientational correlations. Depending on the magnitude of λ in comparison with $\sqrt{\mathbf{M}} \equiv \omega_i$, one distinguishes the cases (*a*) of fast relaxation, $\lambda > \omega_i$ and (*b*) of slow relaxation, $\lambda < \omega_i$. In case (*a*), the inelastic spectrum exhibits a soft-mode Brillouin doublet of frequencies $\pm\Omega_i$ on top of a broad central peak which disappears in the experimental background. This is the situation found experimentally[6] in KCN at temperatures above 168 K (orientationally disordered cubic phase). On the other hand, in the slow-relaxation case (*b*), one has a well defined three-peak structure, where in addition to a narrow central peak, a non-soft phonon doublet is located at $z = \pm\omega_i$. This situation is experimentally realized in NH_4Br.[2,14]

We now turn to the more general situation and take into account the frequency dependence of $\Sigma_r(z)$. We are interested in the correlation function

$$\psi_{ss}^{ii}(\omega) = -\lim_{\varepsilon \to 0} \frac{1}{2\pi i} [\Phi_{ss}^{ii}(\omega + i\varepsilon) - \Phi_{ss}^{ii}(\omega - i\varepsilon)] \tag{4.6}$$

which enters directly into the neutron or Brillouin scattering law.

We shall discuss eqn (4.1) as a function of the coupling $\beta_r'^2$ and of the moment $\langle\omega^2\rangle_r$. It is easy to consider several limiting cases:

1. $\beta_r'^2 \to 0$, vanishing coupling between translations and rotations. Then we see from eqn (3.18*b*) and (3.19*b*) that Ω_r^2 reduces to ω_r^2, and eqn (4.1) becomes

$$\Phi_{ss}^{ii}(z) = \frac{1}{\omega_r^2}\left(\frac{z}{z^2 - \omega_r^2}\right), \tag{4.7a}$$

and

$$\Psi_{ss}^{ii}(\omega) = \frac{1}{2\omega_r^2}[\delta(\omega - \omega_r) - \delta(\omega + \omega_r)]. \tag{4.7b}$$

Consequently we have a Brillouin doublet with bare phonons at $\omega = \pm\omega_r$.

2. $\Omega_r^2 \to 0$, or equivalently [see eqn (2.13*a* and *b*)]$\chi_r^R = \xi^{-1}$, where ξ stands for γ or δ, depending on whether $r = E_g$ or T_{2g}. This situation corresponds to the phase transition, the coupling to the rotation leads to an instability in the lattice.[15,5] Then

$$\Phi_{ss}^{ii}(z) = \frac{1}{z}\frac{1}{\Omega_r^2}, \tag{4.8a}$$

$$\Psi_{ss}^{ii}(\omega) = \frac{1}{\Omega_r^2}\delta(\omega) \tag{4.8b}$$

All scattering intensity is concentrated in an infinitely sharp central peak.

3. $\langle\omega^2\rangle_r \to 0$, the second-order orientational moment vanishes.
Then

$$\Phi_{ss}^{ii}(z) = \frac{1}{\Omega_r^2}\left(\frac{z^2 - \beta_r'^2}{z^2 - \omega_r^2}\right)\frac{1}{z}, \tag{4.9a}$$

and a partial fraction decomposition yields:

$$\Phi_{ss}^{ii}(z) = \frac{1}{\Omega_r^2}\left[\left(1 - \frac{\beta_r'^2}{\omega_r^2}\right)\frac{z}{z^2 - \omega_r^2} + \frac{\beta_r'^2}{\omega_r^2}\frac{1}{z}\right]. \tag{4.9b}$$

The correlation function reads

$$\Psi_{ss}^{ii}(\omega) = \frac{1}{\Omega_r^2}\left\{\left(1 - \frac{\beta_r'^2}{\omega_r^2}\right)\tfrac{1}{2}[\delta(\omega - \omega_r) - \delta(\omega + \omega_r)] + \frac{\beta_r'^2}{\omega_r^2}\delta(\omega)\right\}. \tag{4.9c}$$

In addition to a Brillouin doublet at $\omega = \pm\omega_r$, there is a sharp central resonance due to orientational relaxation.

4. $\langle\omega^2\rangle_r \gg \omega_r^2$. Here we find

$$\Phi_{ss}^{ii}(z) = \frac{1}{\Omega_2^r}\frac{z}{(z^2 - \Omega)_r^2}. \qquad (4.10a)$$

The rotational frequencies are too fast, the dynamics of translations and rotations are decoupled but there is a shift of the phonon frequencies from ω_r to Ω_r. We find

$$\Psi_{ss}^{ii}(\omega) = \frac{1}{2\Omega_r^2}[\delta(\omega - \Omega_r) - \delta(\omega + \Omega_r)]. \qquad (4.10b)$$

Let us now discuss some intermediate cases, where we have to evaluate $\Psi_{ss}^{ii}(\omega)$ numerically. Using (3.20) and (3.21) we can write

$$\Phi_{ss}^{ii}(z) = \cfrac{\Omega_r^{-2}\omega_r^{-1}}{u - \cfrac{1 - B_r}{u - \cfrac{B_r}{u - \cfrac{C_r}{u + \mathrm{i}(B_r + C_r)^{\frac{1}{2}}}}}} \qquad (4.11)$$

with

$$u = z/\omega_r \qquad (4.12a)$$

$$B_r = \delta\chi_{33}^{\mathrm{R}} \quad \text{for} \quad r = T_{2g} \qquad (4.12b)$$

$$B_r = \gamma\chi_{11}^{\mathrm{R}} \quad \text{for} \quad r = E_g \qquad (4.12c)$$

$$C_r = \langle\omega^2\rangle_r/\omega_r^2. \qquad (4.12d)$$

In eqn (4.11) the frequency is scaled by the bare phonon frequency ω_r. We see that the shape of $\Phi_{ss}^{ii}(z)$ only depends on the two parameters, B_r and C_r. If we specify the parameters of the microscopic Hamiltonian [eqn (2.1)-(2.4)] and if we take an explicit form for the one particle potential V^{R} [eqn (2.5)], we can determine B_r and C_r.[5,7] Instead of doing this, we merely discuss the shape of the correlation function for some realistic values of B_r and C_r. The physical meaning of the parameters is clear: B_r close to zero means that we are far away from the phase transition; B_r close to one means that one is in the neighbourhood of the phase transition. C_r is simply a measure of the rotational frequency in terms of the phonon frequency. From figs. 1-3 we draw the following main conclusions: The weight of the central resonance, which has its origin in the translation–rotation coupling, becomes more important as we approach the phase transition. A two-peak structure is observed for some values of the parameters: in particular when the phonon frequency is comparable with the rotational frequency this phenomenon is likely to occur. In these cases the two resonances are always broad. This, however, might be an artefact of our approximation. It is easy to verify that a four-pole approximation such as eqn (4.11) can never give two sharp resonances at non-zero frequencies. As far as we know, two sharp resonances have not been observed experimentally until now. Theoretically one could answer the question, whether such two sharp peak structures can occur, if one could go one step further in the continued fraction. This, however, is not possible at the moment.

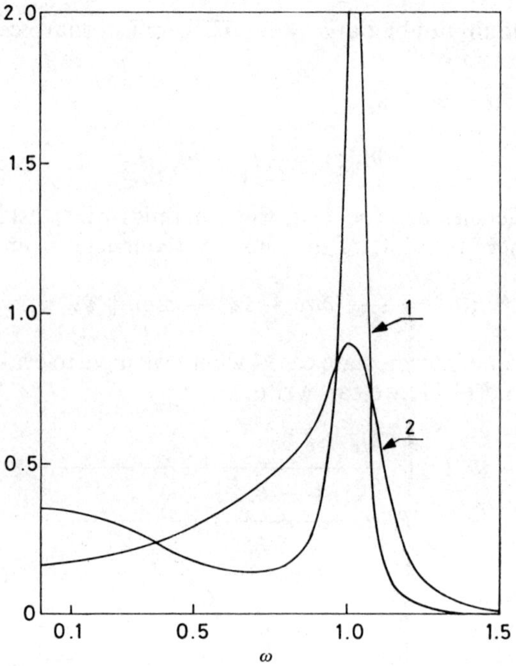

FIG. 1.—Normalized dynamic phonon correlation function. The frequency is measured in units of the bare phonon frequency. (1) $B_r = 0.3$, $C_r = 0.3$; (2) $B_r = 0.3$, $C_r = 0.9$.

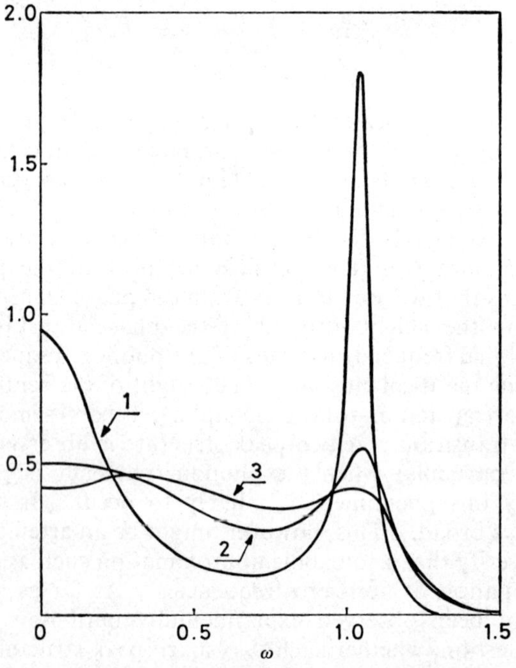

FIG. 2.—See fig. 1. (1) $B_r = 0.5$, $C_r = 0.3$; (2) $B_r = 0.5$, $C_r = 0.7$; (3) $B_r = 0.5$, $C = 0.9$.

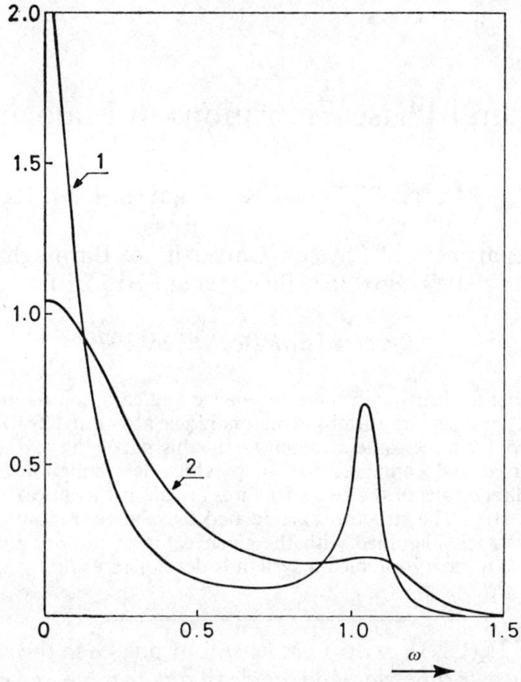

FIG. 3.—See fig. 1. (1) $B_r = 0.7$, $C_r = 0.3$; (2) $B_r = 0.7$, $C_r = 0.9$.

[1] For a review, see, *e.g.*, B. J. Berne and R. Pecora, *Dynamic Light Scattering* (Wiley-Interscience, New York, 1976).
[2] Y. Yamada, H. Takatera and D. Humber, *J. Phys. Soc. Japan*, 1974, **36**, 641.
[3] E. Courtens, *J. Physique Lettres*, 1976, **37**, L-21.
[4] K. H. Michel and J. Naudts, *J. Chem. Phys.*, 1978, **68**, 216.
[5] K. H. Michel and J. Naudts, *Phys. Rev. Letters*, 1977, **39**, 212; *J. Chem. Phys.*, 1977, **67**, 547.
[6] J. M. Rowe, J. J. Rush, N. J. Chesser, K. H. Michel and J. Naudts, *Phys. Rev. Letters*, 1978, **40**, 455.
[7] B. De Raedt and K. H. Michel, *Phys. Rev.*, 1979, **B19**, 767.
[8] A. F. Devonshire, *Proc. Roy. Soc. A*, 1936, **53** 601; P. Sauer, *Z. Phys.*, 1966, **194**, 360.
[9] H. De Raedt and B. De Raedt, *Phys. Rev.*, 1977, **B15**, 5379.
[10] C. J. Bradley and A. P. Cracknell, *The Mathematical Theory of Symmetry in Solids* (Clarendon, Oxford, 1972).
[11] R. Kubo, *J. Phys. Soc. Japan*, 1957, **12**, 570; L. P. Kadanoff and P. C. Martin, *Ann. Phys. (N.Y.)*, 1963, **24**, 419.
[12] H. Mori, *Progr. Theor. Phys.*, 1965, **33**, 423; Here we follow the formulation by W. Götze and K. H. Michel, in *Dynamical Properties of Solids*, ed. G. K. Horton and A. A. Maradudin (North-Holland, Amsterdam, 1974).
[13] H. Mori, *Progr. Theor. Phys.*, 1965, **34**, 399.
[14] Y. Yamada, Y. Noda, J. D. Axe and G. Shirane, *Phys. Rev.*, 1974, **B9**, 4429.
[15] S. Haussühl, *Solid State Comm.*, 1973, **13**, 147.

Structural Phase Transitions in Malononitrile

By Martin T. Dove and Alastair I. M. Rae

Department of Physics, University of Birmingham,
P.O. Box 363, Birmingham B15 2TT

Received 3rd *December*, 1979

Previous studies using nuclear quadrupole resonance and calorimetric methods have shown that there are two second-order phase transitions in malononitrile at 141 and 294.7 K and that the low- and high-temperature phases have the same symmetry. In this paper the crystal structure of the intermediate phase is reported and compared with a previous determination of the high-temperature phase. The unit cell dimensions of the two structures are almost identical and the space groups are $P\bar{1}$ and $P2_1/n$, respectively. The structures are related by relative translations and rotations of the molecules and this motion is identified with the eigenvector of the soft mode associated with the transition. The basis of a theory for such a system is developed using the quasi-harmonic approximation.

Malononitrile, $CH_2(CN)_2$, exists in at least four phases in the solid state. Previous investigations using calorimetric[1] and nuclear quadrupole resonance[2] (n.q.r.) techniques have demonstrated the existence of continuous (second-order) transitions at 141 and 294.7 K and a very slow, probably first-order, transition at 260 K. The n.q.r. data are exemplified in fig. 1 from which it is seen that in the low-temperature

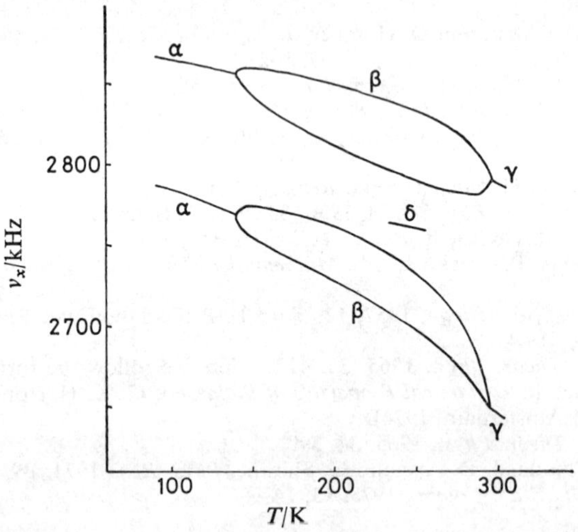

FIG. 1.—Frequency of one of the ^{14}N nuclear quadrupole lines (ν_x) as a function of temperature (T) in malononitrile as measured in ref. (2). The splittings in the different phases (α, β, γ, δ) relate to the number of distinguishable nitrogen sites in the crystal.

(α) phase and the high-temperature (γ) phase there are two distinct nitrogen sites in the crystal while in the intermediate (β) phase there are four. The similarity of the n.q.r. spectra in the α and γ phases and the fact that the frequencies change con-

TABLE 1.—CELL DIMENSIONS OF THE β AND γ PHASES OF MALONONITRILE

	β phase (present work)	γ phase [from ref. (4)]
a/nm	0.782(1)	0.784
b/nm	0.763(1)	0.763
c/nm	0.615(1)	0.618
α/°	89.68(5)	90
β/°	96.8(1)	96.2
α/°	90.23(5)	90

Estimated errors of measurement are given in brackets. No corrections have been made for systematic error.

tinuously with temperature between these two phases strongly imply that the α and γ phases are crystallographically identical.

Heat capacity measurements[1] on malononitrile show a very small anomaly at the α–β transition and no detectable anomaly at the β–γ transition although the accuracy of the latter measurement is adversely affected by the proximity of the melting point (304.9 K). These measurements also demonstrated the existence of a fourth (δ) phase: when a sample of malononitrile which had been held at a temperature around 250 K for several days was heated, a large specific heat anomaly was observed at 260 K indicating a first-order transition at this point. N.q.r. measurements confirmed the existence of the δ phase and it was found that the frequencies from all the nitrogen atoms were identical, indicating that the symmetry of the δ phase is higher than that of any of the other three. The δ phase is clearly the equilibrium phase below 260 K, the α and β phases being metastable in this temperature region; however, the time constant for a transition into the δ phase is very long (although the reverse δ to β change is fast) so that the α and β phases are both accessible for experimental investigation.

Previous investigations of the crystal structure of malononitrile have resulted in a

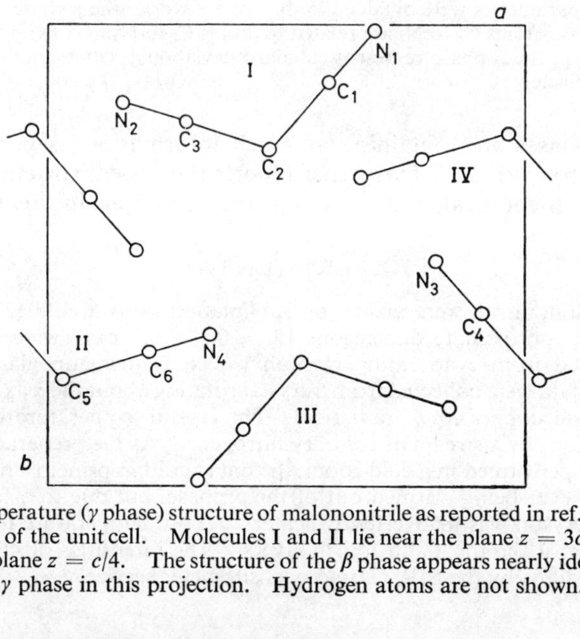

FIG. 2.—Room temperature (γ phase) structure of malononitrile as reported in ref. (4) shown in projection down the c axis of the unit cell. Molecules I and II lie near the plane $z = 3c/4$ and molecules III and IV lie near the plane $z = c/4$. The structure of the β phase appears nearly identical to that of the γ phase in this projection. Hydrogen atoms are not shown.

determination of the complete structure of the γ phase at room temperature[3,4] and a report of the unit-cell dimensions and space group of the β phase.[3] The γ phase has been shown to be monoclinic with space group $P2_1/n$ and cell dimensions as listed in table 1. The crystal structure is illustrated in fig. 2 and the atomic coordinates are listed in table 2. The space group of the β phase was reported to be $P2_1$,[3] but this does not agree with the results reported below.

The present investigation has the aim of determining the crystal structures of all

TABLE 2.—COORDINATES OF THE CARBON AND NITROGEN ATOMS IN THE α AND γ PHASES OF MALONONITRILE

		β phase	γ phase			β phase	γ phase
	x/a	0.595(4)	0.590		x/a	−0.099(4)	−0.090
C_1	y/b	0.120(4)	0.124	C_4	y/b	0.630(4)	0.624
	z/c	0.737(4)	0.726		z/c	0.793(4)	0.774
	x/a	0.464(4)	0.468		x/a	0.029(4)	0.032
C_2	y/b	0.274(4)	0.269	C_5	y/b	0.775(4)	0.769
	z/c	0.700(5)	0.691		z/c	0.816(4)	0.809
	x/a	0.283(3)	0.292		x/a	0.195(5)	0.208
C_3	y/b	0.212(3)	0.210	C_6	y/b	0.719(4)	0.710
	z/c	0.706(4)	0.720		z/c	0.773(5)	0.780
	x/a	0.681(4)	0.687		x/a	−0.196(4)	−0.187
N_1	y/b	0.009(5)	0.012	N_3	y/b	0.516(4)	0.512
	z/c	0.766(5)	0.746		z/c	0.773(4)	0.754
	x/a	0.139(3)	0.157		x/a	0.327(4)	0.343
N_2	y/b	0.178(4)	0.170	N_4	y/b	0.674(4)	0.670
	z/c	0.706(4)	0.733		z/c	0.751(4)	0.767

Notes: The β phase parameters were obtained in the present work; the γ phase parameters are from ref. (4). In the γ phase atoms C_4 to N_4 are related to atoms C_1 to N_2 by a two-fold screw axis. The numbers in brackets in the β phase represent standard deviations; corresponding quantities in the γ phase were not published.

the other three phases and obtaining as much information as possible concerning the nature of the transitions. This paper reports the crystal structure of the β phase and a preliminary theoretical study of the phase transitions in this unusual system.

EXPERIMENTAL

Crystals of malononitrile were grown by sublimation onto a cold (273 K) surface and suitable crystals of approximate dimensions $1.0 \times 0.3 \times 0.3$ mm^3 were mounted on glass fibres and prevented from evaporating by thin walled Lindemann glass capillary tubes. Zero to sixth layer data were collected from a crystal rotating about the a axis using a Weissenberg goniometer and copper $K\alpha$ X-radiation. The crystal temperature was maintained at 273 ± 1 K throughout by a stream of cold dry nitrogen.[5] As the preparation and mounting of the crystals were performed in a cold room, the data could in principle have been collected without the crystal ever being warmed out of the β phase, but due to a failure of the cooling apparatus, the crystal was inadvertently heated to room temperature for a short time between the recording of the third and fourth layers. The intensities of 468 reflections were measured from the films by the S.R.C. microdensitometer service.

RESULTS

Inspection of the films showed that the cell dimensions of the β phase are very nearly equal to those of the α phase, but that the condition $h + l = 2n$ in the $(h0l)$ zone no longer holds; on the other hand reflections with k odd along the $(0k0)$ reciprocal axis are absent, apparently implying that the space group is $P2_1$ as previously reported.[3] Careful examination of the intensity data, however, showed that the intensities of the hkl and $h\bar{k}l$ reflections were not in general equal. It follows that the symmetry of the crystal must be triclinic rather than monoclinic and the absence of the $0k0$ reflections with k odd must therefore be accidental rather than systematic. This possibility had also been noted by earlier workers.[6] The space group was taken to be $P\bar{1}$ as this is the only triclinic space group which can be generated from $P2_1/n$ by a second-order change obeying the Landau rules[7] and as the n.q.r. measurements clearly indicate that the number of symmetry elements in the β phase should be half that in the γ phase.

The unit-cell dimensions measured from the Weissenberg films are shown in table 1 where it is seen that they are indeed very near to the room temperature values: in particular, the α and γ angles are extremely close to the monoclinic values of 90°. This close relationship between the two unit cells raises the possibility that the triclinic crystals will be twinned, as crystallites with their b axes in opposite directions generate the same monoclinic cell when heated above the transition. The X-ray reflection from the hkl planes of one twin will be at almost exactly the same place as that from the $h\bar{k}l$ planes of the other, so the observed intensities will in general be a linear combination of the true intensities, corresponding to these two points in reciprocal space. The presence of twinning was confirmed by a careful inspection of the films: a few spots, mostly at very high Bragg angles, were resolved into two components and the values of the α and γ cell angles given in table 1 were in fact obtained from measurements of these splittings. Methods for analysing the structures of such pseudo-merohedrally twinned crystals have been discussed by Britton[8] and Murray-Rust.[9] Using the statistical methods described in these papers the twinning fraction was estimated to be 0.18; no significant difference was detected between the value of this parameter appropriate to the layers $h = 0$ to 3 and that for the layers $h = 4$ to 6 despite the fact that the crystal had inadvertently been heated into the γ phase between the recording of these two data subsets. The true intensities were then recovered from the measured intensities of pairs corresponding to the hkl and $h\bar{k}l$ planes. Unfortunately, this was not possible for all the reflections as the importance of these pairs had not been realised when the data were collected, and as a result only 298 independent intensities were included in the final data set.

Structure factors derived from these intensities were compared with those calculated on the basis of the atomic coordinates of the high temperature phase and an overall temperature factor exponent of 0.02 nm^2. The R-factor (defined as $\Sigma||F_O|-|F_C||/\Sigma|F_O)|$ was 35%. Ten cycles of least squares refinement were performed in which the coordinates and individual temperature factors of the carbon and nitrogen atoms were refined along with a set of layer scales. These were followed by three cycles of refinement of the anisotropic temperature factors of the non-hydrogen atoms; the hydrogen atoms were included at positions derived from the molecular stereochemistry with isotropic temperature factors the same as those of the central carbon atoms, but these parameters were not varied in the refinement. The R-factor was now 18%. The final atomic coordinates are listed in table 2 and the intramolecular bond lengths and angles are given in table 3.

TABLE 3.—INTRAMOLECULAR BOND LENGTHS AND ANGLES IN MALONONITRILE AS MEASURED IN THE β AND γ PHASES

atoms	β phase	γ phase
C_1—C_2	0.156(4) nm	0.148 nm
C_2—C_3	0.149(4)	0.146
C_1—N_1	0.108(5)	0.115
C_3—N_2	0.116(4)	0.110
C_4—C_5	0.149(4)	0.148
C_5—C_6	0.142(5)	0.146
C_4—N_3	0.115(4)	0.115
C_6—N_4	0.111(5)	0.110
C_1—C_2—C_3	111(2)°	111°
N_1—C_1—C_2	177(3)	177
N_2—C_3—C_2	175(3)	180
C_4—C_5—C_6	112(2)	111
N_3—C_4—C_5	179(3)	177
N_4—C_6—C_5	176(3)	180

The β phase quantities were obtained in the present work; the γ phase quantities are from ref. (4). Standard deviations (where known) are in brackets.

DISCUSSION

The final R factor of 18% is relatively high and this is attributed to inaccuracies in the intensity data, particularly those associated with the crystal twinning; improvements could probably have been made if the twinning fraction had been refined as a parameter, but routines to do this were not readily available. Nevertheless, the final atomic positions are believed to be reliable and free from systematic error. This is confirmed by the good agreement between the measured values of the intramolecular bond lengths and angles and those determined in the γ phase (*cf.* table 3) which in turn agree well with standard values. The final values of the temperature factors, on the other hand, have such large standard deviations that they are of only qualitative significance and are not reported in this paper.

The relationship between the β and γ phases is illustrated in fig. 3 where two molecules, which are related by a screw axis in the γ phase, are shown in projection down the b axis. The transition to the β phase is accomplished by translating the centres of

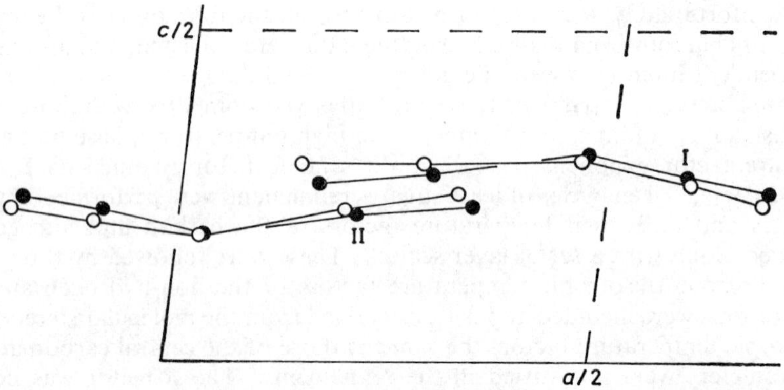

FIG. 3.—Structure of the β (open circles) and γ (filled circles) phases of malononitrile shown in projection down the b axis. Hydrogen atoms are not shown.

mass of each molecule 0.007 nm in the x direction and rotating the molecules in opposite senses by 3.5° about axes parallel to the b axes which pass through these centres. The other two molecules in the unit cell undergo similar changes so that the centre of symmetry is retained. Such displacements leave the y coordinates of all the atoms unchanged which explains why the $(0k0)$ reflections with k odd are apparently systematically absent and why the α and γ unit cell angles remain so close to 90° in the β phase. All the significant differences between the structures at room temperature and 273 K are contained in the above symmetry-breaking motions; if the atomic coordinates of one molecule in the β phase are transformed by the two-fold screw axis of $P2_1/n$ and then averaged with the corresponding coordinates of the other, the result is identical to the high-temperature structure within experimental error.

The β–γ transition in malononitrile is therefore a structural transition driven by a soft mode whose eigenvector can be characterised by the molecular motions described in the previous paragraph. This mode is clearly a zone-centre optic mode and its symmetry is that of the B_g irreducible representation of the point group C_{2h} which is the point group associated with the monoclinic space group $P2_1/n$. Although the malononitrile molecule is polar, the space groups of both phases are centrosymmetric, implying that neither is ferroelectric.

Although the crystal structure of the α phase has yet to be determined, it is almost certainly the same as that of the γ phase. This statement is based on the fact that the n.q.r. measurements show the nitrogen atom environments in the α and γ phase to be very similar and the α–β and β–γ transitions to be both continuous, which proves that the α and γ phases have the same symmetry. If the α-phase structure were to be significantly different from that of the γ phase, while maintaining the symmetry, then differences between the γ-phase structure and the average structure of the β phase would be expected, but these were not observed (see above). Examples of substances which change from a high-symmetry, low-temperature phase to an intermediate low-symmetry phase and back again at high temperature are rare. Rochelle salt[10–12] is probably the best known example of such a system, but in this case the transitions have been shown to be of the order–disorder type. Malononitrile appears to be the only case where such a sequence of *structural* phase transitions has been observed. The basis of a theory to describe such a system has been developed and will now be described.

The form of the temperature dependence of the n.q.r. frequencies close to the β–γ transition [2] indicates that fluctuation effects are negligible and that the system should therefore be capable of being described by mean-field theory. Landau[7] has described such a theory which provides a phenomenological explanation of the properties of many phase transitions: the free energy (F) is expressed as a power series in the order parameter (P), *i.e.*

$$F = AP^2 + CP^4. \tag{1}$$

If A is positive, there is only one minimum in F and this is at the point $P = 0$, while if A is negative, the equivalent configuration corresponds to $P = \pm(-A/2C)^{\frac{1}{2}}$. A second-order phase change is therefore produced if A is temperature dependent and changes sign at the transition temperature (T_C). That is

$$A = A'(T - T_C) \tag{2}$$

where A' is a constant or slowly varying function of temperature. To extend Landau theory to the present case we replace eqn (2) by the expression

$$A = A''(T - T_1)(T - T_2) \tag{3}$$

where $T_2 > T_1$. A is now positive if $T < T_1$ or $> T_2$ and negative for T between T_1 and T_2. We thus have

$$\left.\begin{aligned} P &= 0, T \leqslant T_1 \text{ or } T \geqslant T_2 \\ P &= [-A''(T - T_1)(T - T_2)/(2C]^{\frac{1}{2}}, T_1 \leqslant T \leqslant T_2 \end{aligned}\right\}. \qquad (4)$$

In the case of malononitrile, the order parameter, P, is the amplitude of the " frozen in " part of the soft mode in the β phase, and this has been measured directly at only one temperature. However, the n.q.r. frequency splittings are expected to be proportional to P, at least to first order, and these have been measured over a wide range of temperature.[2] Fig. 4 shows the average splitting as a function of temperature

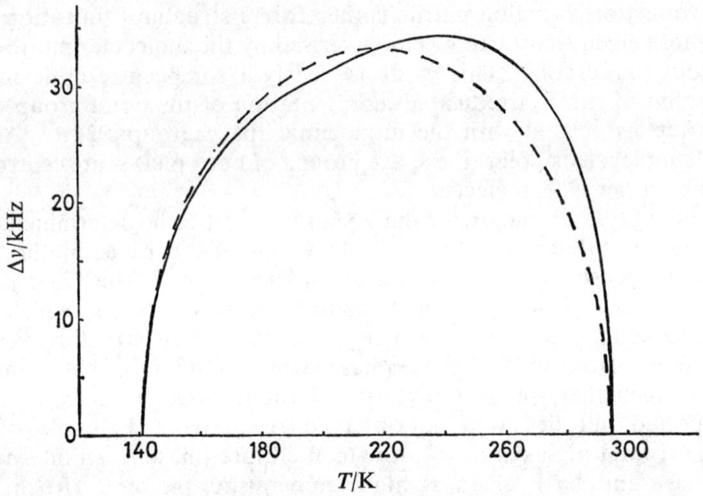

FIG. 4.—Average of the n.q.r. frequency splittings ($\Delta\nu$) in malononitrile as a function of temperature (T) (continuous line) along with a quadratic curve predicted by the version of Landau theory described in the text (broken line).

along with a curve of the form of eqn (3). The agreement is reasonably good, the discrepancies being about what would be expected from the neglect of higher-order effects such as the weak temperature dependence of A'' and C.

Landau theory is, however, a phenomenological theory and does not provide a microscopic explanation for a phase transition. In the case of a structural phase change from a low-temperature, low-symmetry structure to a high-temperature, high-symmetry phase, a more fundamental explanation can be given in terms of anharmonic interactions between phonons.[13] The temperature-independent term in eqn (2) corresponds to the harmonic contribution to the energy of the soft mode giving

$$A'T_C = -\tfrac{1}{2}\omega_0{}^2(\boldsymbol{q}, \lambda) \qquad (5)$$

where $\omega_0(\boldsymbol{q}, \lambda)$ is the harmonic frequency of the soft mode (whose wavevector is $\boldsymbol{q}$ and polarisation λ) in the high-symmetry phase and is imaginary because this mode is unstable at low temperatures. The term proportional to T in eqn (2) arises from the anharmonicity and if the quasi-harmonic approximation is applied,[13] it gives

$$A' = k_B \sum_{\boldsymbol{k}, \mu} g^{(4)}_{\lambda\mu}(\boldsymbol{q}, \boldsymbol{k})/\omega_0^2(\mu, \boldsymbol{k}) \qquad (6)^*$$

* In this paper, Boltzmann's Constant is printed k_B, to avoid confusion with wavevector $\boldsymbol{k}$.

where $g^{(4)}_{\lambda\mu}(\boldsymbol{q}, \boldsymbol{k})$ is the fourth-order anharmonic coupling constant between the soft mode and another phonon of wavevector $\boldsymbol{k}$ and polarisation μ.

To extend this theory to malononitrile, comparison of eqn (2) and (3) shows that three changes from the above are required: the temperature independent term must be positive, the term proportional to T must be negative, and there must be an additional positive term proportional to T^2. To fulfil the first condition, the only requirement is that the soft mode be stable at low temperatures and its harmonic energy consequently positive. For the term proportional to T to be negative, the right-hand side of eqn (6) must be negative which can be achieved provided some or all of the fourth-order anharmonic coupling constants are negative. There would seem to be no physical reason why this should not be the case, and it seems surprising that structural phase transitions where the low-temperature phase has a higher symmetry than the high-temperature phase are not more common. Finally the required term proportional to T^2 will result from a consideration of higher-order anharmonic interactions. If we apply the quasi-harmonic approximation in the usual way,[13] but extended to include sixth order as well as fourth order terms, a contribution to A which is proportional to T^2 is obtained which corresponds to an expression for A'' in eqn (3) as

$$A'' = k_B{}^2 \sum_{\substack{\boldsymbol{k}, \boldsymbol{k}' \\ \mu, \mu'}} g^{(6)}_{\lambda\mu\mu'}(\boldsymbol{q}, \boldsymbol{k}, \boldsymbol{k}')/[\omega_0^2(\mu, \boldsymbol{k})\omega_0^2(\mu', \boldsymbol{k}')] \tag{7}$$

where $g^{(6)}_{\lambda\mu\mu'}(\boldsymbol{q}, \boldsymbol{k}, \boldsymbol{k}')$ is the sixth-order anharmonic coupling constant connecting the soft mode with the two phonons whose wavevectors and polarisations are $(\boldsymbol{k}, \mu)$ and $(\boldsymbol{k}', \mu')$.

The basis of a microscopic theory for a series of phase transitions such as those observed in malononitrile has therefore been obtained and it can be concluded that the occurrence of such a pattern of phase transitions is evidence for the importance of sixth-order three-phonon interactions in this case.

CONCLUSION

The crystal structure of the β phase of malononitrile has been successfully obtained from X-ray diffraction measurements. The unit cell is almost identical to that of the γ phase, but the symmetry has been lowered from monoclinic $P2_1/n$ to triclinic $P\bar{1}$. The eigenvector of the soft mode associated with the transitions consists mainly of opposite rotations of the two molecules, which are related by the screw axis in the high-symmetry structure, about axes parallel to the b axis of the unit cell. On the assumption that the α phase is identical to the γ phase, this system of phase transitions can be understood on the basis of the quasi-harmonic approximation, provided the contribution from sixth-order anharmonic three-phonon interactions is included.

Further work planned on malononitrile includes an extension of the measurements on the β phase to improve the accuracy of the analysis and to study the temperature dependence of the order parameter, and the determination of the crystal structures of the α and δ phases. It is hoped to measure the changes in other physical quantities associated with the transition and to develop the theoretical analysis further by the calculation of some relevant quantities on the basis of an intermolecular force model.

Technical assistance by Mr. R. Pflaumer is gratefully acknowledged as is help from the staffs of the S.R.C. microdensitometer service and the computer centre of the

University of Birmingham. Financial support was received from the S.R.C. for M.T.D.

[1] H. L. Girdhar, E. F. Westrum Jr and C. A. Wulff, *J. Chem. Eng. Data*, 1968, **13**, 239.
[2] A. Zussman and S. Alexander, *J. Chem. Phys.*, 1968, **49**, 3792.
[3] N. Nakamura, S. Tanisaki and K. Obatake, *Phys. Letters*, 1971, **34A**, 372.
[4] K. Obatake and S. Tanisaki, *Phys. Letters*, 1973, **44A**, 341.
[5] R. A. Young, *J. Sci. Instr.*, 1966, **43**, 449.
[6] S. Tanisaki, personal communication, 1978.
[7] L. D. Landau and E. M. Lifschitz, *Statistical Physics* (Pergamon, Oxford, 1969).
[8] D. Britton, *Acta Cryst*, 1972, *A***28**, 296.
[9] P. Murray-Rust, *Acta Cryst*, 1973, *B***29**, 2559.
[10] B. C. Frazer, M. McKeown and R. Pepinsky, *Phys. Rev.*, 1954, **94**, 1435.
[11] T. Mitsui, *Phys. Rev.*, 1958, **111**, 1259.
[12] B. Žekš, G. C. Shukla and R. Blinc, *Phys. Rev*, 1971, *B*3, 2306.
[13] R. Blinc and B. Žekš, *Soft Modes in Ferroelectrics and Antiferroelectrics* (North Holland, Amsterdam, 1974).

Theoretical Studies of Phase Transitions in Tetracyanoethylene

BY ROBERT W. MUNN

Department of Chemistry,
University of Manchester Institute of Science and Technology,
Manchester M60 1QD

AND TADEUSZ LUTY *

Institute of Organic and Physical Chemistry,
Technical University, 50-370 Wrocław, Poland

Received 15*th November*, 1979

The equilibrium structure and harmonic lattice dynamics of monoclinic tetracyanoethylene under imposed symmetry-preserving strains are calculated from a 6-exp atom–atom potential function. Some lattice frequencies become imaginary for extensions and shears above ≈10%. Calculations of the eigenvalues of the dielectric function for the strained structures show the absence of dielectric singularities, but a dielectric mechanism for the observed transitions is not completely excluded.

1. INTRODUCTION

Theories of molecular crystals seek to explain the crystal properties in terms of the properties of the molecules and their arrangement in the crystal structure. To explain the temperature and pressure dependence of the crystal properties, the corresponding dependence of the crystal structure must be known. For the unit cell parameters, such information is fairly readily obtainable, but for the arrangement of molecules within the unit cell it is not. One may be able to deduce changes of molecular arrangement from crystal properties [1] or from the unit cell parameters through empirical rules,[2] but in general another source of structural information is needed. If an intermolecular potential is available, the molecular arrangement can be determined from the condition that it must minimize an appropriate energy function for given lattice parameters.[3]

In this paper, we first report calculations of the equilibrium structure of the tetracyanoethylene (TCNE) crystal under imposed external symmetry-preserving strains. An improved 6-exp potential function derived from lattice-dynamical calculations [4] is used, with the molecules treated as rigid. TCNE is of interest because under atmospheric pressure it forms monoclinic crystals above 292 K[5] and cubic crystals below.[6] The monoclinic phase is readily supercooled, and appears to transform to another monoclinic phase around 280 K.[7]

Secondly we report calculations of the lattice dynamics in the harmonic approximation for the strained structures. For sufficiently large strains, certain frequencies become imaginary, indicating a possible lattice instability.

TCNE acts as the acceptor in many donor–acceptor complexes, and is highly

* Work performed while on leave of absence at Institute of Theoretical Chemistry, University of Nijmegen, The Netherlands.

polarizable. We have shown elsewhere[8] that for crystals composed of molecules with a sufficiently high and anisotropic polarizability, the inverse dielectric function can become singular for some wavevector. This causes the polarization energy to diverge, leading to static and dynamic instabilities. In the third part of this paper we therefore report calculations of the dielectric function for the strained structures to investigate whether there are any singularities which can be related to the instabilities given by the empirical potential.

Finally we discuss our results, their inter-relation and possible extensions. Our treatment concentrates on dynamic instabilities and so is better suited to elucidating the smooth transition to the other monoclinic phase of TCNE than that to the cubic phase. We do not consider thermodynamic stability.

2. STRUCTURE OF STRAINED CRYSTAL

In a homogeneous deformation, a point $\boldsymbol{x}$ in the reference configuration is displaced by $\boldsymbol{u}$ to $\boldsymbol{x}'$. The strain can be described by the symmetric tensor with cartesian components

$$l_{\alpha\beta} = \tfrac{1}{2}(\partial u_\alpha/\partial x_\beta + \partial u_\beta/\partial x_\alpha), \tag{2.1}$$

although for finite strains this may include a contribution from rotation.[9] We take the reference configuration as the unstrained monoclinic structure with lattice parameters a, b, c and β. Under symmetry-preserving strains, the lattice parameters become a', b', c' and β' and the strain components are[10]

$$l_{11} = a'/a - 1, \tag{2.2}$$
$$l_{22} = b'/b - 1, \tag{2.3}$$
$$l_{33} = c' \sin\beta'/c \sin\beta - 1, \tag{2.4}$$
$$l_{13} = \tfrac{1}{2}(c' \cos\beta'/c \sin\beta - a' \cos\beta/a \sin\beta). \tag{2.5}$$

In the monoclinic TCNE structure there are two molecules in the unit cell at centres of symmetry,[5] so that under symmetry-preserving strains the molecular arrangement can be specified by three Euler angles, φ, ψ and θ.

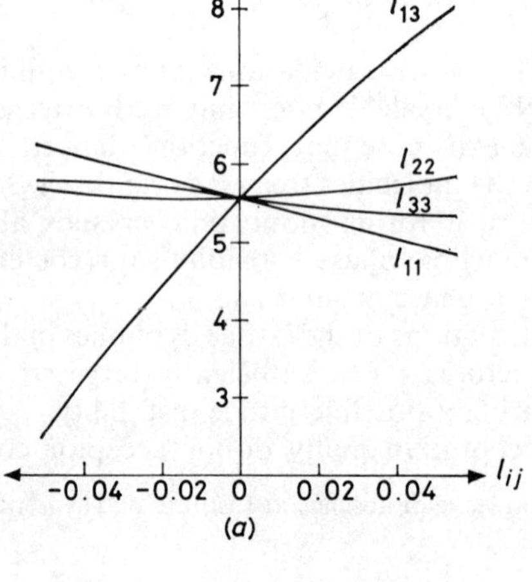

(a)

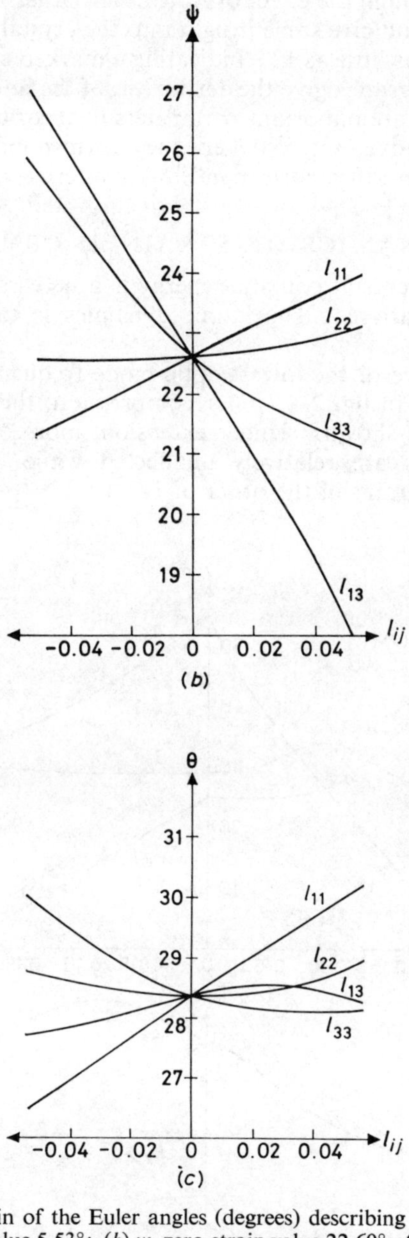

FIG. 1.—Variation with strain of the Euler angles (degrees) describing the molecular orientation in TCNE: (*a*) φ, zero-strain value 5.53°; (*b*) ψ, zero-strain value 22.60°; (*c*) θ, zero-strain value 28.36°.

The crystal potential energy is calculated from an improved parameter set for the 6-exp atom–atom potential function for crystals containing cyano groups.[4] Deformed lattice parameters corresponding to a single non-zero strain component are imposed, and the potential energy is then minimised with respect to changes in the three Euler angles. The results are shown in fig. 1.

The diagonal strain components represent compression for negative values and expansion for positive values. As can be seen, the orientations change very differently with different strains, though the effect of l_{22} is never large. These differences reflect the molecular packing, and give some insight into the crystal potential. Some of the graphs curve for strains as little as 1%, indicating a marked curvature in the potential. The initial slopes of the graphs give the derivatives of the internal strains with respect to external strain; these are important parameters in theories of elasto-optic properties,[2] generalized thermodynamics[13] (where they correspond to the bridging matrix $R_{k\lambda}$) and elasticity[11] (where they correspond to the internal strain tensor A^{λ}_{ij}).

3. DYNAMICS OF STRAINED CRYSTAL

Minimisation of the crystal potential energy is a necessary standard first step in lattice dynamical calculations. The lattice dynamics is calculated as for the unstrained structure.[4]

The strain dependence of the lowest optic mode frequencies at the centre of the Brillouin zone is shown in fig. 2. Under compression, the frequencies all increase as expected and are not shown. Under extension, more complicated behaviour is observed. Some modes are relatively unaffected while others soften markedly, becoming unstable for strains of the order of 12%.

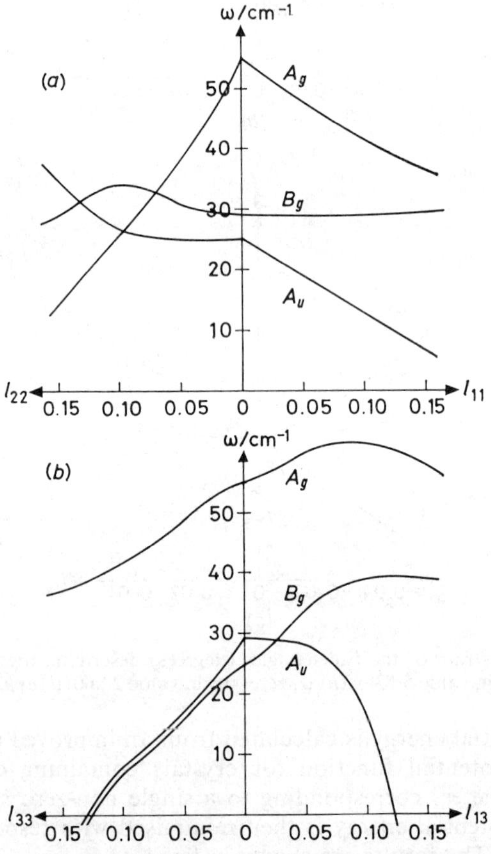

FIG. 2.—Variation with strain of the lowest zone-centre optic mode frequencies in TCNE. (*a*) Strains l_{22} (left-hand side) and l_{11} (right-hand side); (*b*) strains l_{33} (left-hand side) and l_{13} (right-hand side). The modes are labelled by their symmetry species.

The dispersion curves for modes propagating along high-symmetry directions have also been calculated. The most strongly strain-dependent dispersion curves are shown in fig. 3, again only for positive strains but this time including acoustic modes.

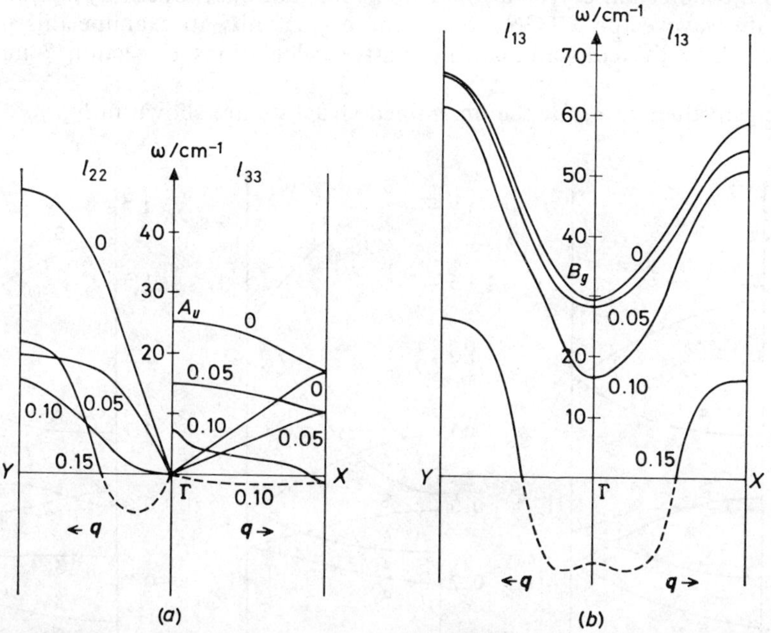

FIG. 3.—Variation with strain of selected dispersion curves along high-symmetry directions in TCNE. The point Γ is the Brillouin zone centre (0, 0, 0), X is (0.5, 0, 0) and Y is (0, 0.5, 0). (*a*) Strains l_{22} (left-hand side) and l_{33} (right-hand side); (*b*) strains l_{13}. The value of the strain is marked on each curve. Broken lines indicate unstable regions corresponding to imaginary frequencies.

Various instabilities are seen: with respect to longitudinal acoustic modes in the Y direction for the strain l_{22}; with respect to transverse acoustic and translational optic modes in the X direction for the strain l_{33}; and with respect to librational optic modes at the zone centre and in the X and Y directions for the strain l_{13}.

4. DIELECTRIC FUNCTION OF STRAINED CRYSTAL

The dielectric function relates the electric field applied to a crystal to the local field it produces. In a molecular crystal containing Z molecules labelled k in the primitive unit cell, it is convenient to describe the applied and local fields by their Fourier transforms $\boldsymbol{e}_q^0(k)$ and $\boldsymbol{f}_q(k)$, where $\boldsymbol{q}$ is the wavevector. The molecules are viewed as being polarized by the local fields, which are given by[8]

$$\boldsymbol{f}_q(k) = \sum_{k'} \boldsymbol{\epsilon}_q^{-1}(kk') \cdot \boldsymbol{e}_q^0(k'). \tag{4.1}$$

Here the dielectric function $\boldsymbol{\epsilon}_q$ is a $3Z \times 3Z$ matrix with elements given by

$$\boldsymbol{\epsilon}_q(kk') = \mathbf{1}\delta_{kk'} - \mathbf{t}_q(kk') \cdot \boldsymbol{\beta}(k'), \tag{4.2}$$

where $\mathbf{t}_q(kk')$ is a modulated lattice dipole sum and $\boldsymbol{\beta}(k')$ is a dimensionless reduced polarizability.

If the molecular polarizability is sufficiently large and anisotropic, an eigenvalue of ϵ_q may become zero so that ϵ_q^{-1} is singular. The inverse dielectric function determines the crystal polarization energy[8,12] and through it a mechanism of electron–phonon coupling. A singularity in ϵ_q^{-1} thus implies an instability of the electronic system of the molecular crystal against dipole fluctuations induced by a phonon of the appropriate wavevector. TCNE offers the opportunity to examine this prediction in a case where the conventional alternative calculations in section 3 indicate an instability.

The eigenvalues of ϵ_q for the unstrained structure are shown in fig. 4. They are

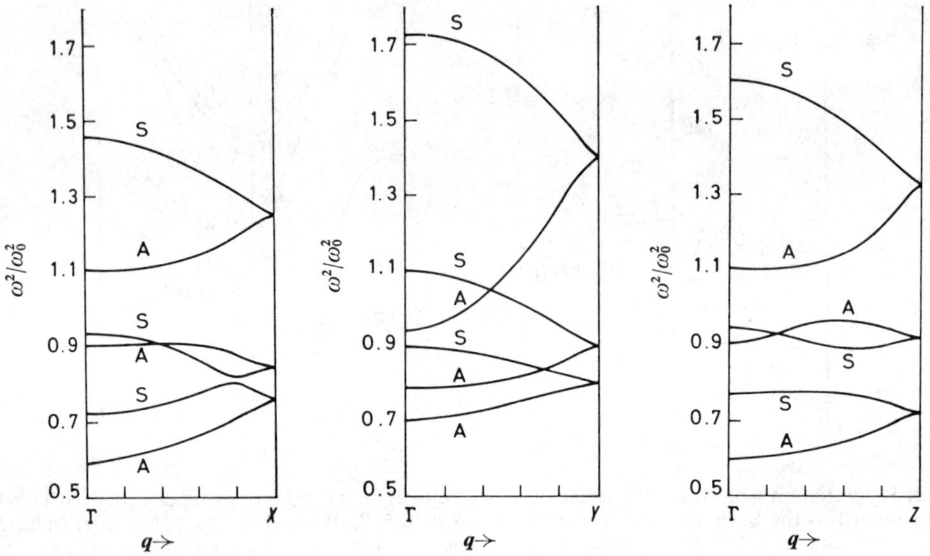

FIG. 4.—Eigenvalues of the dielectric function in TCNE as a function of wavevector at zero strain. Symmetric and antisymmetric branches are labelled S and A. Points Γ, *X* and *Y* are as in fig. 3, and *Z* is (0, 0, 0.5).

expressed in the usual way as the square of a frequency scaled relative to the corresponding frequency for the free molecule. The curves are labelled as symmetric or antisymmetric as for phonon dispersion curves.[13] Some of the eingenvalues depend quite markedly on $\boldsymbol{q}$, but none falls below ≈ 0.6.

The lowest pair of symmetric and antisymmetric eigenvalues of ϵ_q for positive and negative strains (l_{11} and l_{22}, as an illustration) are shown in fig. 5. For the diagonal strain components, compression lowers the lowest eigenvalue and extension raises it. Thus extension tends to stabilize the crystal against dielectric singularities, and although compression tends to destabilize the crystal the eigenvalues remain above 0.5. This weak tendency towards dielectric instability on compression contrasts with the dynamic instability on extension.

5. DISCUSSION

The results presented here fall into two distinct groups which must be compared cautiously.

The results in Sections 2 and 3 explore the detailed consequences of the 6-ex-

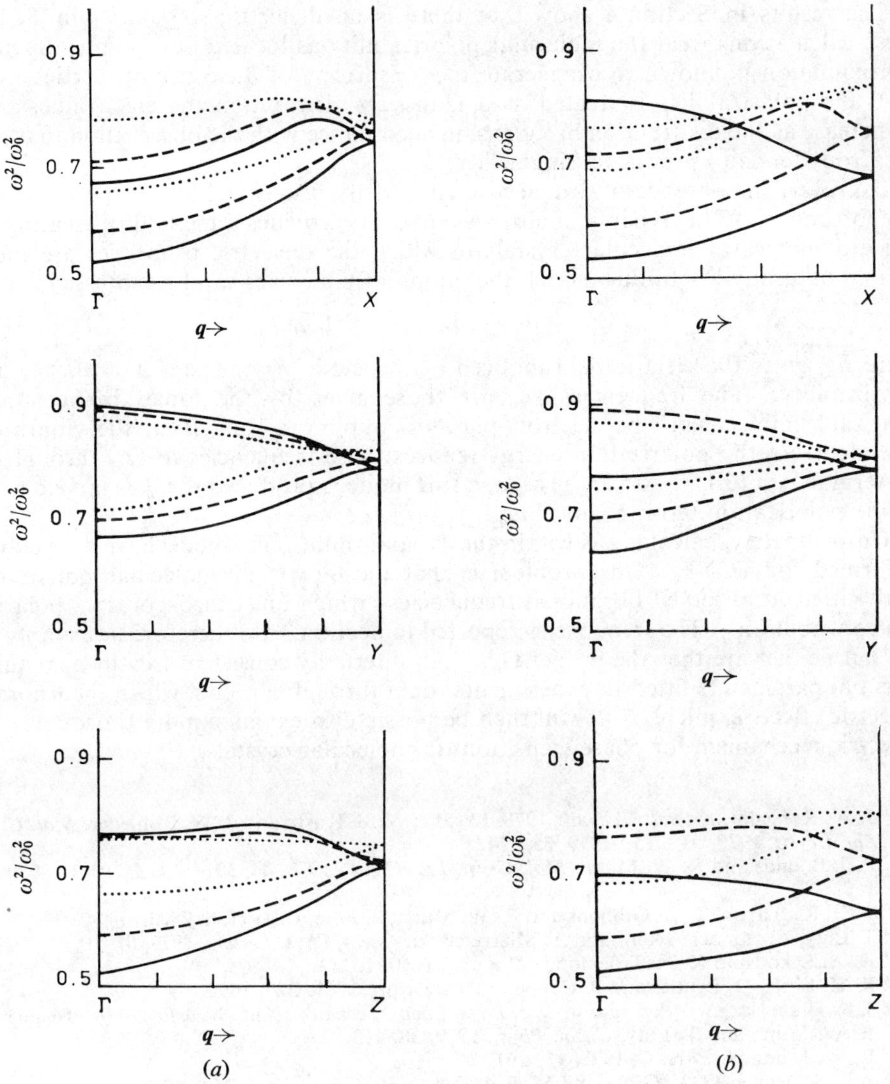

FIG. 5.—Variation with strain of the lowest pair of eigenvalues of the dielectric function in TCNE. (*a*) Strains l_{11}; (*b*) strains l_{22}. The solid lines denote a strain of −0.10, the broken lines 0, and the dotted lines +0.15.

potential function and parameters for molecules containing cyano groups. The structural changes under strain are of use in theoretical treatments of the various experiments carried out on TCNE [see references in ref. (4)]. The dynamical instabilities under strain imply a phase transition like that predicted[14] and observed[15] in HCN. However, the HCN calculations did take explicit account of electrostatic interactions not treated in section 3. The predicted instabilities in TCNE should therefore be regarded with some scepticism. In any case, it is unlikely that the required extension could be applied mechanically in order to seek the instabilities, although the thermal expansion is markedly and increasingly anisotropic above room temperature.[7]

The results in Section 4 show that there is no dielectric singularity in TCNE. These calculations treat the molecular polarizability as located at a point. Such an approximation is known to exaggerate the anisotropy of dielectric properties[16] and when the polarizability is treated as a Gaussian distribution the eigenvalues of $\boldsymbol{\epsilon}_q$ do increase as the distribution broadens, in accordance with the observation that high anisotropy tends to produce a singularity.

However, the absence of a dielectric singularity does not exclude the possibility of a soft mode. The dynamic multipoles model[17] provides a method of treating the lattice dynamics of molecular crystals in which the dielectric properties are incorporated naturally.[18] In this model, the phonon frequencies can be expressed as

$$\Omega_{qj}^2 = \omega_{qj}^2 - \boldsymbol{E}_{qj} \cdot \boldsymbol{\alpha} \cdot \boldsymbol{\epsilon}^{-1} \cdot \boldsymbol{E}_{qj}^T. \quad (5.1)$$

where $\boldsymbol{E}_{qj}$ gives the electric field induced by the mode $\boldsymbol{q}j$ and $\boldsymbol{\alpha}(kk') = \boldsymbol{\alpha}(k)\delta_{kk'}$ is a polarizability. The frequencies ω_{qj} are those given by the forces between non-polarizable molecules; the electron–phonon coupling arising from the vibrational dependence of the polarization energy reduces these frequencies to Ω_{qj} through the last term in eqn (5.1). As can be seen, a soft mode is produced if $\boldsymbol{\epsilon}_q^{-1}$ is large enough for the polarization term to cancel ω_{qj}^2.

Unfortunately, calculations within the dynamic multipoles model have not yet been performed for TCNE. One problem is that the 6-exp intermolecular potential is partly fitted to observed librational frequencies[4] which must incorporate a polarization contribution. The frequencies reported in Section 3 are therefore not simply the ω_{qj}, but neither are they the proper Ω_{qj}. An internally consistent treatment requires potential parameters fitted to experimental data through a model which incorporates dielectric effects explicitly. It will then be possible to examine more thoroughly the dielectric mechanism for phase transitions in molecular crystals.

[1] J. W. Rohleder, *Kristall. Technik*, 1978, **13**, 517; W. J. Kusto and J. W. Rohleder, *Mol. Cryst. Liq. Cryst.*, 1979, **51**, 215; 1979, **55**, 151.

[2] P. J. Bounds and R. W. Munn, *Mol. Cryst. Liq. Cryst.*, 1978, **44**, 301; *Chem. Phys.*, 1979, **39**, 165.

[3] T. H. K. Barron, T. G. Gibbons and R. W. Munn, *J. Phys. C*, 1971, **4**, 2805.

[4] T. Luty, A. van der Avoird and A. Mierzejewski, *Chem. Phys. Letters*, 1979, **61**, 10.

[5] D. A. Bekoe and K. N. Trueblood, *Z. Krist.*, 1960, **1**, 113.

[6] R. G. Little, D. Pautler and P. Coppens, *Acta Cryst.*, 1971, B**27**, 1493.

[7] J. Świątkiewicz, *Sci. Papers Inst. Org. Phys. Chem. Technical Univ. Wrocław*, 1978, **16**, 305.

[8] R. W. Munn and T. Luty, *Chem. Phys.*, 1979, **38**, 413.

[9] R. W. Munn, *J. Phys. C*, 1978, **11**, L61.

[10] J. L. Schlenker, G. V. Gibbs and M. B. Boisen, *Acta Cryst.*, 1978, **34**A, 52.

[11] C. S. G. Cousins, *J. Phys. C*, 1978, **11**, 4867.

[12] P. J. Bounds and R. W. Munn, *Chem. Phys.*, 1979, 44, 103.

[13] G. S. Pawley, *Acta Cryst*, 1974, *A***30**, 585.

[14] A. I. M. Rae, *J. Phys. C*, 1972, **5**, 3309.

[15] O. W. Dietrich, G. A. Mackenzie and G. S. Pawley, *J. Phys. C*, 1975, **8**, L98; G. A. Mackenzie and G. S. Pawley, *J. Phys. C*, 1979, **12**, 2717.

[16] P. J. Bounds and R. W. Munn, *Chem. Phys.*, 1977, **24**, 343; J. H. Meyling, P. J. Bounds and R. W. Munn, *Chem. Phys. Letters*, 1977, **51**, 234.

[17] T. Luty, *J. Chem. Phys.*, 1977, **66**, 1231.

[18] T. Luty and R. W. Munn, *Chem. Phys.*, 1979, **43**, 295.

Some Novel Symmetry Aspects of Phase Transitions

BY L. L. BOYLE, J. R. WALKER† AND ANNE C. WANJIE‡

University Chemical Laboratory,
Canterbury, Kent

Received 11*th January*, 1980

Two aspects of phase transitions in the solid state are discussed, namely the structure of the symmetry group of a plastic crystal and the prediction of plausible pathways for first-order phase transitions. It is explained how group extension theory may be used for the former and how space group representation theory is an essential adjunct of crystallographic considerations in the latter.

THE SYMMETRY OF PLASTIC CRYSTALS

It is the authors' contention that the description of the symmetry properties of the plastic phases of crystalline solids may well be improved by the use of symmetry groups which are rather more complicated in structure than the familiar Federov space groups. The case for a new group-theoretical description depends on how one regards the relative motions of the molecules in the plastic phase. If one accepts the conclusion of Finbak[1] that in the high-temperature cubic phases of CBr_4, C_2Cl_6 and camphor, the rotation of the molecules is a co-operative phenomenon since the sphere of rotation of any one molecule penetrates those of its neighbours, one can describe the rotational motion of the molecules as a lattice vibration of a non-plastic phase. In molecules such as tin tetraethyl, however, two or three polymorphs can co-exist,[2] and a new model should at least be able to cater for the case where different polymorphic forms can be interconverted by the rotational motion. To achieve this one has to get away from the idea of treating the lattice vibration as spanning an irreducible representation of a space group, which is really only valid for small amplitudes of vibration, and to find a description in which the reorientational motion is an integral part of the description. The symmetry elements of a group corresponding to such a description would not only contain the translations (which interrelate the co-ordinate axes of different unit cells) and the point group operations (which interrelate equivalent settings, relative to the lattice vectors, of the co-ordinate axes of a given unit cell) but also the molecular rotations (which interrelate equivalent settings of the molecular axes relative to the co-ordinate axes in a given unit cell.) By not requiring the molecular rotations to conserve the periodicity of the original lattice one can include the orientationally-disordered structures as well as those highly ordered structures which actually correspond to non-plastic solid phases. Thus the enlarged symmetry group can have more than one classical space group as a subgroup: further, librational and translational motions can be included by use of the irreducible representations of this group.

The problem now arises as to how one might construct such a group: extensive details may be found in Dr. Walker's thesis,[3] but an attempt will be made here to

† Present address: Chemistry Laboratory, University College, 20 Gordon Street, London W.C.1.
‡ *née* Philpot.

summarize the structure of the solution. The mathematical technique needed is group extension theory, a formal technique developed by Schur[4] in 1907 and especially by Schreier in 1924[5] and 1926.[6] (An account for chemists has recently been published by Boyle and Walker.)[7] Group extension theory uses two ingredients: an invariant (or normal) subgroup and a factor group. The invariant subgroup is a genuine subgroup of the extension and in physical terms we might regard it as a subgroup which is unaffected by all the permutations due to the other elements of the group. Thus, the point group D_{2h} is an invariant subgroup of O_h because, as a group, it is unaltered by permutations caused by other elements of O_h such as C_3. In rigorous mathematical terms, an invariant subgroup is invariant to all the inner automorphisms of the group, *i.e.*, if we choose to permute the elements N_i of the invariant subgroup by conjugation with an operation G_j of the group, then, whichever element G_j is chosen, the set of permuted elements $\{G_j^{-1}N_iG_j\}$ is in 1 : 1 correspondence with the set $\{N_i\}$. In other words the conjugation just changes the order in which we write the elements down.

The factor group is a little more difficult to explain because it is not a subgroup of the original group, although it may be isomorphic to one. Its elements are cosets of the original group and their multiplication law simply states that the product of two representative elements is an element representative of the product coset. A familiar example may help: a space group is an extension of the lattice (*i.e.*, translation group) by the point group. The lattice is the invariant subgroup because, as a group, it is unchanged by the point operations. The point group is not a genuine subgroup, because the space group elements act on the lattice as well as on the co-ordinate axes while the point group operations act on co-ordinate axes alone. The coset of space group operations corresponding to a point group operation will contain all those space group operations having the same rotational part, but differing in translational element. Such cosets are therefore infinite, although the point group itself is finite. The multiplication law of the cosets defines a group which is isomorphic to the point group. In mathematical terms, one says that the point group is the quotient of the space group with the translational group and is hence called the factor group or quotient group. Some familiar examples of group extensions are listed in table 1.

TABLE 1.—SOME EXAMPLES OF GROUP EXTENSIONS

invariant subgroup	isomorph of factor group	group extension
lattice	point group	space group
C_1'[a]	point group	double group
torsional symmetry	frame symmetry	non-rigid molecule symmetry group
reorientations	space group	plastic crystal
non-magnetic	time-reversal	magnetic symmetry
Schur multiplicator	point group	representation group

[a] C_1' is the group consisting of the identity and R, the element which reverses the signs of one-electron spin functions.

It should be noted that since the factor group is actually a group of cosets, the point group, where quoted in the preceding description, is only isomorphic to the factor group. In cases where the groups concerned all act on the same basis functions, *e.g.*, the resolution of point groups as direct or semi-direct products,[8,9] this distinction becomes unimportant but in solid-state problems it is essential to bear it in mind.

The final stage in setting up the mathematical machinery to establish these group extensions is to decide exactly what permutations of the invariant subgroup are induced by the various elements of the factor group. This may not be easy for various reasons. On the mathematical side, one must be sure that the " extension " really is a group and know whether it is or is not isomorphic to any other extensions obtained by this process. On the physical side, one must know what the physical consequences are of any proposed extension, *i.e.*, is the specification of the action of the generators of the factor group on the generators of the invariant subgroup in accordance with experiment and geometric considerations? In the case of the plastic-crystal problem this is not a simple question to answer. Recognizing the rotations of the molecules as generating the invariant subgroup and the space group (or something closely related to it) as generating the factor group, the problem boils down to asking us to specify in advance how the rotations are correlated in the " unit cell ", however big this may be, of the plastic crystal. Once one has agreed on this the group can be produced by computation.

In such a problem it is always useful to know what are the possible mathematically meaningful ways of permuting the elements of a group. These ways form a group called the outer automorphism group of the group being permuted. Dr. Walker's thesis contains both the elegant mathematical and computational derivations of the outer automorphism groups of all point groups as well as an indication of the possible automorphisms of the molecular rotational groups required in the plastic-crystal problem.

SYMMETRY PROPERTIES OF FIRST-ORDER PHASE TRANSITIONS

To contribute a new approach to the elucidation of the mechanism of first-order phase transitions necessarily involves various assumptions which of their essence will rule out its application to numerous observed systems, at least in so far as they are presently understood. Following the definition of Ehrenfest,[10] second-order phase transitions are those involving continuous changes of thermodynamic variables with temperature while first-order phase transitions are those involving discontinuity. From the group-theoretical point of view, second-order phase transitions have been successfully interpreted as involving a reduction in space-group symmetry as one cools the crystal through the transition point. Second-order transitions usually involve a *zellengleich* reduction in symmetry, *i.e.*, the translational elements are unaffected by the change and the volume of the primitive unit cells remains constant. By contrast, we have been able to propose mechanisms for numerous first-order phase transitions on the basis of a multi-stage process which involves an intermediate space group which is a subgroup (or, occasionally, a supergroup) of the space groups of both high- and low-temperature phases. First-order phase transitions often involve changes in the volume of the primitive unit cell.

Our interest, then, has been to find feasible pathways which the atoms would follow in the transition from one structure to another. The existence of such a pathway is difficult to justify in the all-too-frequent case where there is no " crystallographic continuity ", *i.e.*, the crystal breaks up into a powder on being taken through the transition. It may well be that the kinetics of the transition are so sluggish that taking the crystal through the transition temperature at an experimentally reasonable speed sets up stresses within the crystal which disrupt its macroscopic form. Certainly the reproducibility of phase transitions argues in favour of the idea that the atoms know where they are going, at least within a given grain. To produce an absolutely random arrangement of the atoms at the point of transition is probably more difficult energetically than to allow them to follow rather more symmetrical pathways. Even in a theory which depends implicitly on the use of symmetry, the rôle of dislocations and defects in providing the " breathing space " which catalyses the transition is clearly recognized. Such features, however, only need to be taken into account when comparing the relative energetics of different plausible pathways.

One of the key requirements of the present theory for predicting plausible pathways is that the motion of the atoms is determined by lattice vibrations and that there is a continuity between the lattice vibrations in both phases and through the intermediate space groups postulated. This is essentially an extension of one of Landau's criteria[11] concerning which space-group representations were allowed to generate a given distorted structure. Its consequences, however, were more far-reaching than were expected as far as the actual prediction of plausible pathways was concerned. Rather than work out the symmetries of the lattice modes at all the possible different types of special points in the Brillouin zone of the high-symmetry phase and then examine the distortions to which they corresponded, it seemed a very much simpler task to use some extremely reliable tables of maximal subgroups of the space groups which had been produced in 1969 by Neubüser and Wondratschek,[12] and which will be incorporated in the forthcoming Volume A of the International Tables for Crystallography. Coupled with some other tables of Ascher[13] and Boyle and Lawrenson,[14,15] it is a relatively simple matter to find chains of space groups which link the two structures involved in a phase transition. However, the application of the extended Landau criterion mentioned above really highlights the need for these crystallographic tables to be supplemented by some indication of which space-group representation is capable of generating the distortion.

One of the simplest examples of this difficulty arises in the attempt to switch from a face-centred cubic (*Fm*3*m*, O_h^5, no. 225) structure to a simple cubic structure (*Pm*3*m*, O_h^1, no. 221). According to the Neubüser–Wondratschek tables, *Pm*3*m* is a maximal subgroup (of index 4) of *Fm*3*m*. One can draw this rather easily by colouring the atoms on the centres of the faces of a conventional unit cell a different colour from those at the apices of the cube. However, there is no space-group representation which would allow this process which we can draw so readily! To confirm this, one merely has to glance at the list of symmetries of the special points of the Brillouin zone of the face-centred cubic lattice which may be conveniently found in Bradley and Cracknell's book.[16] This list shows that at no point outside the centre of the Brillouin zone can a vibration produce a cubic structure; whilst at the centre, no change of lattice type is possible.

Another pitfall encountered in the use of maximal subgroups is that one may well insert intermediate space groups into a pathway which are such that they do not correspond to the instantaneous positions of the atoms at any given moment. An example of this is the cubic (*Pm*3*m*, O_h^1, no. 221) to tetragonal (*I*4/*mcm*, D_{4h}^{18}, no. 140) phase transition[17] which is found in various perovskites, *e.g.*, in $SrTiO_3$ at 110 K and in $KMnF_3$ at 184 K. Now D_{4h}^{18} happens to be a subgroup of O_h^1, but is of the *allgemein* or general type, and hence according to a theorem of Hermann,[18] is non-maximal and can be reached in stages from O_h^1 which are either purely *zellengleich* (cell-preserving) or *klassengleich* (crystal class-preserving). A possible route is therefore

$$Pm3m \longrightarrow P4/mmm \longrightarrow I4/mcm$$

i.e.

$$O_h^1 \longrightarrow D_{4h}^1 \longrightarrow D_{4h}^{18}$$

and the Wyckoff positions of the atoms correlate between the different phases.[19] However, the perovskite is incapable of undergoing a *zellengleich* distortion to the primitive tetragonal structure since the atoms involved cannot generate an E_g-type lattice mode at the centre of the Brillouin zone. Indeed, since all of the atoms are on centrosymmetric sites at room temperature all of the zero wave-vector vibrations are anticentrosymmetric and hence would lead to polar structures. It is, of course, possible for those simple cubic, O_h^1, crystals having atoms occupying Wyckoff's *e, f, g, h, i, j, k, l* or *m* sites[20,21] to undergo an E_g-distortion and produce a D_{4h}^1-structure.

The mode which is thought[22] to be responsible for the phase transition in the perovskites is a triply-degenerate mode at the R-point of the Brillouin zone. On descent to D_{4h}^{18} symmetry, this yields a totally symmetric Γ-point mode in accordance with the Landau theory. It is interesting to note that it correlates with A-point modes in D_{4h}^1 which in turn correlate with the same Γ-point modes in D_{4h}^{18}: this is, however, no indication that the D_{4h}^1 step has any real existence.

Once a set of lattice modes capable of producing the transition has been identified, taking full account of these representation-theoretic restrictions, one must decide whether there are alternative plausible routes and, if so, how one should choose between them. In principle some spectroscopic or other criterion (*e.g.*, measurement of certain components of a physical property tensor) should be suggested which is capable of diagnosing the intermediate steps in a phase transition. In practice, there is remarkably little choice of routes for many substances: it would therefore lend added credibility to a proposed pathway if the crystal could be pumped with radiation of the soft-mode frequency near the transition temperature to see if the crystal could be taken through the transition (and perhaps back) with less physical damage. This would lend support to the damage being caused by the simultaneous thermal excitation of other lattice modes, rather than by the soft mode, and hence strengthen the case for considering preferred pathways.

The authors thank the S.R.C. for the award of a research studentship to J. R. W. for his work on plastic crystals and the University of Kent Graduate Studies Support Fund for the award of a research studentship to A. C. W. for her work on phase transitions.

[1] C. Finbak, *Arch. Math. Naturvidenskab*, 1938, (*B*), **42**, 71.
[2] L. A. K. Staveley, J. B. Warren, H. P. Paget and D. J. Dowrick, *J. Chem. Soc.*, 1954, 1992.
[3] J. R. Walker, *Automorphisms and Extensions in Chemical Physics, Ph.D. Thesis* (University of Kent at Canterbury, 1977), chap. 5.
[4] I. Schur, *J. reine angew. Math.*, 1907, **132**, 85.
[5] O. Schreier, *Abh. math. Sem. Hamburg. Univ.*, 1924, **3**.
[6] O. Schreier, *Monatsh. Math. Phys.*, 1926, **34**, 165.
[7] L. L. Boyle and J. R. Walker, *Int. J. Quantum Chem.*, 1977, **12**, 157.
[8] S. L. Altmann, *Phil. Trans.*, *A*, 1963, **255**, 216.
[9] S. L. Altmann, *Rev. Mod. Phys.*, 1963, **35**, 641.
[10] P. Ehrenfest, *Proc. Acad. Sci.* (*Amsterdam*), 1933, **36**, 153.
[11] L. D. Landau, *Phys. Z. Sowjetunion*, 1937, **11**, 26 and 545.
[12] J. Neubüser and H. Wondratschek, *Maximal Subgroups of the Space-groups* (privately published, Kiel and Karlsruhe, 2nd edn, 1969).
[13] E. Ascher, *Lattices of Equi-translation Subgroups of the Space Groups* (Internal report, Battelle Institute Advanced Studies Centre, Geneva, 2nd edn, 1968).
[14] L. L. Boyle and J. E. Lawrenson, *Acta Cryst*, 1972, **28***A*, 485.
[15] L. L. Boyle and J. E. Lawrenson, *Acta Cryst*, 1972, **28***A*, 489.
[16] C. J. Bradley and A. P. Cracknell, *The Mathematical Theory of Symmetry in Solids* (Clarendon Press, Oxford, 1972), p. 118.
[17] H. Unoki and T. Sakudo, *J. Phys. Soc. Japan*, 1967, **23**, 546.
[18] C. Hermann, *Z. Krist.*, 1929, **69**, 533.
[19] J. E. Lawrenson, *Theoretical Studies in Crystallography, Ph.D. Thesis* (University of Kent at Canterbury, 1973).
[20] L. L. Boyle, *Acta Cryst*, 1972, **28***A*, 172.
[21] L. L. Boyle, *Spectrochimica Acta* 1972, **28***A*, 1355.
[22] P. A. Fleury, J. F. Scott and J. M. Worlock, *Phys. Rev. Letters*, 1968, **21**, 16.

GENERAL DISCUSSION

Prof. G. M. Schneider (*University of Bochum*) said: The phase diagram of deuterated *p*-terphenyl (fig. 4 of the paper of Cailleau *et al.*) is remarkable in two respects:

(1) The $T(p)$ transition curve has a negative slope; that would mean that ΔV is negative (since ΔH is positive; see page 11). This is an unusual pressure dependence since normally $T(p)$ transition curves with positive slopes are found in similar cases.

(2) No break in the transition line is found at the triple point. The behaviour shown is unlikely in the context of phase theory.

Mr. H. Cailleau (*University of Rennes*) said: Concerning point (1), the negative slope of the $T(p)$ transition curve appears as a confirmation of the proposed transition mechanism in polyphenyls (competing intramolecular and intermolecular forces). This mechanism is different from the orientational ordering of numerous rigid molecules where $T(p)$ transition curves with positive slopes are found.

With respect to point (2), within the precision of our measurements we cannot say that no break in the transition line is found at the triple point. Above the triple point the shift of the transition temperature with pressure is clearly not linear, but below this, the shift is closely linear.

Prof. R. M. Pick (*Université Pierre et Marie Curie, Paris*) said: In an order–disorder phase transition, the ordering temperature is not related to the height of the energy barrier separating the various minima, but to the difference of interaction energy between molecules when they are in different relative positions. The value of T_c in *para*-terphenyl is thus entirely related to intermolecular interactions. Furthermore, the low-temperature structure may result from a balance of the interactions between different pairs of molecules; the effect of pressure on the strength of these various interactions may differ from one to another. It is thus not clear that the sign of the change in the ordering temperature with pressure may be easily predictable.

Prof. A. J. Leadbetter (*University of Exeter*) said: Can Mr. Cailleau tell us something about the phase transition in quaterphenyl?

Mr. H. Cailleau (*University of Rennes*) said: In *p*-quaterphenyl, superlattice reflections appear at the $C(\frac{1}{2}, \frac{1}{2}, 0)$ reciprocal point of the high-temperature phase as in *p*-terphenyl, and the two low-temperature structures are similar. The average torsion angles are 17.1° between an external and an internal ring and 22.7° between two internal rings. These torsion angles, like the thermal motion in the high-temperature phase, lead us to suppose that the phase transition is order–disorder as in *p*-terphenyl, although no inelastic neutron scattering investigations have been performed.

Dr. G. S. Pawley (*University of Edinburgh*) said: We have recently discovered what we believe to be similar incommensurate behaviour in *p*-dibromotetrafluorobenzene. The space group is the same as that for biphenyl and superlattice reflections are observed at points associated with a wave-vector $(0, \eta, 0)$ for $\eta \simeq 0.45$.

Does Mr. Cailleau think that in biphenyl the incommensurate phase transition is in any way related to the easy torsion about the central C—C bond?

Mr. H. Cailleau (*University of Rennes*) said: The possibility of incommensurate phases is related to the occurrence of a minimum in the soft-phonon branch away from a high-symmetry point in the Brillouin zone. In biphenyl, this is realized with the existence of a Lifshitz invariant at the B $(0, \frac{1}{2}, 0)$ zone boundary point. The easy torsion about the central C—C bond is only related to the mechanism of the soft mode and numerous different mechanisms have been observed in numerous organic and inorganic compounds.

Dr. N. Boden (*University of Leeds*) said: Ewen *et al.* state in their paper that " phase transitions can generally be regarded as reactions of the crystal lattice to the onset of specific types of molecular motion ". The results presented for n-$C_{33}H_{68}$ certainly seem to substantiate this assertion. However, I wonder whether this is really true in general?

My reason for asking this question is based upon some recent n.m.r. measurements of ours (measurements made by Mr. Gene Peterson) on the neighbouring homologue n-$C_{32}H_{66}$. The different relative orientations of the terminal CH_3 groups of the two molecules impart different crystal structures and sequences of phase transitions on the two solids. Thus, n-$C_{32}H_{66}$ exhibits a monoclinic structure at low temperatures and undergoes a transition to a rotor phase at 339 K before finally melting at 343 K. Yet measurements of the contribution of the proton second moment M_2 from the intra-CH_2 interaction (fig. 1) reveal four distinct temperature intervals I-IV for motion,

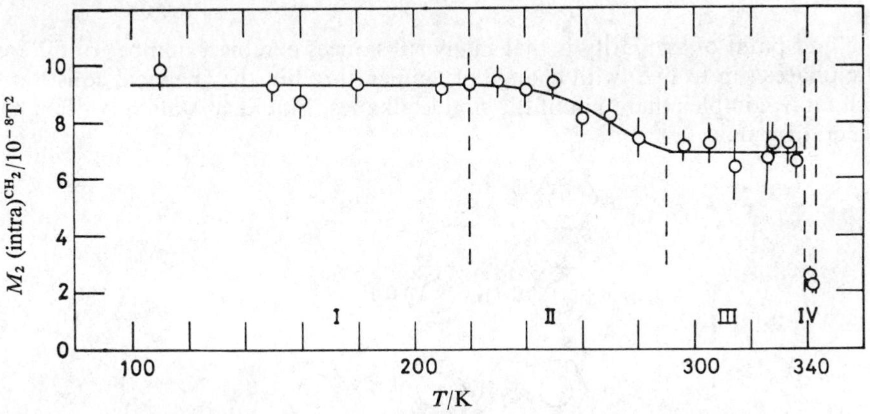

FIG 1.—Contribution to the proton second moment M_2 from the intra-CH_2 interaction.

with I-III corresponding to the monoclinic phase. In interval I, M_2(intra)^{CH_2} changes very little with temperature, whilst the total M_2 (not shown) exhibits a more marked decrease due to a reduction in the extramolecular interaction caused by lattice expansion. The value of M_2(intra)^{CH_2} implies 25° angular oscillations. In the interval II, M_2(intra)^{CH_2} decreases more rapidly with temperature due to more extensive intramolecular oscillations. The behaviour in interval III is more exciting. M_2(intra)^{CH_2} is constant whilst M_2(extra)^{CH_2} decreases smoothly up to the transition to the rotor phase. This requires 180° reorientational jumps of the molecules about their long axes and is the " α-process " region as described by Olf and Peterlin[1]. Here, the onset of the α-process is not accompanied by a phase transition as for n-$C_{33}H_{68}$.

Our measurements suggest that the fraction of molecules undergoing such jumps increases linearly from 0 at 290 K to 0.5 at 339 K, the temperature of transition to the rotor phase. Thus, the α-process in n-$C_{32}H_{66}$ seems to involve a translational defect mechanism; the defect is thermally generated and its concentration increases strongly with temperature. Interval IV is the rotor phase: two types of molecules are again identified, one undergoing reorientational motion similar to the molecules in the rotor phase of n-$C_{19}H_{40}$ and the other reorientational and translational diffusion.

To conclude, it appears that n-$C_{32}H_{66}$ exhibits similar though not exactly corresponding motional processes as Ewen *et al.* have so elegantly demonstrated for n-$C_{33}H_{68}$, but without undergoing a parallel sequence of phase transitions. It is, of course, possible, though unlikely, that the difference in the behaviour of these two molecular solids could originate from impurities in the former case.

[1] H. G. Olf and A. Peterlin, *J. Polymer Sci.*, 1970, **8**, 791.

Prof. G. Chapuis (*University of Lausanne*) said: In their model Ewen *et al.* assume that phase B is accompanied by a pure 180° rotational jump. From geometrical considerations, one sees also that a 90° rotational jump leaves the cell dimensions and symmetry invariant. Such a motion has been observed in, *e.g.*, $(C_{10}H_{21}NH_3)_2CdCl_4$ (paper by Blinc *et al.*, this Discussion). Is there any evidence suggesting that a 90° rotational jump can be excluded from the model?

Prof. A. J. Leadbetter (*University of Exeter*) said: I wish to draw attention to similarities and differences between the behaviour of the long-chain paraffin systems and that of smectic liquid crystals. It seems to me that it might be fruitful to pursue this comparison more deeply and I would be interested to have comments from the authors.

The first point of similarity is that many substances exhibit a number of different smectic phases (up to five) with change of temperature but the chemical constitution is much more complex than that of the simple alkanes. Liquid crystals very often have the general formula,

e.g.,

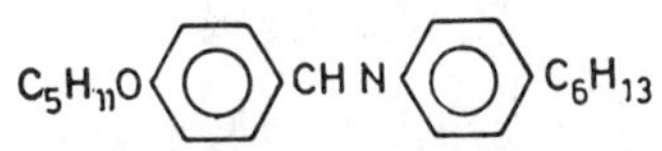

It seems inevitable that some differences in detailed behaviour must arise from the loss of the simple chain structure of the alkanes. However, a major difference is in the crystallinity: the transitions A → B → C → D all have lower entropy than D → Liquid whereas for the smectic phases the first transition from ordered crystal (equivalent to the A phase) to the lowest-temperature (most ordered) smectic phase generally has a greater entropy than that of all the other transitions combined. Nevertheless, the more ordered smectic phases (below) are now generally considered to be true 3-dimensional crystals, although the diffraction spots, while sharp, are confined to relatively low scattering vectors and there is very strong diffuse scattering: both indicate considerable disorder. For example, the reciprocal lattice of the ordered smectic phases generally shows *hk*1 or *hk*0 reflections only for *h* or $k \lesssim 2$ and not usually more than 5 orders of the 00*l* layer reflections. How does this compare with diffraction patterns of phases B-D of the alkanes?

A brief summary of the main smectic phases is given in table 1. The molecular long axes may be tilted or orthogonal to the smectic layer planes and the symmetry is given of the molecular packing in planes perpendicular to the molecular long axes and of the lattice.

TABLE 1.—SUMMARY OF THE MAIN SMECTIC TYPES

orthogonal name	molecular packing	lattice type	tilted (tilt angle name typically *ca.* 25°)
SmA	liquid		
	liquid		SmC
?	hexagonal (short range order)	hexagonal	SmF
SmB	hexagonal	hexagonal	—
—	hexagonal	monoclinic	SmG
SmE	chevron	orthorhombic	—
—	chevron	monoclinic	SmH

In addition to the structural similarities there are remarkable similarities in the dynamic behaviour, except for the special effects arising from chain kinking. The observations with a particular technique depend on the relation of the time-scale of the motion to that of the experiment but, as for the alkanes, the molecular dynamics in the smectic phases are dominated by reorientation about the long axis together with a longitudinal motion of this axis. In the most ordered smectic phases E and H the orientational motion on a time-scale of *ca.* 10^{-10} s or less is restricted with, respectively, bipolar and polar distribution functions. This motion becomes less restricted in the smectic B and H phases, tending toward a random 6-fold distribution, which tends to a uniform distribution in smectic A and C phases. In all phases there exists a restricted longitudinal motion of the molecules which in the E and H phases is of amplitude *ca.* 1 Å (on a time-scale of $\lesssim 10^{-10}$ s) and not strongly coupled to the reorientations. In the B (G) and A (C) phases the amplitude increases to $\lesssim$ 2 Å and the strength of the coupling also increases. Except that the amplitude of the longitudinal motion is not related to the $>CH_2$ repeat unit as for the alkanes, the existence of this motion seems to be a characteristic feature of all the disordered phases of long rod-like molecules.

Dr. A. Hüller (*KFA Jülich*) said: Prof. Leadbetter says that he cannot explain his EISF data with just one type of motion of the molecules. Even if one has only one type of motion, say a rotation around the threefold symmetry axis of the molecules, one may observe more than one correlation time. The rapidly rotating molecules will feel a much lower frictional force than the slowly rotating molecules.[1] This leads to long relaxation times for the rapidly rotating molecules and to short relaxation times for the slow molecules.[2] I want to demonstrate this by using a model contributed by Werner Press and myself. It consists of magnetic quadrupoles sitting on pins which allow a rotation of the magnets around a vertical axis. If I set one of them into rapid rotation the neighbours cannot respond to the high-frequency a.c. field which results. Therefore the rotation is almost undamped. The small friction one sees is due to the bearing. As the magnet gradually slows down its neighbours begin to respond. The energy is dissipated over the whole system, the rotating particle now loses its kinetic energy very quickly and the rotation comes to an end. Slowly rotating molecules feel much higher frictional forces.

[1] R. Gerling and A. Hüller, to be published.
[2] R. G. Gordon, *J. Chem. Phys.*, 1965, **43**, 1307.

Prof. A. J. Leadbetter (*University of Exeter*) said: I agree with Dr. Hüller's remarks to the effect that rotational motion about even a single axis might have complex dynamics and I congratulate him on his beautiful model demonstration. However, the EISF gives direct *geometric* information on the average proton trajectory in space, and this shows that the protons do more than simply move in a circle, irrespective of the dynamical details. In other words there are certainly at least two types of motion occurring and the overall dynamics (as revealed by the quasielastic scattering component) are indeed quite complex. We have proposed in the paper an interpretation in terms of a centre-of-mass relaxation triggered by simultaneous rotations of adjacent molecules and have modelled this assuming a superposition of local structures similar to that of the low-temperature modification, such as to give the correct high-temperature symmetry. This is the most novel aspect of our work and I wonder if anyone has any comments on this model: I know that Prof. Pick for example has thought about this kind of problem.

Dr. D. N. Batchelder (*Queen Mary College, London*) said: In our paper we have suggested that the asymmetry in the intensities of certain Friedel pairs, as illustrated in the Bijvoet ratio Δ, could be due to anomalous scattering. We have now used a simple model to estimate the maximum possible value of Δ for the (5,0,1)-($\bar{5}$,0,$\bar{1}$) pair which could result from anomalous scattering and found it to be an order of magnitude smaller than the experimental result. The model assumed that only the sulphur atoms contribute to the anomalous scattering and that the structure factor F_{hkl} can be separated into two parts, F^{s}_{hkl} due to sulphur atoms in the unit cell and F^{o}_{hkl} due to the rest. F^{o}_{hkl} is real while F^{s}_{hkl} has an imaginary component which, in a vector diagram, is anti-parallel to the imaginary component of $F^{s}_{\bar{h},\bar{k},\bar{l}}$. By adjusting the real and imaginary components in the vector diagram one can obtain a non-centrosymmetric " structure " which maximizes the difference between the structure factors of the (5,0,1) and ($\bar{5}$,0,$\bar{1}$) reflections.

Since anomalous scattering thus appears to be ruled out as the cause of the asymmetry of the intensities of Friedel pairs, we are now investigating the possibility that there is a variation of transition temperature throughout the sample. With an absorption coefficient of 3 mm^{-1} for Cu $K\alpha$ radiation, different parts of the crystal could contribute more to the intensity of one reflection than its reciprocal.

It is unlikely that the asymmetry is due to residual strain resulting from the solid-state polymerization in the crystal. Although the initial polymer which forms in the crystal is under strain, with the completion of polymerization all that strain should be relieved. The only residual strain should be that due to lattice defects present in the starting monomer crystal. Indeed these defects must be very important as the transition temperature appears to vary considerably from one crystal to another.

We accept that any order parameters obtained for this phase transition must be treated with caution. It is certainly not yet possible to say that the transition has no first-order component and, in addition, it is still far from clear what the behaviour of a " perfect " crystal would be.

Prof. A. J. Leadbetter (*University of Exeter*) said: Would Prof. Blinc expand a little on just what he means by the " melting of chains"?

Prof. R. Blinc (*University of Ljubljana*) said: In the low-temperature phase the chains are rigid, well ordered and parallel to a given preferred direction. In the high-temperature phase, on the other hand, a chain may have several kinks which can be created and annihilated and which can move up and down the chain. Each configuration of kinks (defined as a series of *gauche* configurations of C—C bonds between which an odd number of trans C—C bonds must be present) decreases the overall chain length by an integer multiple of 1.26 Å. Taking into account the increase in the c lattice constant by 3 Å and a reorientation of the chains from an initially tilted position with respect to the layer normal, to a position parallel to this normal one finds that the average number of $gt\bar{g}$ kinks per chain is $1.5/1.26 = 1.2$ in the high-temperature phase. The chains thus retain a significant amount of " nematic " order in the high-temperature phase above the " melting " transitions.

Prof. R. M. Pick (*Université Pierre et Marie Curie, Paris*) said: Other transitions take place with the same type of material $CdCl_4(C_nH_{2n+1}NH_3)_2$ for $n < 10$. Is it possible to relate, at least partly, the transitions with $n \approx 10$, with those with $n = 1$, 2 or 3?

Prof. R. Blinc (*University of Ljubljana*) said: In layer-structure perovskites there are two driving mechanisms which induce the transitions: the melting of the chains and the disordering of the keys, *i.e.* the NH_3 groups in the cavities formed by the $CdCl_6$ octahedra. In long-chain compounds the melting of the chains is dominant, whereas in compounds with $n = 1$, 2 or 3 it is the disordering of the keys and the transitions between the various NH—Cl hydrogen-bonding schemes which are responsible for the observed phase transitions.

Dr. A. Novak (*CNRS, Thiais, France*) said: The compound of the general formula $(NH_3(CH_2)_3NH_3)MnCl_4$ shows four different phases in the temperature range 10-350 K. The phase below 110 K is completely ordered while the three high-temperature phases are more or less disordered. Infrared and Raman spectra show discontinuities in the vicinity of T_c in frequency, intensity and half-width for several bands due to internal and lattice vibrations, NH_3 torsional motion in particular. The mechanism of the phase transitions is believed to be due mainly to rotational jumps of NH_3 groups which couple with the librational motions of the $(NH_3(CH_2)_3NH_3)^{2+}$ ion about its long axis when the temperature is raised.[1]

[1] C. Sourisseau and G. Lucazeau, *J. Raman Spectrosc.*, 1979, **8**, 311.

Dr. D. N. Batchelder (*Queen Mary College, London*) said: A number of diacetylenes also form lipid bilayers in a crystalline matrix which exhibit phase transitions. These materials have the general chemical formula $C_mH_{2m+1}—C{\equiv}C—C{\equiv}C—(CH_2)_n—COOH$ and they have been prepared as polycrystalline samples and, in a few cases, as multilayers using a Langmuir trough. Samples which contain *ca.* 1% diacetylene polymer show optical and Raman spectra which are very sensitive to phase transitions which are dominated by the lipid bilayer. The lithium salt of tricosa-10,12-diynoic acid is particularly convenient to study as there is a wide temperature range between the onset of the phase transition and the melting point. In this salt the optical and Raman spectra begin to shift and broaden above 50 °C indicating the onset of disorder. Provided the temperature is kept below 120 °C the changes are reversible. Research into these materials has been carried out by Prof. G. Wegner and Dr. B. Tieke at the University of Freiburg and Dr D. Bloor and Mr. C. Hubble at Queen Mary College.

Prof. R. M. J. Cotterill (*Technical University of Denmark, Lyngby*) said: I am currently carrying out a Monte Carlo study of the gel–fluid transition in a lipid monolayer, the preliminary results of which appear to endorse those obtained by Prof. Blinc. Each molecule in the computational model has two fully-saturated acyl chains, each having sixteen carbon and thirty-three hydrogen atoms. Rotation is possible about each carbon–carbon bond, the angular potential being taken to be the same as that for the central bond in n-butane. This has a barrier of 3.5 kcal mol^{-1} between the *trans* and *gauche* positions, and the *gauche* and *cis* angular energies are 0.6 and 5.8 kcal mol^{-1}, respectively, relative to the *trans* state. Intra- and inter-molecular potentials for all carbon–carbon, hydrogen–hydrogen, and carbon–hydrogen non-bonding interactions are assumed to have the Lennard-Jones form,

$$V(r) = \varepsilon\{(\sigma/r)^{12} - 2(\sigma/r)^{6}\},$$

and the individual parameters are $\sigma_{CC} = 0.51$ nm, $\sigma_{HH} = 0.36$ nm, $\sigma_{CH} = 0.44$ nm, $\varepsilon_{CC} = 0.120$ kcal mol^{-1}, $\varepsilon_{HH} = 0.123$ kcal mol^{-1} and $\varepsilon_{CH} = 0.107$ kcal mol^{-1}. The first carbon atoms of pairs of chains are at fixed separation distances, simulating the glycerol backbones of (dipalmitoyl) phospholipid molecules. Periodic boundaries are used and the irreducible cell contains 24 (= 6 × 4) molecules, *i.e.*, 48 paired acyl chains. A typical elevational view of the model, during the order–disorder process, at 40 °C, is given in fig. 2.

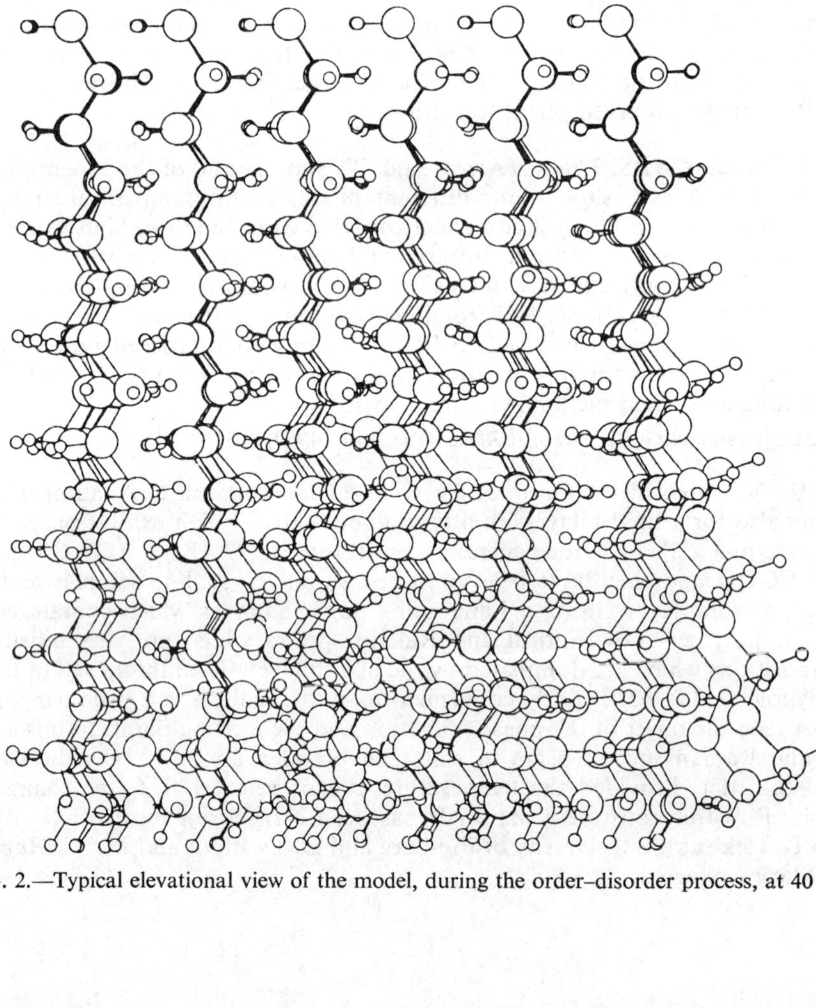

FIG. 2.—Typical elevational view of the model, during the order–disorder process, at 40 °C.

An interesting feature of such an intermediate stage of the disordering process is the fact that the disorder increases with increasing proximity to the CH_3 terminal ends (seen at the bottom of fig. 2). This is in agreement with what has been found by Prof. Blinc. A particularly significant difference between his system and my own lies in the constraints imposed on the other ends of the acyl chains: these ends are fixed in his case but relatively free in my own. In his system the disorder might lead to chain–chain interdigitation, while in mine the most important effect is probably the increase in free volume, which promotes the transition.

Dr. A. Dworkin (*Université Paris Sud, Orsay*) said: I wish to emphasize the conclusions of Dr. Hüller about using heat capacity measurements on molecular crystals in order to get evidence for second-order (continuous) phase transitions and consequently to determine their critical exponents.

First, I draw attention to the paramount importance of the resolution in temperature, by giving the example of $SnCl_2.2H_2O$, whose transition appears continuous when looked at with temperature increments of 0.15 K[1] but shows a discontinuity when studied at 0.02 K intervals by the same authors.[2] The very popular a.c. calorimetry works with such small temperature intervals,[3] but this method does not give absolute values and its results should always be suspected, because of a lack of thermal equilibrium.

Secondly the critical exponent α is obtained when plotting $\ln\Delta C_p$ against $\ln (T - T_c)/T_c$, which implies that a " normal " heat capacity has to be determined in order to get ΔC_p. This is done either by a phenomenological treatment, [3,4] by a comparison with a similar substance without phase transition,[1] or by an extrapolation of the results obtained at lower or higher temperatures. Therefore it is clear that the critical exponent depends sensitively on the choice of the base line, as was shown previously.[5] Many other criteria are nevertheless discussed by Kornblit and Ahlers on this subject.[4]

[1] T. Matsuo, M. Oguni, H. Suga and S. Seki, *Bull. Chem. Soc. Jpn*, 1974, **47**, 57.
[2] T. Matsuo, M. Tatsumi, H. Suga and S. Seki, *Solid State Comm.*, 1973, **13**, 1829.
[3] F. L. Lederman, M. B. Salamon and L. W. Shacklette, *Phys. Rev B*, 1974, **9**, 2981.
[4] A. Kornblit and G. Ahlers, *Phys. Rev. B*, 1973, **8**, 5163.
[5] H. Cailleau and A. Dworkin, *Mol. Cryst. Liq. Cryst.*, 1979, **50**, 217.

Dr. A. Hüller (*KFA Jülich*) said: I agree with your comment—one has to be very careful with published critical exponents on apparently second-order phase transitions. In many cases a close look reveals a small discontinuity. One of the best safeguards against a misinterpretation of experimental data is to see if the order parameter extrapolates to zero at the same temperature where the susceptibility extrapolates to infinity. If they do not, then the transition is discontinuous.

Prof. G. M. Schneider (*University of Bochum*) said: With respect to fig. 1 of Hüller's paper, from which the existence of a phase transition at negative pressures can be deduced, I would like to mention a similar phenomenon found for the phase separation in liquid binary mixtures in spite of the fact that there is no perfect analogy with respect to generalized theory. In some binary liquid mixtures, *e.g.*, 2-methylpyridine + H_2O, closed immiscibility loops have been found to exist at negative pressures that can be shifted to positive pressures by the substitution of H_2O by D_2O or by adding outsalting salts to the 2-methylpyridine + water mixture.[1]

[1] G. M. Schneider, in *Chemical Thermodynamics*, ed. M. L. McGlashan (Specialist Periodical Report, The Chemical Society, London, 1978), vol. 2, chap. 4.

Prof. J. S. Rowlinson (*University of Oxford*) said: Prof. Schneider has suggested an analogy between the pressure maximum in Stevenson's generalised phase diagram for the ammonium halides, which can lie at positive or negative pressures according to the composition, and the pressure extrema in liquid–liquid immiscibility curves which can also be changed from positive to negative pressures by changes of composition. There is, however, an important difference between the two phenomena; in his representation of the second, Schneider used mole fraction (a generalised density) as his independent variable and so it follows that the extrema are always (ordinary) critical points. In contrast, the extremum in Stevenson's diagram, as drawn in Hüller's paper, is merely a turning point on a line of first-order transitions—there is no critical point.

Dr. A. Hüller (*KFA Jülich*) said): I thank Prof. Rowlinson for his comment, but I would like to remark that one can really do something to NH_4Cl which has the same effect as applying negative pressures. If you replace some of the ammonium ions by Cs ions the lattice constant becomes larger (as with negative pressure) and the triple point is reached.[1]

[1] P. Brauer and A. Bußmann, *Z. Naturforsch. Teil A*, 1976, **31**, 213.

Prof. H. Suga (*Osaka University*) said: I also would like to make a comment on the generalized phase diagram of ammonium halide crystals reported by Dr. Stevenson that Dr. Hüller cited as fig. 1 in his paper. The phase diagram led us to suppose that we have an interesting sequence of phase transitions at constant high pressure just above the triple point. We questioned this interesting idea and studied the equilibrium coexistence curves by use of a high-pressure calorimeter recently constructed in our laboratory.

Fig. 3 shows the heat capacities and molar volumes of ND_4Br measured at 0.1, 50 and 111 MPa. These data were obtained along the constant pressure in the appropriate temperature range. At low pressures, the transition takes place in two stages; from ordered cubic to tetragonal, and then to disordered cubic. At 111 MPa, the transition proceeds in a single step; from ordered cubic to disordered cubic.

Fig. 4 gives the entropy of transition as a function of temperature and pressure around the triple point. The total entropy change is *ca.* $R\ln 2$ irrespective of pressure, and corresponds to the two possible orientations of NH_4^+ ion in the disordered cubic phase.

Fig. 5 is the phase diagram obtained by plotting the temperature of the heat capacity peak as a function of pressure. There is a complicated situation around the triple point owing to the extremely slow rate of the lower phase transition, but the triple point is represented essentially by an intersection of the three coexisting lines. There seems to be no maximum in one branch of the coexisting curves. The dotted line in fig. 5 corresponds to that given by Dr. Stevenson. Although the present result is only for ND_4Br crystals, we believe that it is generally true for the other ammonium halides.

Dr. A. Hüller (*KFA Jülich*) said: I fully agree with Dr. Suga's valuable comment on the shape of the Stevenson plot. On heating, the phase diagram of NH_4Cl is exactly as represented in fig. 5 shown by Dr. Suga. I have shown the generalized Stevenson plot including the deuterated species which has first been published by Press *et al.* [ref. (14) of my paper]. It was my intention to stress the effect of deuteration and I was not so much interested in the detailed shape of the diagram in the vicinity of the triple point.

I would like to remark, however, that all the transition lines shown in Dr. Suga's

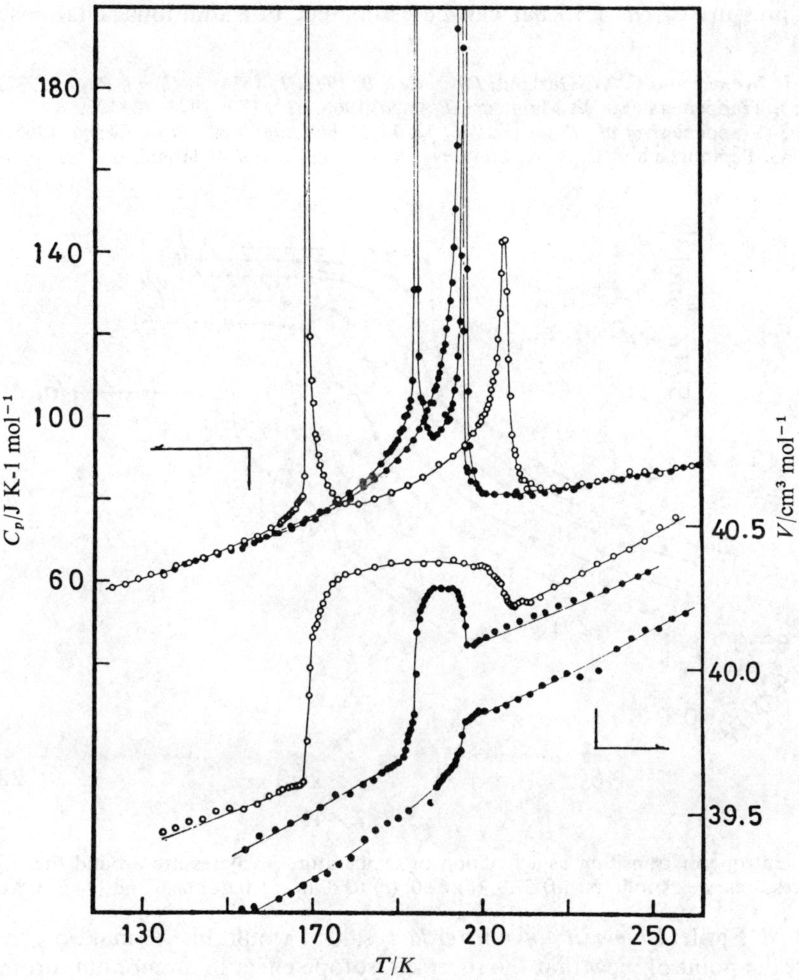

FIG. 3.—Heat capacities and molar volumes of ND_4Br measured at 0.1, 50 and 111 MPa.

plot represent first-order phase transitions. There is an appreciable hysteresis for the δ to γ transition and the transition temperature is ill defined. Extensive measurements by Jahn[1] on mixtures of NH_4Br and NH_4Cl confirm Dr. Suga's point on heating. On cooling the shape of the diagram is rather close to Stevenson's original suggestion.

[1] I. R. Jahn, personal communication and to be published.

Prof. G. M. Schneider (*University of Bochum*) said: With respect to fig. 1 of Hüller's paper I would like to mention that the literature values of temperature and pressure for the tricritical point of NH_4Cl differ considerably.

Weiner and Garland[1] found 255.95 K and 1492 bar from capacitance measurements of the change in length of a single crystal. Trappeniers and Mandema[2] determined 251.3 K and *ca.* 950 bar from spin–lattice relaxation time experiments; older d.t.a. measurements of this group[3] are probably erroneous since no corrections for the pressure dependence of the heat conductivity of the pressurizing gas were made. From preliminary d.s.c. experiments of M. Wirths in our Institute at Bochum a tricritical pressure of *ca.* 1.5 kbar can be estimated. For additional citations see *e.g.*, ref. (4).

[1] B. B. Weiner and C. W. Garland, *Phys. Rev. B*, 1971, **3**, 1634; *J. Chem. Phys.*, 1972, **56**, 155.
[2] N. J. Trappeniers and W. Mandema, *Physica*, 1966, **32**, 1170; 1974, **76**, 123.
[3] N. J. Trappeniers *et al.*, *Physica*, 1966, **32**, 1161; *Ber Bunsenges. Phys. Chem.*, 1966, **70**, 1080.
[4] N. G. Parsonage and L. A. K. Staveley, *Disorder in Crystals* (Clarendon Press, Oxford, 1978).

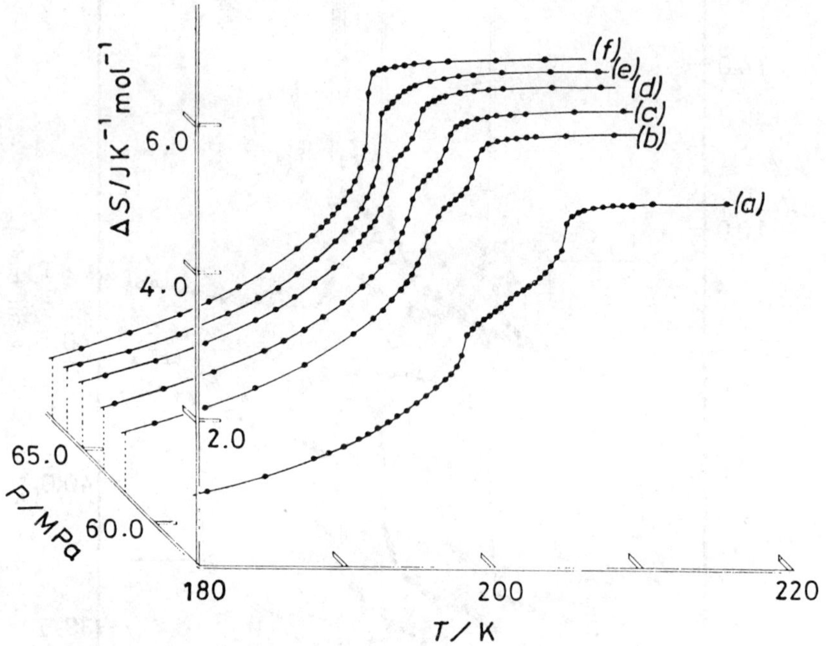

FIG. 4.—Entropy of transition as a function of temperature and pressure around the triple point. Pressures are as follows: (*a*) 57.0, (*b*) 62.0, (*c*) 63.5, (*d*) 65.0, (*e*) 66.0 and (*f*) 67.0 MPa.

Mr. M. Sprik (*University of Amsterdam*) said: I would like to make a comment to support the point of view that the reversed isotope effect in ammonium bromide may be understood without the introduction of a pseudo spin–phonon coupling or an isotope-dependent hydrogen bonding. In a first approximation, the stress induced in the lattice by the orientational system can be represented by an isotropic contribution to the total pressure, the so-called orientational pressure. In an ordered phase, the orientational pressure is generally lower than in a disordered phase because of the negative anisotropic interaction energy due to the ordering. On the other hand, the rotational kinetic energy will, in general, be larger in an ordered phase as a consequence of the confinement of rotational or librational motion by the molecular field. For discontinuous phase transformations these general considerations may be applied in approximate Clausius–Clapeyron relations for the derivatives of transition temperature and pressure with respect to the inverse inertial moment.[1]

$$\left(\frac{\partial T_t}{\partial B}\right)_p = \frac{T\,\Delta K^\circ}{B\,\Delta Q}$$

$$\left(\frac{\partial p_t}{\partial B}\right)_T = \frac{B_T^L\ \Delta K^\circ}{B\ V\Delta p^\circ}$$

$$\left(\frac{\partial p_t}{\partial T}\right)_B = \frac{B_T^L\ \Delta Q}{T\ V\Delta p^\circ}$$

where p° is the orientational pressure; K° is the rotational kinetic energy; B_T^L is the isothermal bulk modulus of the stable lattice and $B = \hbar^2/I$ is a rotational parameter.

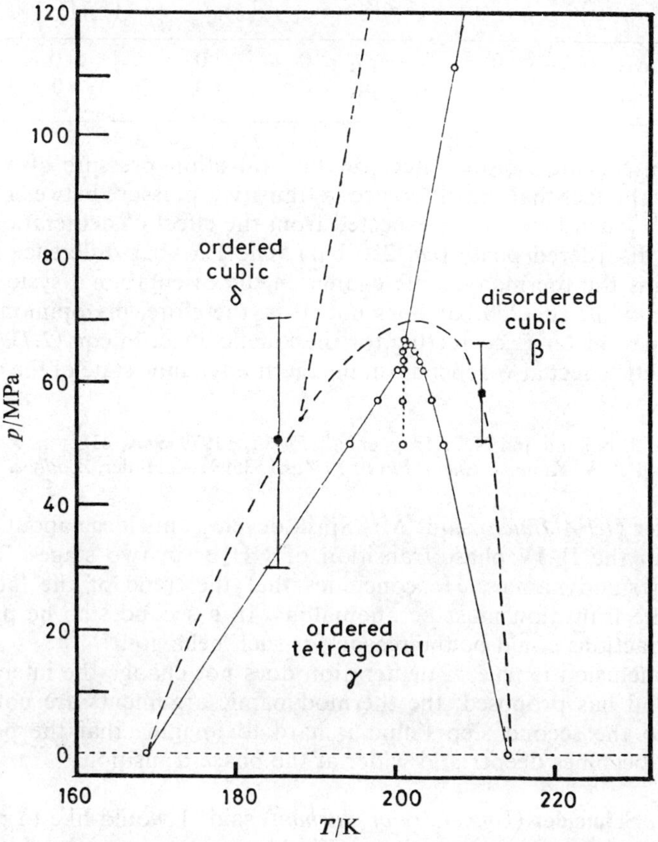

FIG. 5.—Phase diagram obtained by plotting the temperature of the heat-capacity peak as a function of pressure: (—O—) present data, (— — —) R. Stevenson (1961).

The first relation implies that, in general, the heavier isotope has the higher transition temperature. The II–IV phase transition in ammonium bromide shows this common behaviour, as is summarized in table 2. The second relation predicts that the transition pressure is reduced by deuteration. The resulting slope of the transition line in the pT phase diagram of a particular isotope is given by the third relation.

These values for the signs of the derivatives are not a thermodynamic necessity. I would like to propose that an interaction competing with the ordering interaction [for ammonium bromide, see ref. (2)] may be able to decrease the kinetic energy in the ordered phase with respect to the disordered phase and raise the orientational pressure, while at the same time the total Gibbs free energy is still reduced by the ordering. One may imagine that the competing interaction, by opposing the ordering potential, renders the angular dependence of the potential well more square in comparison to the simple crystal-field potential, confining the librational motion in the disordered phase. This allows the potential well in the ordered phase to be both deeper and wider. A similar argument can be given for the change in orientational pressure [see ref. (1)]. In this case, the Clausius–Clapeyron relations yield signs for the derivatives as is observed for the II → III transition (see table 2).

TABLE 2.—SUMMARY OF THE II–IV PHASE TRANSITION IN AMMONIUM BROMIDE

	ΔK°	Δp°	$\left(\frac{\partial T_t}{\partial B}\right)_p$	$\left(\frac{\partial p_t}{\partial B}\right)_T$	$\left(\frac{\partial p_t}{\partial T}\right)_B$
II→ IV	$K_{IV}^\circ - K_{II}^\circ > 0$	$p_{IV}^\circ - p_{II}^\circ < 0$	<0	>0	>0
II → III	$K_{III}^\circ - K_{II}^\circ < 0$	$p_{III}^\circ - p_{II}^\circ > 0$	>0	>0	<0

The relation for the isotope effect on the transition pressure also suggests an explanation for the fact that the difference in transition pressure between two isotopes (*ca.* 1.3 kbar) is much larger than expected from the effect of deuteration in a stable phase, say the disordered phase (*ca.* 250 bar). The 1.3 kbar difference in transition pressure contains the thermodynamic change in the orientational system pertaining to the ordering, while the 250 bar does not. It is, therefore, my opinion that the renormalization parameter ξ, correcting for the kinetic effect in eqn (2.7), is confusing because the kinetic effect also depends on the thermodynamic state of the orientational system itself.

[1] M. Sprik, A. J. Nijman and N. J. Trappeniers, *Physica*, 1979, **98A**, 231.
[2] A. Hüller and J. W. Kane, *J. Chem. Phys.*, 1974, **61**, 3599; A. Hüller, *Z. Phys.*, 1974, **270**, 343.

Dr. A. Hüller (*KFA Jülich*) said: Mr. Sprik develops his ideas about the negative isotope effect in the II–IV phase transition of NH_4Br in two stages. The first one is based on thermodynamics. He concludes that the trend of the " orientational pressure " at the transition must be anomalous. In a second step he proposes that competing interactions could be the reason for such behaviour.

The first conclusion is true, if deuteration does not change the interactions. If it does, as Garland has proposed, the thermodynamic arguments are not conclusive. With respect to the second step I find it hard to imagine that the potential well simultaneously becomes deeper and wider at the phase transition.

Prof. G. M. Schneider (*University of Bochum*) said: I would like to present some results on mesomorphic phase transitions in liquid crystals at normal and high pressures that are related to the general scope of this Discussion. Phase-transition temperatures T_{tr} and phase-transition enthalpies ΔH_{tr} were determined up to 3 kbar using differential thermal analysls (d.t.a.) and differential scanning calorimetry (d.s.c.) as experimental methods. Measurements were performed especially on some nematic liquid crystals (*e.g.*, MBBA, EBBA) and on some homologues of the bis(4,4′-n-alkoxybenzylidene)-1,4-phenylenediamine series with 4 to 13 carbon atoms in the n-alkoxy chains where up to six smectic phases exist and triple points involving smectic phases only are found. Some results have already been published.[1] Most of the $T_{tr}(p)$ transition curves measured run essentially parallel, thus exhibiting rather similar slopes for very different phase transitions (solid–solid, solid–smectic, smectic–smectic, smectic–nematic, smectic–liquid isotropic, nematic–liquid isotropic). This fact provides evidence that the ratio $\Delta V/\Delta S$ (or to a better degree of approximation $\Delta V/\Delta H$) is nearly constant for all these very different phase transitions, whereas the absolute values of ΔV and ΔS or ΔH, respectively, vary over orders of magnitude, $\Delta V/\Delta H$ being *ca.* 0.0013 $cm^3\ J^{-1}$. Approximately the same numerical value of $\Delta V/\Delta H$ has been found in our paper of this Discussion for 1,3-dimethyladamantane.

[1] W. Spratte and G. M. Schneider, *Ber. Bunsenges. Phys. Chem.*, 1976, **80**, 886; W. Spratte and G. M. Schneider, *Mol. Cryst. Liq. Cryst.*, 1979, **51**, 101; J. Hermann, J. Quednau and G. M. Schneider, *Mol. Cryst. Liq. Cryst.*, submitted for publication.

Prof. A. J. Leadbetter (*University of Exeter*) said: Can Dr. More say whether there is any relationship between the short-range structural correlations observed in the high-temperature disordered phase and structure of the low-temperature ordered phase?

Dr. M. More (*University of Lille*) said: There is no direct relation between the two phases in CBr_4. In the high-temperature phase, the observed diffuse scattering has been attributed to the effect of correlations *via* a hard-core potential but no critical effect has been observed when lowering the temperature. However, the low-temperature phase shows a small distortion from a pseudo-cubic lattice with parameter equal to the high-temperature lattice parameter. Then one can distinguish between " centre-of-mass reflections " and the other ones which are characteristic of long-range orientational order. Because of the translation rotation coupling in phase I, the orientational order is coupled with the lattice distortion in a complicated way, leading through a first-order transition to a new phase where the translation–rotation coupling vanishes.

Prof. R. M. Pick (*Université Pierre et Marie Curie, Paris*) said: Concerning the problem of linewidths in neutron scattering, I would like to mention that in our laboratory, we have performed Raman scattering experiments on the internal lines of CBr_4 in its plastic phase. For the ν_2 and ν_4 lines, which are affected by molecular reorientations but not by induced-dipole–induced-dipole interactions, the half-width at half-maximum is *ca.* 3 cm^{-1} (*i.e.*, 0.1 THz), out of which 0.5 or 1 cm^{-1} might be related to a vibrational lifetime, and these lines stand on a rather flat background.

This 0.1 THz value is within a factor of 2 that obtained for Γ_2 (*cf.* fig. 2 of More's paper); it is also in the same frequency range as the one at which the longitudinal sound velocities saturate (*cf.* fig. 6). It seems reasonable to relate this 0.2 THz value to the reorientational motion of a CBr_4 molecule from one potential well to another. Indeed (*a*) in a plastic crystal, the saturation of the sound velocity should be related to the quenching of the coupling between a macroscopic distortion and a molecular reorientation. (*b*) If the reorientations represent 90° rotations [as may be suggested by the p.d.f. P_0 (Ω)] and if they occur with an angular velocity of the order of the free-rotation angular velocity $\omega_R = (kT/I)^{\frac{1}{2}}$ one also obtains a half-width at half-maximum of *ca.* 0.1 THz.

If this interpretation is correct, the difficulty is transferred to the explanation of the Γ_3 value of 0.5 THz obtained in the neutron coherent scattering experiments throughout the entire Brillouin zone. In this respect one can note that, as far as librations are concerned, the disordered nature of a plastic crystal also affects strongly the force constants (and the eigenvectors which are no longer plane waves) as the detection mechanism, which takes place only through the disordered Br atoms. The difference between a collective (or extended) librational mode and an individual (or local) one might then be difficult to detect even in a coherent scattering experiment: the presence of a given peak in the whole Brillouin zone could not be a decisive argument against its assignment to collective librational modes.

Dr. G. S. Pawley (*University of Edinburgh*) said: The dynamic aspects of the plastic phase with translation rotation coupling included can be attacked by computer simulations, always providing that a good potential function is available. Such a function exists for SF_6 which gives instabilities for the lattice dynamics in the cubic phase but which gives a stable model in the simulation. In this model the molecules are oriented on average in the way X-ray diffraction suggests, but the molecules undergo occasional reorientational jumps. These jumps are about the (100) axes, and a jump by one molecule is usually followed by a jump of a neighbouring molecule about a different (100) axis. As yet insufficient calculations have been done to know whether there is any cooperative motion which localizes the excess jump energy for greater than a few picoseconds. The model used for the calculation was oversimplified as translational motions were not included.

Mr. M. Sprik (*University of Amsterdam*) (*communicated*): Prof. Michel's choice of the operators in eqn (3.12) as the set of dynamical variables in the Mori formalism is, in my opinion, justified, provided the orientational system behaves classically. If quantum mechanics is no longer negligible in the dynamics of the orientational system, additional quasiconstants of motion have to be taken into account. This makes the validity of the expressions (4.1) and (4.2) in the NH_4Br case questionable since, if the damping parameter λ for NH_4Br is still given by the expression (4.4), it may be expected that λ is of the order of a typical phonon frequency instead of being smaller, as is stated on page 94. Unless the coupling coefficient β' is extremely large, expression (4.4) implies that λ will be comparable with the square root of the second moment of the spectral density of a second-order irreducible tensor [Raman spectrum, see eqn (3.18) and (3.19)]. The latter quantity is approximately equal to the librational frequency of the NH_4^+ ions in the crystal field potential and, hence, of the order of an optical phonon frequency. On the other hand, NH_4Br is certainly an instance of the slow relaxation case $\lambda < \omega_i$. But instead of identifying the damping parameter λ with expression (4.4), it should, in my opinion, be interpreted as the rate of re-orientation between the two available orientations with respect to the cubic lattice.

Prof. K. H. Michel (*University of the Saar*) (*communicated*): Our first aim in the present approach is to get a microscopic expression of the transport coefficient λ, based on the Hamiltonian (2.1). The calculation is indeed classical and the choice of variables is the most simple non-trivial one. As a major approximation, the second moments (4.5*a*) and (4.5*b*) are calculated with a single-particle potential in the classical limit. As has not been mentioned explicitly in the present paper, subsequent investigations show that the present theory enables us to understand the temperature dependence of the phonon linewidth measured by Brillouin scattering in KCN.[1] In addition, for KCN we understand from eqn (4.4) why the λ measured by neutron coherent scattering [ref. (6)] has about the same value as the single-particle relaxation frequency measured by Raman scattering.[2]

As far as NH_4Br is concerned, the Hamiltonian (2.1) with the coupling to acoustic phonons is not relevant. Consequently the calculation of the moments (A_n^i, A_n^j), eqn (3.13*a*)–(3.13*c*) would give different results. Nevertheless I believe that a calculation according to the lines of the present approach, eventually with consideration of additional dynamic variables, would also be of great interest in the case of NH_4Br.

[1] M. Boissier, R. Vacher, D. Fontaine and R. M. Pick, *J. Physique*, 1978, **39**, 205.
[2] D. Fontaine and R. M. Pick, *J. Physique*, 1979, **40**, 1105.

Dr. A. I. M. Rae (*University of Birmingham*) said: Since submitting our paper, we have collected single-crystal X-ray data from the *h0l* zone of malononitrile at a temperature of *ca.* 100 K, from which we have confirmed that the space-group of the α phase is $P2_1/n$ identical with that of the γ phase. We have also collected X-ray powder data at a temperature of *ca.* 85 K and have refined this using Dr. Pawley's EDINP program with rigid-body constraints. The resulting structure of the α phase is very similar to that of the γ phase and will be reported in detail elsewhere.

We have given some further consideration to the implications of the Landau theory discussed in our paper [eqn (1)-(4)] which describes the temperature variation of the free energy and order parameter in an α–β–α series of phase transitions. The resulting contributions to the entropy and heat capacity are:

$$S = -\left(\frac{\partial F}{\partial T}\right)_P = \frac{A''^2}{2C}(2T - T_1 - T_2)(T - T_1)(T - T_2)$$

and

$$C = T\frac{\mathrm{d}S}{\mathrm{d}T} = \frac{A''^2 T}{2C}[(T - T_1)^2 + 4(T - T_1)(T - T_2) + (T - T_2)^2].$$

(Note: P is the order parameter not the pressure.)

It follows that $C = 0$ when $T = \frac{1}{2}[(T_1 + T_2) \pm (T_1 - T_2)/\sqrt{3}]$ and is negative between these two temperatures. Such a negative contribution to the specific heat is a necessary consequence of the assumption that the entropies of the high- and low-temperature phases are identical, and does not depend on the details of the model used. In practice other contributions (*e.g.*, vibrational) are positive and much larger than this and the total heat capacity is therefore always positive. However, this property is not an intrinsic requirement of the model and it would appear to be possible to imagine a hypothetical ensemble of oscillators whose coupling coefficients were such that the ratio $A''^2/2C$ was very large and the total heat capacity consequently negative over some temperature range.

Prof. G. J. Hills (*University of Southampton*) said: In his verbal introduction of his paper, Dr. Rae touched on the possibility of systems exhibiting negative heat capacities. Quite apart from any ensuing damage that such a possibility might inflict on the laws of thermodynamics it needs to be restated (with surprising regularity) that *isobaric* enthalpy changes in condensed media always contain a temperature-dependent–density-dependent contribution which is invariably large because the internal pressure of condensed phases is high. Moreover, the contribution to the specific heat may be positive or negative depending on the sign of the volume change. Thus,

$$(\Delta H)_p = (\Delta E)_V + \frac{\alpha T}{\beta}(\Delta V)_T$$

and

$$C_p = \left(\frac{\partial \Delta H}{\partial T}\right)_p = \underbrace{\left(\frac{\partial \Delta E}{\partial T}\right)_V}_{\text{necessarily positive}} + \underbrace{\frac{\alpha T}{\beta}\left(\frac{\mathrm{d}\Delta V}{\mathrm{d}T}\right)_p + \frac{\alpha}{\beta}(\Delta V)_T}_{\text{possibly negative}},$$

where the terms have their usual thermodynamic meaning (p is the pressure, α the molar expansivity and β the molar compressibility). The question of apparently negative specific heats therefore needs to be examined in a wide framework of related quantities.

Dr. R. M. Lynden-Bell (*University of Cambridge*) said: In introducing this paper Dr. Rae remarked that his model gives a negative contribution to the specific heat and that he sees no reason to stop such a model from having a total heat capacity that is negative.

There *is* no theoretical reason why an isolated system should not show such behaviour. Systems with negative specific heat are well known in astrophysics; both self-gravitating gas spheres[1] and black holes[2] can show this phenomenon. We can see the reason for this from the Virial theorem for an r^{-1} potential:

$$E + T = 0$$

where T is the kinetic energy. As the temperature is proportional to the kinetic energy, an increase in energy, E, leads to a decrease in temperature.

Although systems with negative specific heats can exist in microcanonical ensembles (that is when isolated), they are unstable in contact with a heat bath or with a number of duplicate systems. The specific heat of a Gibbs canonical ensemble is always positive as it depends on the square of the energy fluctuations

$$C_v = k\beta^2 (E - \bar{E})^2.$$

A system which has a negative specific heat when isolated shows a first-order phase transition in a canonical ensemble.[3] Thus the negative specific heat regions of Dr. Rae's models might be difficult to observe.

[1] D. Lynden-Bell and R. Wood, *Mon. Not. R. Astr. Soc.*, 1967, **136**, 101.
[2] S. W. Hawking, *Phys. Rev. D*, 1976, **13**, 191.
[3] D. Lynden-Bell and R. M. Lynden-Bell, *Mon. Not. R. Astr. Soc.*, 1977, **181**, 405; P. Hertel and W. Thirring, *Comm. Math. Phys.*, 1972, **28**, 159.

Dr. A. Novak (*CNRS, Thiais, France*) said: I have three comments on phase transitions in malononitrile. They are based mainly on an infrared and Raman study of this compound. The first one concerns the space-group symmetries of α, β, γ phases. The infrared and Raman spectra of α and γ phases are very similar and consistent with C_{2h} factor group symmetries and four molecules in the unit cell. In the spectral region corresponding to internal vibrations we observe mainly doublets and their infrared and Raman frequencies generally do not coincide. The spectra of the intermediate phase β, on the other hand, are different. Triplets and even quadruplets are frequently observed in both Raman and infrared. We would thus prefer a non-centrosymmetric space group such as $P2_1 \equiv C_2^2$ as initially suggested by Nakamura *et al.*[1] in order to interpret our spectra. Some dielectric or other measurements of the phase β would be useful. The second comment is about the soft mode which has been observed in the Raman spectra as the lowest lattice frequency at 12 cm^{-1} at 15 K. It has been assigned to a predominantly R_y' libration, *i.e.*, to a rotational vibration about the C_2 molecular axis which is in very good agreement with the X-ray diffraction work given in the discussed paper (Dove and Rae). The most remarkable feature of the soft mode, however, is its behaviour with temperature: its frequency decreases continuously to zero at the $\alpha \rightarrow \beta$ transition temperature ($T_c = 141$ K) and then increases to a maximum at 16 cm^{-1} in phase β at 215 K, *i.e.*, zero frequency is reached when T_c is approached from above and below. When the temperature is raised above 215 K the frequency decreases again to zero at $T_c = 296$ K corresponding to $\beta \rightarrow \gamma$ transition. No equivalent of this mode has been observed in the γ phase and there is no anomalous broadening of the Rayleigh line either. The soft-mode behaviour is strikingly similar to the ^{14}N n.q.r. results of Zussman and Alexander[2] and reminds one of that of the Rochelle salt.[3] The vibrational data confirm thus the suggestion that the $\alpha \rightarrow \beta \rightarrow \gamma$ transitions are of second order and of the displacive type. The β exponent can be extracted from a log ν against T plot and its value amounts to 1/2. The last comment deals with the δ phase which is quite different in many aspects. The low-temperature δ phase is stable with respect to β phase below 260 K. It contains, unlike α, β, γ phases, intermolecular C—N···N hydrogen bonds as indicated by a low-frequency shift (*ca.* 60 cm^{-1}) of CH stretching bands and by a high-frequency shift of CN stretching and CH_2 scissoring bands in going from α to δ phase at 90 K. There are fewer Raman lattice bands with respect to any of α, β, γ phases, showing a higher symmetry and/or smaller Z of the δ phase crystal in agreement with the n.q.r. results. Finally, the lowest two Raman lattice frequencies at 95.5 and 59.6 cm^{-1} (110 K) decrease strongly with increasing temperature but they never reach zero. This shows that the δ → β transition is a first-order one which is consistent with the normal specific-heat anomaly.[4] The fact that there is a considerable volume change during the transformation and that the δ phase alone contains hydrogen bonds would suggest that δ → β phase change is of reconstructive type.

[1] N. Nakamura and K. Obatake, *Phys. Letters, A*, 1971, **34**, 3721.
[2] A. Zussman and S. Alexander, *J. Chem. Phys.*, 1968, **49**, 372.
[3] J. Habluetzel, *Helv. Phys. Acta*, 1939, **12**, 489.
[4] H. L. Girdhar, E. F. Westrum, and C. A. Wulf, *J. Chem. Eng. Data*, 1968, **13**, 239.

Dr. A. I. M. Rae (*University of Birmingham*) said: The X-ray evidence for the β-phase of malononitrile being triclinic is set out in our paper and is, we believe, incontrovertible. If the structure were also to be non-centrosymmetric, the space group would have to be $P1$ rather than $P\bar{1}$ as assumed by us. However, this is highly improbable, both because our structure refinement assuming $P\bar{1}$ was quite successful and also because a continuous transition from $P2_1/n$ to $P1$ is forbidden by the Landau rules. We believe that the multiplet structures of the Raman and infrared spectra reported by Dr. Novak and also by Savoie *et al.*[1] are in fact consistent with the structure of the β phase reported in our paper, when it is remembered that this contains two independent malononitrile molecules in the asymmetric unit.

The temperature dependence of the soft mode reported by Dr. Novak is in good agreement with what would be expected, both from our experimental results and from the extended form of Landau theory reported in our paper.

The spectroscopic evidence concerning the nature of the δ phase is also very interesting and we are currently carrying out experimental and theoretical investigations into the structure of this modification.

[1] R. Savoie *et al.*, *Can. J. Chem.*, 1976, **54**, 3293.

Prof. R. M. Pick (*Université Pierre et Marie Curie, Paris*) said: The Raman scattering experiments by Le Calvé and Novak presented at the poster session on malononitrile, show the existence of a soft mode both above and below the α–β transition. As the Dove and Rae experiments show that the α phase has a higher symmetry than the β one, the existence of a Raman active soft mode in the α phase implies a coupling of this mode with an elastic constant, and it is this latter quantity which should go to zero at the α–β transition.[1] The same effect should also exist at the β–γ transition as the α and γ phases are identical.

It would be interesting to look for those elastic constant softenings. It also means that the thermodynamics of these phase transitions has to be partly remodelled to take into account these effects which are required by symmetry considerations.

[1] Axe–Miller theorem: see P. B. Miller and J. Axe, *Phys. Rev.*, 1967, **163**, 924.

Dr. R. W. Munn (*UMIST, Manchester*) **and Dr. T. Luty** (*Technical University, Wroclaw*) (*partly communicated*): Prof. Strobl asked whether from the calculations of the lattice dynamics under different strains one could calculate microscopic

Grüneisen parameters and hence the equation of state in the quasiharmonic approximation.

The microscopic Grüneisen parameters are defined as[1]

$$\gamma_{ij}^{(r)} = -(\partial \ln \omega^{(r)}/\partial l_{ij})_{l'}, \tag{1}$$

where $\omega^{(r)}$ is the frequency of mode r, l_{ij} is a strain component and l' denotes that other strain components are held constant. These quantities are simply related to the slopes of the curves in fig. 2 of our paper. We have calculated Grüneisen parameters for the zone-centre optic modes under symmetry-preserving strains in the limit of zero strain. The results are given in table 3.

TABLE 3.—MICROSCOPIC ZERO-STRAIN GRÜNEISEN PARAMETERS CALCULATED FOR ZONE-CENTRE OPTIC MODES IN TETRACYANOETHYLENE

symmetry of mode r	$\omega^{(r)}/\text{cm}^{-1}$	$\gamma_{11}^{(r)}$	$\gamma_{22}^{(r)}$	$\gamma_{33}^{(r)}$	$\gamma_{13}^{(r)}$
A_g	54.8	3.72	6.79	1.84	−0.36
A_g	77.5	3.81	1.16	4.22	−1.66
A_g	100.8	4.05	3.02	2.00	1.10
B_g	29.1	0.34	−0.68	11.26	−2.71
B_g	91.8	2.76	4.00	2.65	1.76
B_g	105.7	4.05	3.07	1.91	0.85
A_u	25.3	5.27	0.40	8.66	−5.38
A_u	89.4	6.23	1.80	1.69	1.80
B_u	73.3	1.93	5.04	1.51	1.37

Calculation of the equation of state would require Grüneisen parameters at points throughout the Brillouin zone for all strains, including those which lower the crystal symmetry. Macroscopic Grüneisen functions could then be obtained in the quasi-harmonic approximation as averages of the microscopic Grüneisen parameters weighted with the heat capacity contribution $c^{(r)}$ of the corresponding mode:

$$\gamma_{ij} = (1/C_l)(\partial S/\partial l_{ij})_{Tl'}, \tag{2}$$

$$= \sum_r c^{(r)}\gamma_{ij}{}^{(r)}/\sum_r c^{(r)} \tag{3}$$

where C_l is the heat capacity at constant strain.[1] With the elastic constants, derived from the slopes of the dispersion curves for the acoustic modes at the zone centre, the Grüneisen functions yield the thermal expansion. In this way the equation of state could be obtained for arbitrary strain (ignoring constant-strain anharmonicity). However, we have not attempted this rather complicated procedure.

[1] T. H. K. Barron and R. W. Munn, *Philos. Mag.*, 1967, **15**, 85.

Calorimetric Investigations of Plastic Crystals at Low Temperatures and High Pressures with Differential Scanning Calorimetry

By Horst Arntz and Gerhard M. Schneider

Department of Chemistry,
University of Bochum, 4630 Bochum, West Germany

Received 3rd December, 1979

A new high-pressure low-temperature differential scanning calorimeter (d.s.c.) has been developed. The calorimeter has been designed for the investigation of some plastic crystals over wide ranges of temperature and pressure, *e.g.*, between 100 and 450 K at normal pressure and 100 and 300 K at 4 kbar. This d.s.c. apparatus has been applied to the determination of transition temperatures and transition enthalpies of cyclohexane, 1,3-dimethyladamantane, and 1,3,5-trimethyladamantane at normal and elevated pressures from which the transition entropies were also determined. By means of these values and the slopes of the $T(p)$ transition curves transition volumes could be additionally calculated.

Quantitative determinations of enthalpies of transformation at high pressures were recently performed by Kamphausen[1–3] using the d.s.c. principle. The high-pressure microcalorimeter developed was also applied to the investigation of phase transitions of some liquid crystals up to 2.5 kbar.[4] The results show that the d.s.c. technique is capable of yielding accurate values, *e.g.*, of the enthalpy changes for solid–solid and solid–liquid transitions at high pressures.

Since the use of this microcalorimeter was limited to temperatures above 250 K a new high-pressure low-temperature d.s.c. microcalorimeter has been developed to make possible high-pressure investigations in the low-temperature region down to 100 K.

EXPERIMENTAL

The main parts of the high-pressure low-temperature d.s.c. microcalorimeter are two identical cylindrical autoclaves, each of them closed by two Bridgman pistons. They are installed in a thermostatted jacket that can be cooled with liquid nitrogen and also heated with linear heating rates from 0.1 to 2.5 K min^{-1}. The sample and reference holders of the high-pressure d.s.c. measuring head are mounted in the vessels on the top of the lower piston, respectively; the control unit is a Perkin–Elmer DSC-2 regulator.

In order to get quantitative values for the transition enthalpies one has to prevent heat transport by convection. Therefore both holders are carefully insulated with glass fibre, Teflon, and steatite.

The pressure is transmitted by highly purified helium and measured with a Heise bourdon gauge and a strain gauge. The substances are enclosed into a capsule made of indium or stainless steel in order to prevent the pressure-transmitting gas from being dissolved in the samples.

The temperature was calibrated with the melting point of indium (99.999%) and the transition points of highly purified cyclopentane. Energy calibrations were based on the fusion of indium and the transitions of cyclopentane. At high pressures mercury was used as an internal standard.

SUBSTANCES

Cyclohexane was obtained from Merck AG, Darmstadt, with a purity of 99.7%. 1,3-Dimethyladamantane (>99%) was obtained from EGA-Chemie, Steinheim/Albuch. 1,3,5-Trimethyladamantane was prepared from the appropriate bromide using the Grignard coupling procedure described by E. Osawa *et al.*[5] Both adamantanes were purified by preparative gas chromatography; the final purity was found to be better than 99.8%.

RESULTS AND DISCUSSION

CYCLOHEXANE

In fig. 1 five original d.s.c. traces obtained experimentally are presented for cyclohexane at 1, 270, 500, 1000 and 1300 bar; only the peaks of the solid–solid transitions are plotted.

At normal pressure cyclohexane exhibits a solid–solid transition at 186.2 K. The plastic crystalline phase s_I is face-centred cubic[6] and stable between 279.8 and 186.2 K; phase s_{II} is monoclinic and stable below 186.2 K. At elevated pressures a new solid

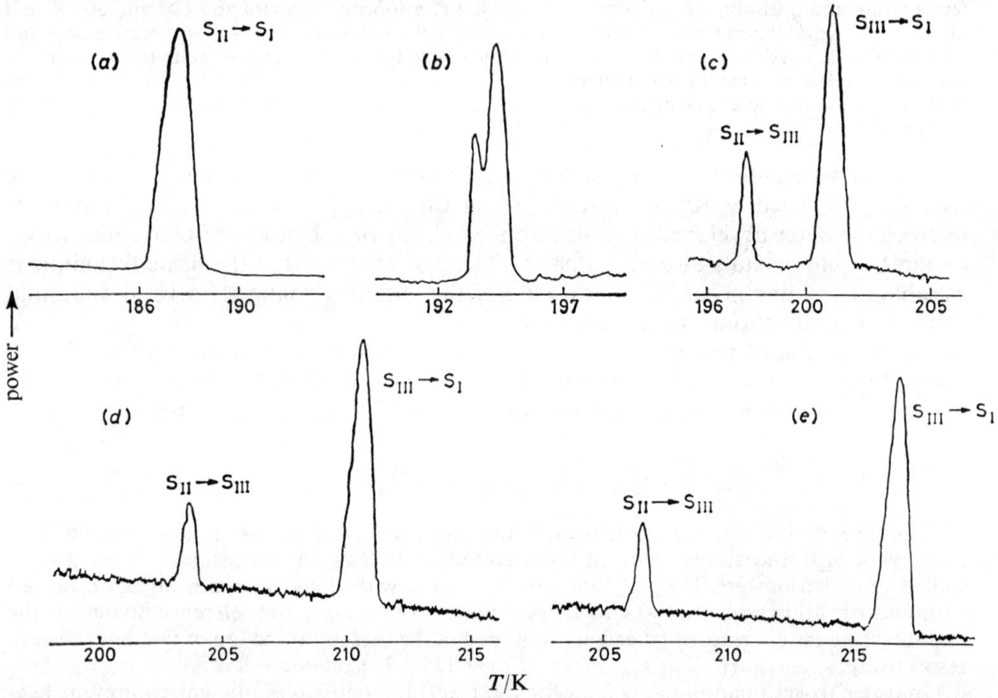

FIG. 1.—D.s.c. peaks of cyclohexane at different pressures (see text), (*a*) 1, (*b*) 270, (*c*) 500,(*d*) 1000, (*e*) 1300 bar.

phase, s_{III}, which has been reported by Würflinger,[7] was confirmed; its structure is not yet known.

Fig. 2 shows the $T(p)$ phase diagram of cyclohexane (without the melting curve) obtained from measurements of this work; it is in good agreement with the values found by Würflinger.[7] The triple point where the three solid phases s_I, s_{II} and s_{III} coexist is at *ca.* 200 bar and 191 K.

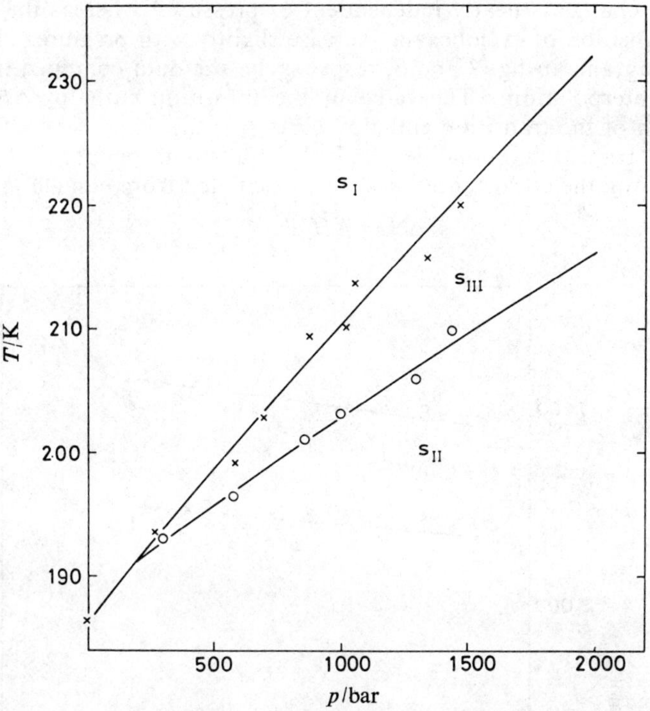

FIG. 2.—*T(p)* phase diagram of cyclohexane (without the melting curve).

In fig. 3 the enthalpies of the solid–solid transitions $\Delta H(s_{II} \rightarrow s_{III})$ and $\Delta H(s_{III} \rightarrow s_I)$ of cyclohexane along the coexistence lines as well as the sum $\Delta H(s_{II} \rightarrow s_{III}) + \Delta H(s_{III} \rightarrow s_I)$ of both transitions are plotted as a function of pressure. The enthalpy of melting of mercury (internal standard substance) $\Delta H(s \rightarrow 1)$ is additionally given.

TABLE 1.—PHASE TRANSFORMATIONS OF CYCLOHEXANE AS A FUNCTION OF PRESSURE. TRANSITION TEMPERATURES, *T*, ENTHALPY CHANGES, ΔH, ENTROPY CHANGES, ΔS, AND VOLUME CHANGES, ΔV, ALONG THE COEXISTENCE LINES.

p/bar	*T*/K			ΔH/kJ mol^{-1}		
	$s_{II} \rightarrow s_I$	$s_{II} \rightarrow s_{III}$	$s_{III} \rightarrow s_I$	$s_{II} \rightarrow s_I$	$s_{II} \rightarrow s_{III}$	$s_{III} \rightarrow s_I$
1	186.2	—	—	6.72	—	—
500	—	195.6	198.7	—	1.18	5.76
1000	—	202.7	210.2	—	1.21	5.85
1500	—	209.4	221.2	—	1.25	6.00

p/bar	ΔV/cm^3 mol^{-1}			ΔS/J K^{-1} mol^{-1}		
	$s_{II} \rightarrow s_I$	$s_{II} \rightarrow s_{III}$	$s_{III} \rightarrow s_I$	$s_{II} \rightarrow s_I$	$s_{II} \rightarrow s_{III}$	$s_{III} \rightarrow s_I$
1	9.7	—	—	36.1	—	—
500	—	0.87	7.1	—	6.03	29.0
1000	—	0.81	6.3	—	5.97	27.8
1500	—	0.77	5.8	—	5.97	27.1

This enthalpy change is nearly independent of pressure,[8] whereas the values for the solid–solid transition of cyclohexane increase slightly with pressure. From the $T(p)$ and $\Delta H(p)$ diagrams in fig. 2 and 3, respectively, the data compiled in table 1 were obtained by interpolation. The value of the transition enthalpy $\Delta H(s_{II} \rightarrow s_{III})$ is about one fifth of the transition enthalpy $\Delta H(s_{III} \rightarrow s_{I})$.

From the transition enthalpies and the transition temperatures the entropy changes ΔS along the coexistence lines were calculated from eqn (1)

$$\Delta S = \Delta H/T \, . \tag{1}$$

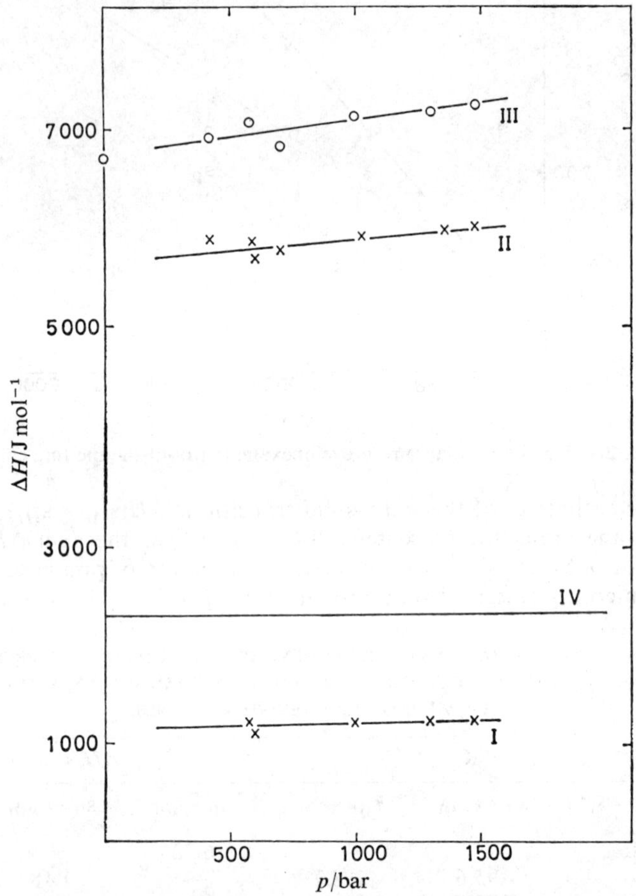

FIG. 3.—Transition enthalpies ΔH of cyclohexane and mercury (standard substance) as a function of pressure along the coexistence lines [I, $\Delta H(s_{II} \rightarrow s_{III})$; II, $\Delta H(s_{III} \rightarrow s_{I})$; III, $\Delta H(s_{II} \rightarrow s_{III}) + \Delta H(s_{III} \rightarrow s_{I})$; IV, $\Delta H(s \rightarrow l)$ of Hg].

The corresponding volume changes ΔV could be calculated using Clapeyron's equation, eqn (2)

$$(\mathrm{d}p/\mathrm{d}T)_{\mathrm{coex}} = \Delta H/T\Delta V \, . \tag{2}$$

The data for the entropy changes and the volume changes accompanying the phase transformations are also summarized in table 1. It follows that $\Delta V(s_{II} \rightarrow s_{III})$ is only $\approx 13\%$ of $\Delta V(s_{III} \rightarrow s_{I})$.

From the precision of the transition enthalpies (better than 3%) and the transition temperatures ($\approx \pm 1$ K) and taking into account the error of $(dp/dT)_{coex}$; an error of $\pm 10\%$ for the ΔV values was estimated.

The precision of the results can be tested by the condition that at the triple point the sum of the corresponding transition enthalpies $\Delta H(s_{II} \rightarrow s_{III})$ and $\Delta H(s_{III} \rightarrow s_I)$ must be equal to $\Delta H(s_{II} \rightarrow s_I)$ [see eqn (3*a*)]. The corresponding condition also holds for the transition volumes [see eqn (3*b*)]

$$\Delta H(s_{II} \rightarrow s_{III}) + \Delta H(s_{III} \rightarrow s_I) = \Delta H(s_{II} \rightarrow s_I) \qquad (3a)$$

$$\Delta V(s_{II} \rightarrow s_{III}) + \Delta V(s_{III} \rightarrow s_I) = \Delta V(s_{II} \rightarrow s_I) \qquad (3b)$$

These equations only hold at the triple point.

According to eqn (3*a*) and (3*b*) deviations of $<5\%$ result from the sums of the volume changes and enthalpy changes, respectively.

1,3-DIMETHYLADAMANTANE

1,3-Dimethyladamantane ($C_{12}H_{20}$) consists of three condensed cyclohexane rings in the chair configuration with two methyl groups. It has a high melting point ($T =$ 244 K) and a low entropy of melting ($\Delta S = 2.85$ J K^{-1} mol^{-1}) at normal pressure. The plastic crystal undergoes a solid–solid phase transition at 221 K with a high value

TABLE 2.—PHASE TRANSFORMATIONS OF 1,3-DIMETHYLADAMANTANE AS A FUNCTION OF PRESSURE. TRANSITION TEMPERATURES T, ENTHALPY CHANGES ΔH, ENTROPY CHANGES ΔS AND VOLUME CHANGES ΔV ALONG THE COEXISTENCE LINES.

p/bar	T/K		ΔH/kJ mol^{-1}	
	$s_{II} \rightarrow s_I$	$s_I \rightarrow 1$	$s_{II} \rightarrow s_I$	$s_I \rightarrow 1$
1	221.0	244.0	7.65	0.94
500	235.0	258.9	7.23	1.01
800	243.6	268.0	7.08	1.10
1200	255.4	—	6.76	—

p/bar	ΔV/cm^3 mol^{-1}		ΔS/J K^{-1} mol^{-1}	
	$s_{II} \rightarrow s_I$	$s_I \rightarrow 1$	$s_{II} \rightarrow s_I$	$s_I \rightarrow 1$
1	9.9	1.2	34.6	3.85
500	8.8	1.2	30.8	3.9
800	8.3	1.2	29.1	4.1
1200	7.6	—	26.5	—

of the entropy of transition $\Delta S = 34.6$ J K^{-1} mol^{-1}. The phase transition enthalpies $\Delta H(s_I \rightarrow 1) = 0.94$ kJ mol^{-1} and $\Delta H(s_{II} \rightarrow s_I) = 7.65$ kJ mol^{-1} and the above mentioned temperatures determined in this work at atmospheric pressure are in good agreement with those reported by T. Clark *et al.*[9]

In fig. 4(*a*) the $T(p)$ phase diagram and in fig. 4*b* the enthalpies of melting and of the solid–solid transition of 1,3-dimethyladamantane are shown as a function of pressure along the coexistence lines; whereas the values for the solid–solid transition enthalpy decrease slightly with increasing pressure, the enthalpy change on melting increases by ≈16% between 1 and 800 bar.

The slopes of both $T(p)$ transition curves are nearly equal implying that the quotient $\Delta V/\Delta S$ is essentially constant. Nevertheless at atmospheric pressure $\Delta S(s_{II} \rightarrow s_I)$ is about nine times as large as $\Delta S(s_I \rightarrow l)$ and $\Delta V(s_{II} \rightarrow s_I)$ about eight times as large as $\Delta V(s_I \rightarrow l)$ giving evidence for a correlation between ΔV and ΔS or ΔH, respectively. The same has also been observed for liquid crystals [10] where similar $\Delta V/\Delta H$ values

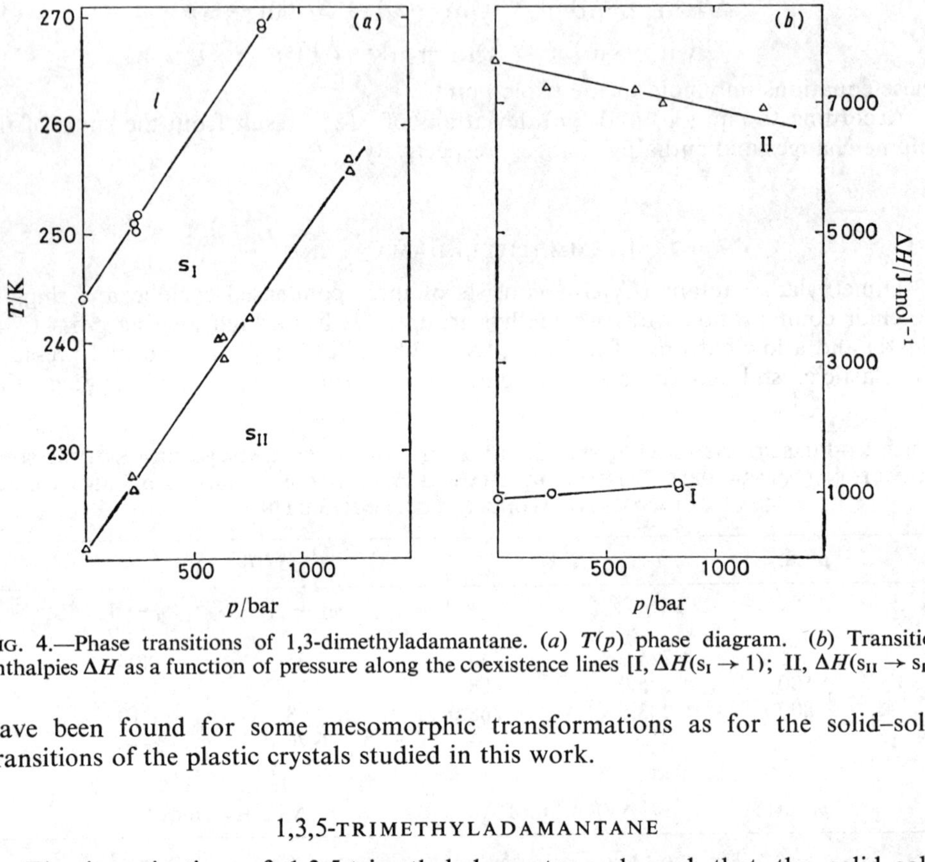

FIG. 4.—Phase transitions of 1,3-dimethyladamantane. (*a*) $T(p)$ phase diagram. (*b*) Transition enthalpies ΔH as a function of pressure along the coexistence lines [I, $\Delta H(s_I \rightarrow l)$; II, $\Delta H(s_{II} \rightarrow s_I)$].

have been found for some mesomorphic transformations as for the solid–solid transitions of the plastic crystals studied in this work.

1,3,5-TRIMETHYLADAMANTANE

The investigation of 1,3,5-trimethyladamantane showed that the solid–solid transition behaviour depends strongly on thermal pretreatment of the sample whereas melting temperature (255 ± 1) K and melting enthalpy (1.92 ± 0.05) kJ mol^{-1} were found to be independent of the pretreatment.

Because of large supercooling effects transition temperatures and transition enthalpies have been determined from heating runs. From heating the solid starting at 180 K a solid–solid transition enthalpy of (4.38 ± 0.06) kJ mol^{-1} at 210.5 K is obtained. When 1,3,5-trimethyladamantane, however, is tempered at 100 K, the solid–solid transition occurs at 234.7 K on heating, the transition enthalpy being (7.75 ± 0.12) kJ mol^{-1}, and sometimes even an exothermic transition is found in the region of 180 K. The transition temperatures and the transition enthalpies obtained experimentally in this work differ from the values reported by Clark *et al.*[9]

At high pressures only the second transition is found. Some results for 1,3,5-trimethyladamantane at elevated pressures are given in Table 3.

TABLE 3.—PHASE TRANSFORMATIONS OF 1,3,5-TRIMETHYLADAMANTANE AS A FUNCTION OF PRESSURE. TRANSITION TEMPERATURES, T, ENTHALPY CHANGES, ΔH, ENTROPY CHANGES, ΔS, AND VOLUME CHANGES, ΔV, ALONG THE COEXISTENCE LINES.

p/bar	T/K		ΔH/kJ mol^{-1}	
	$s_{II} \to s_I$	$s_I \to l$	$s_{II} \to s_I$	$s_I \to l$
200	241.2	263.1	7.62	1.94
300	244.4	267.1	7.56	1.95
400	247.6	271.1	7.49	1.96
500	250.9		7.43	
800	260.6		7.23	
1100	270.3		7.04	

p/bar	ΔV/cm^3 mol^{-1}		ΔS/J K^{-1} mol^{-1}	
	$s_{II} \to s_I$	$s_I \to l$	$s_{II} \to s_I$	$s_I \to l$
200	10.2	3.0	31.6	7.4
300	10.0	2.9	30.9	7.3
400	9.8	2.9	30.3	7.2
500	9.6		29.6	
800	9.0		27.7	
1100	8.4		26.1	

There is only a small temperature range (*ca.* 20 K) for the orientationally disordered phase of 1,3,5-trimethyladamantane. Whereas the melting enthalpy is essentially constant, the solid–solid transition enthalpy decreases with increasing pressure in analogy with the results on 1,3-dimethyladamantane. Accordingly the transition entropy also decreases with pressure here by *ca.* 17% up to 1 kbar.

The authors thank Dr. G. Klaerner for his help in the preparation and purification of the adamantanes. Financial support of the Deutsche Forschungsgemeinschaft (D.F.G.) is gratefully acknowledged.

[1] M. Kamphausen and G. M. Schneider, *Thermochim. Acta*, 1978, **22**, 371.
[2] M. Kamphausen, *Rev. Sci. Instr.*, 1975, **46**, 668.
[3] M. Kamphausen *Dissertation* (University of Bochum W. Germany, 1976).
[4] R. Sandrock, M. Kamphausen and G. M. Schneider, *Mol. Cryst. Liq. Cryst.*, 1978, **45**, 257.
[5] E. Osawa, Z. Majerski and P. von R. Schleyer, *J. Org. Chem.*, 1971, **36**, 205.
[6] R. Kahn, R. Fourme, D. André and M. Renaud, *Acta Cryst. B*, 1973, **29**, 131.
[7] A. Würflinger, *Ber. Bunsenges. phys. Chem.*, 1975, **39**, 1195.
[8] Landolt-Bornstein, *Zahlenwerte und Funktionen* (Springer-Verlag, Berlin, 6th edn, 1960), vol 112*a*.
[9] T. Clark, T. Mc. O. Knox, H. Mackle, M. A. McKervey. *J.C.S. Faraday I*, 1977, **73**, 1224.
[10] W. Spratte and G. M. Schneider, *Ber. Bunsenges. phys. Chem.*, 1976, **80**, 886; *Mol. Cryst. Liq. Cryst.*, 1979, **51**, 101.

High-pressure Phase Transitions in Molecular and Plastic Crystals

By Albert Würflinger

University of Bochum, Institute of Physical Chemistry,
4630 Bochum, West Germany

Received 3*rd December*, 1979

The phase transitions of some molecular and plastic crystals have been studied by pVT and dielectric constant measurements up to 3000 bar. The pVT data show that the density of solid dodecane immediately below the freezing point is comparable with the density of solid undecane immediately below the rotational transition temperature. The phase diagram of acetonitrile exhibits a solid–solid transition (denoted as an α–β solid–solid transition at atmospheric pressure), the pressure dependence of which has not yet been reported. Also cyclohexanone has been studied; it exhibits three solid phases (I, II and III), one of which (II) is only observed at elevated pressures. The dielectric results for cyclohexanone are discussed in terms of the Kirkwood–Fröhlich–Onsager theory. The Kirkwood correlation g-factor is about unity both for liquid and solid (I) cyclohexanone. However, a small increase in the g-factor is observed at the freezing point and a considerable decrease at the I–II solid–solid transition.

In recent years the high-pressure phase behaviour of molecular and plastic crystals has been studied by different experimental methods, *e.g.*, differential thermal analysis,[1–6] dielectric constant,[7,8] and pVT measurements.[9–11] Plastic crystals exhibit at least one solid–solid transition, accompanied by a large enthalpy and volume change. This phase transition is associated with the loss of orientational freedom, leading to a distinct break in the permittivity insofar as polar compounds are considered. Some plastic crystals, moreover, have an intermediate phase transition for which the orientational order is partially lost.[6,12,13] In some cases the intermediate solid–solid transition is only observed at elevated pressures.[6,7]

This paper gives new experimental results for the pVT data of undecane and dodecane, the phase behaviour of which has been studied in earlier investigations.[1] Also the phase behaviour of acetonitrile is studied with pVT and dielectric measurements. New dielectric results of cyclohexanone are discussed, together with recently determined pVT data,[11] in terms of the Kirkwood–Fröhlich–Onsager theory. This theory has been widely used for the discussion of polar fluids,[14,15] but should also be applicable to the plastic phases of polar compounds, where the high value of the dielectric constant is maintained or even increased. Böttcher *et al.*[16] have shown that the Kirkwood–Fröhlich equation is also valid in the case of molecules with isotropic polarizabilities on a cubic lattice, and Hassel and Sommerfeldt[17] found cyclohexanone to have a face-centred cubic lattice in the high-temperature solid phase (denoted here as solid I).

EXPERIMENTAL

Two high-pressure devices have been used for the differential thermal analysis.[1,4] The pVT measurements were carried out with a new high-pressure apparatus developed by Landau and described in detail elsewhere.[9,10] The substance under investigation is enclosed within a steel capsule that is closed by a special moving piston. The piston consists of two parts, separated by an indium seal. A hollow space between these works like a Bridgman's unsupported area yielding a very good sealing of the piston, even after solidification of the substance. The displacement of this moving piston is recorded inductively allowing the calculation of volume changes.

The high-pressure equipment for the dielectric measurements has been described in ref. (7) where details of the experimental procedure and instruments used are also given. Preliminary results on cyclohexanone were obtained with an open capacitor that consisted of two cylinders.[7] A remarkable improvement was achieved in transmitting the pressure with a moving piston similar to that mentioned above for the *pVT* measurements. The improved capacitor was first applied to the dielectric investigation of acetonitrile and will be described elsewhere.[8]

Commercial cyclohexanone (Merck, Darmstadt) was purified as described in ref. (6) and re-examined with the improved capacitor. The frequency used was 1 MHz. No significant frequency dependence or dielectric relaxation was observed. The accuracy in ε (dielectric constant) is believed to be between 0.5 and 1%.

RESULTS

THERMODYNAMIC MEASUREMENTS

Fig. 1 shows former results of the phase behaviour of some even (on the left) and odd (on the right) n-alkanes.[1–3] Most of them exhibit a solid–solid transition

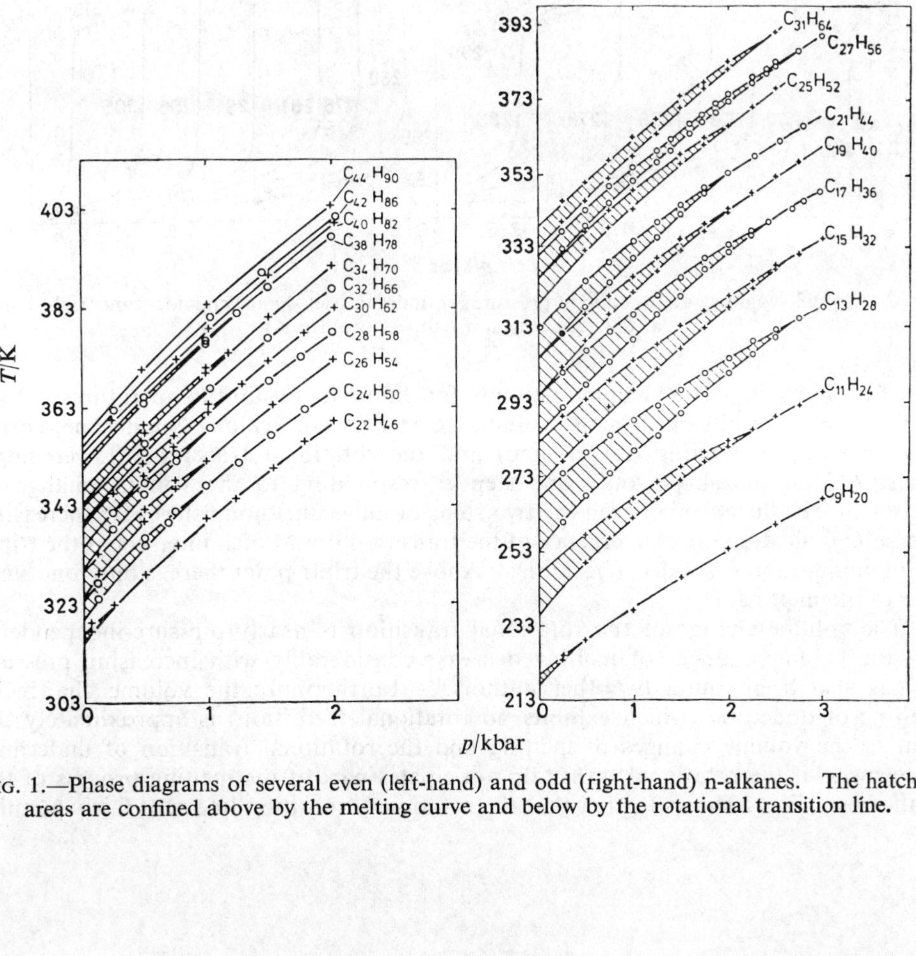

FIG. 1.—Phase diagrams of several even (left-hand) and odd (right-hand) n-alkanes. The hatched areas are confined above by the melting curve and below by the rotational transition line.

(rotational transition) immediately below the melting curve. The high-temperature solid phase is to a certain degree plastic and marked as a hatched area in fig. 1. The rotational transition is observed between nonane and henicosane only for the odd n-alkanes and between docosane and about C_{40} also for the even n-alkanes. A detailed discussion for the odd n-alkanes has already been given in ref. (1). The even n-alkanes reported by Koppitz *et al.*[2] were partially re-examined by Josefiak *et al.*[3] The plot of $C_{36}H_{74}$ is missing from fig. 1 (because it has not yet been redetermined), but is most probably similar to the phase behaviour of the adjacent n-alkanes. In all cases the experimentally observed plastic phase of the alkanes is restricted to a limited pressure range.

Fig. 2 shows recently determined *pVT* data of undecane and dodecane.[9] The

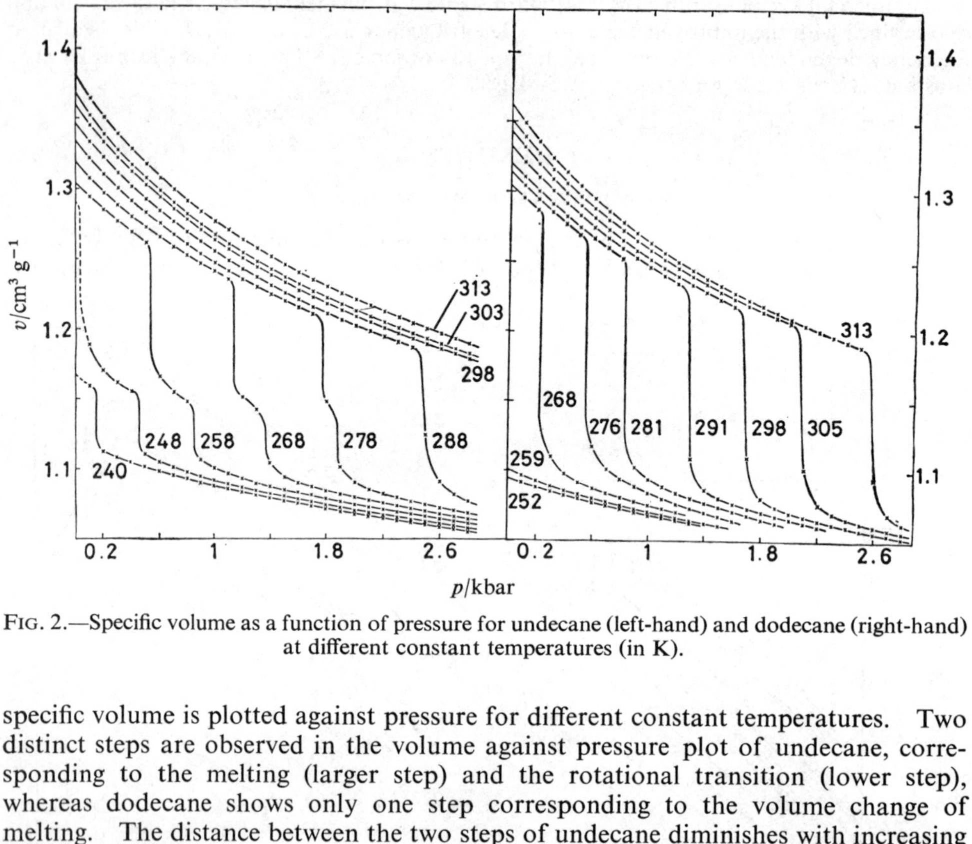

FIG. 2.—Specific volume as a function of pressure for undecane (left-hand) and dodecane (right-hand) at different constant temperatures (in K).

specific volume is plotted against pressure for different constant temperatures. Two distinct steps are observed in the volume against pressure plot of undecane, corresponding to the melting (larger step) and the rotational transition (lower step), whereas dodecane shows only one step corresponding to the volume change of melting. The distance between the two steps of undecane diminishes with increasing pressure, indicating the convergence of the transition lines which intersect at the triple point liquid–solid-I–solid-II (*cf.* fig. 1). Above the triple point there is only one step, due to the melting.

The volume change of the rotational transition is nearly pressure-independent, but the volume changes of melting decrease considerably with increasing pressure as has also been found by other authors.[18] Furthermore the volume change of melting of dodecane (which exhibits no rotational transition) is approximately the sum of the volume changes of melting and the rotational transition of undecane. It seems as if the rotational transition anticipates part of the melting process of the n-alkanes. Fig. 2 shows further that the rotational transition line runs fairly parallel to an isochore, a statement that had already been suggested in earlier investigations.[1]

pVT data of acetonitrile (Landau),[9,10] cyclohexane and cyclohexanone (Wisotzki)[11] have also been determined, both for the liquid and solid states. Fig. 3 shows the

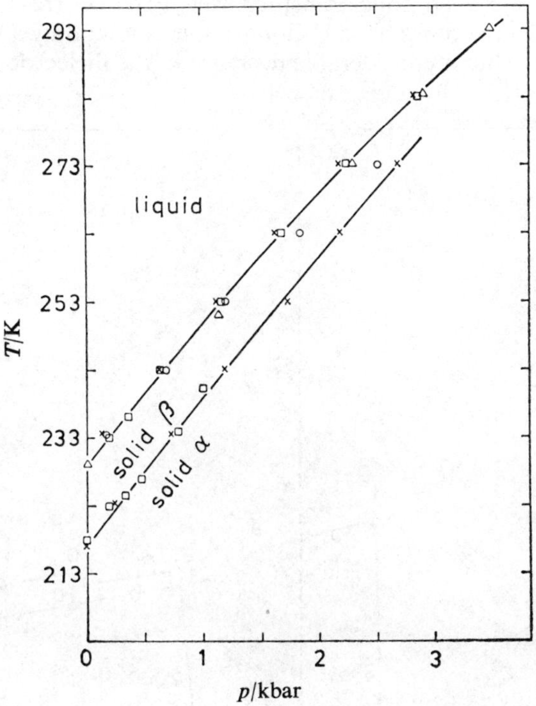

FIG. 3.—Phase diagram of acetonitrile. ×, *pVT* measurements;[9] □, dielectric constant measurements;[8] △, ref. (19); ○, ref. (20).

phase diagram of acetonitrile in comparison with literature data. The melting curve has already been reported by other authors,[19,20] but the pressure dependence of the α–β solid–solid transition has not been found in the literature. However, there is an additional paper of interest by Jakobsen and Mikawa[21] who believe they have found a third solid phase γ, using high-pressure infrared spectroscopy. They report a transition at 30 kbar and room temperature which most probably does not belong to the α–β transition.

The volume changes accompanying the phase transitions allow the calculation of the associated enthalpy changes using the Clausius–Clapeyron equation: Melting of dodecane: 35.4 kJ mol^{-1} at 1 bar rising to 41.4 at 2600 bar. Melting of undecane: 21.5 kJ mol^{-1}, nearly pressure-independent; rotational transition: 6 kJ mol^{-1} at 1 bar rising almost linearly to 12 kJ mol^{-1} at the triple point (2300 bar). Melting of acetonitrile: 8 kJ mol^{-1} at 1 bar rising to 9.5 kJ mol^{-1} at 3000 bar; α–β solid–solid transition: 0.87 kJ mol^{-1} at 1 bar increasing to ≈ 1 kJ mol^{-1} at 2700 bar.

In the case of the n-alkanes the pressure dependence of the enthalpy changes has been determined for related compounds by Kamphausen *et al.*[22] using high-pressure d.s.c. calorimetry. The reported pressure dependence[22] agrees qualitatively with the calculations in this paper.

DIELECTRIC MEASUREMENTS

Fig. 4 shows the dielectric constant as a function of temperature for various substances. Liquid acetonitrile has a large value of the permittivity, which decreases by an order of magnitude at the freezing temperature. There is a further small change ($\Delta\varepsilon \approx 0.05$) at the α–β solid–solid transition (see arrow) corresponding to the small volume change. On the other hand cyclohexanone shows only a small dielectric step at the freezing point, but a considerable decrease in the dielectric constant at a lower

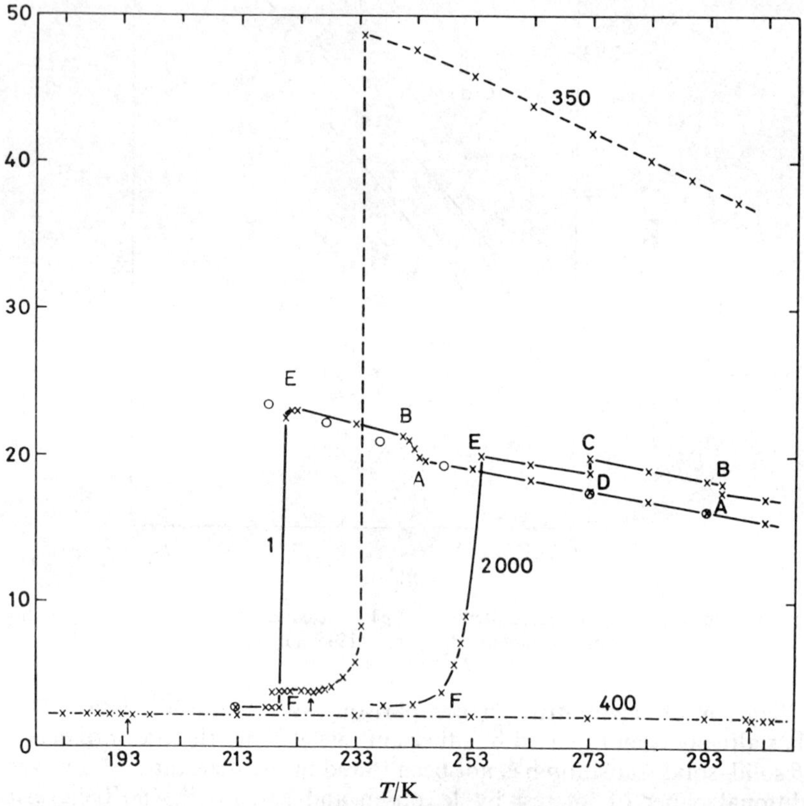

FIG. 4.—Dielectric constant as a function of temperature at different pressures (in bar): cyclohexanone (———), acetonitrile (– – – –), and cyclohexane (– · – · – · –). AB = liquid–solid I, CD, I–II solid–solid and EF, II–III solid–solid transition of cyclohexanone. ○, Corfield and Davies.[12]

solid–solid transition. The completely different dielectric behaviour of these two substances indicates clearly that cyclohexanone must possess orientational freedom within the high temperature solid phase, whereas acetonitrile obviously does not belong to the class of " reorientational solids ". Cyclohexanone is a typical representative of the " plastic crystals " which usually have low enthalpy- and volume-changes of melting, as has been confirmed in high pressure calorimetric[6] and volumetric[11] measurements. Fig. 4 also shows atmospheric pressure data of Corfield and Davies[12] for comparison. The agreement is good at room temperature and at ≈ -50 °C, but some deviations occur near the melting temperature. Even low concentrations of water impurities cause a sluggish melting transition and values of ε that are too low in solid phase I, which might perhaps be the reason for the deviations.

Furthermore fig. 4 shows that cyclohexanone has an additional dielectric step at 2000 bar, due to a second solid–solid transition recently found by differential thermal analysis.[6] Preliminary dielectric measurements of cyclohexanone[7] have now been extended to 3000 bar and are summarized in table 1. The dielectric results of acetonitrile will be discussed in detail elsewhere.[8]

TABLE 1.—STATIC DIELECTRIC CONSTANT, ε, OF CYCLOHEXANONE AS A FUNCTION OF PRESSURE AND TEMPERATURE. THE COLUMNS ARE SUBDIVIDED, REFERRING TO THE LIQUID, SOLID I AND SOLID II, RESPECTIVELY. THE DATA WERE SMOOTHED USING A POLYNOMIAL $\varepsilon = a + bp + cp^2$ AT CONSTANT TEMPERATURE FOR EACH PHASE SEPARATELY.

T/K p/bar	303	293	283	273	263	253	243	233	223
1	15.54	16.17	16.85	17.57	18.35	19.16	20.00	22.10	23.1
250	15.77	16.35	17.05	17.79	18.52	19.32	21.10	22.20	
500	16.00	16.53	17.26	18.00	18.70	20.40	21.30	20.86	
750	16.20	16.72	17.46	18.19	19.78	20.60	21.50		
1000	16.40	16.91	17.65	18.37	19.96	20.80	20.38		
1250	16.59	17.10	17.83	19.30	20.15	19.75	20.5		
1500	16.76	17.29	18.68	19.50	20.34	19.87			
1750	16.92	17.48	18.87	19.68	19.33	20.0			
2000	17.07	18.30	19.04	19.87[a]					
2000				18.84[a]	19.44				
2250	17.21	18.47	19.22	18.94	19.54				
2500	17.94	18.65	18.40	19.05	19.65				
2750	18.15	18.80	18.50	19.15					
3000	18.22	18.03	18.60	19.25					

[a] Both values refer to the I–II solid–solid transition which takes place at 2000 bar.

Cyclohexane is a further representative of the " plastic crystals " and has also been studied by calorimetric[4,23] and pVT measurements.[11] However, this compound does not possess a permanent dielectric dipole moment; therefore, its dielectric constant is small and does not change much ($\Delta\varepsilon \approx 0.05$) at the phase transitions (see arrows). The temperature dependence of ε agrees on the whole with the literature data of Chew and Chan.[24]

Fig. 5 shows another representation of the dielectric results of cyclohexanone. The dielectric constant is plotted as a function of pressure along the phase transition lines of cyclohexanone. The curves A, B, C . . . correspond to the edges A, B, C . . . of fig. 4. The distance AB shows the increase in ε at the freezing temperature and the distance CD refers to the decrease in ε at the solid-I–solid-II transition, respectively. The low value of ε in the solid phase III (edge F) is not shown in fig. 5, because it was rather difficult to perform dielectric measurements in that brittle and hard solid phase. The value at atmospheric pressure is ≈ 2.8, in accordance with data of Corfield and Davies.[12] At higher pressures the accuracy of ε in the solid phase III is less, but ε does seem to be below 3.0.

Fig. 5 further shows that the transition lines E and D diverge with increasing pressure according to the increasing solid phase II region. They intersect at atmospheric

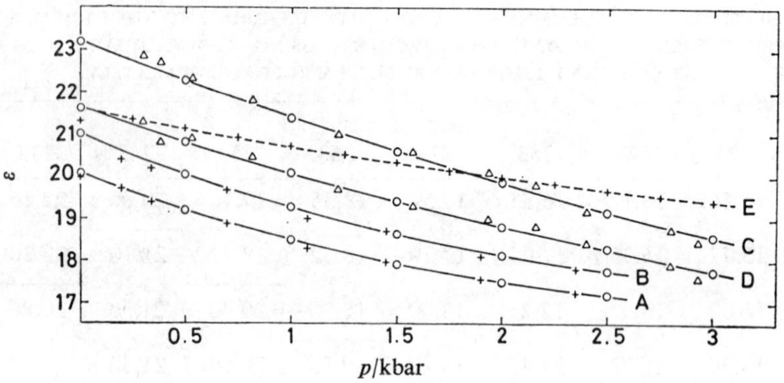

FIG. 5.—Dielectric constant of cyclohexanone as a function of pressure along the phase transition lines. The letters A, B, C... refer to fig. 4. O, smoothed values according to a polynomial $\varepsilon = a + bp + cp^2$; Δ and +, experimental points except curve E which was also smoothed.

pressure at the triple point solid-I–solid-II–solid-III; for details of the phase diagram see ref. (6), where the phase-transition temperatures as a function of pressure are also given.

DISCUSSION

n-ALKANES

One of the striking features of the n-alkanes of higher chain length is the occurrence of the " rotational transition ". The nature of this phase transition has been investigated intensively,[25–28] and will certainly be discussed by Ewen *et al.*[29] Apart from the difficulties of explaining the mechanism of the rotational transition the question arises: why does this phase transition occur at all? Pople and Karasz[30] have suggested a model that allows one to a certain extent to predict the existence of plastic phase transitions. A good review of this model is given by Findenegg.[31] The essential parameters of the model are two energy barriers w and w', concerning the repulsion of a positionally disordered molecule in the lattice and the energy of an orientationally disordered molecule, respectively. Then the ratio w'/w is a measure of the anisotropy of the non-spherical molecule. Assuming a certain volume dependence of w and w' (but such that the ratio w'/w is independent of temperature and pressure) it is possible to predict the pressure dependence of the phase transitions. The main result is that the high-temperature solid phase region should be increased with increasing pressure. This is indeed true for many plastic crystals,[4,6] but is not observed in the case of the n-alkanes (see fig. 1), as has already been pointed out by Kohler.[32]

This different high-pressure phase behaviour of the n-alkanes may perhaps be qualitatively understood if a pressure dependence of w'/w is assumed. According to early X-ray results of Müller[33] the compressibility of tricosane is anisotropic. Therefore the ratio w'/w is most probably pressure-dependent and should increase with increasing pressure. However, an increasing ratio of w'/w is unfavourable for the existence of a plastic phase. According to Karasz and Pople[30] a plastic phase does not occur for $w'/w > 0.66$, occurs only at elevated pressures for $0.325 < w'/w < 0.66$, and at atmospheric pressure for $w'/w < 0.325$. It is possible that, in the case of the n-alkanes, the ratio w'/w is considerably enlarged with increasing pressure and leads to a disappearance of the plastic phase region (hatched areas in fig. 1) for a sufficiently high pressure (triple-point pressure). However, the theory of Pople and Karasz should not be taken too seriously in the case of the n-alkanes, because certainly more than two orientations, as well as more than two energy barriers, are involved with these comparatively long chain molecules, not to mention that more than one solid–solid transition is observed in some cases.[3,24]

CYCLOHEXANONE

Fig. 4 shows that the dielectric constant of cyclohexanone increases with decreasing temperature. There is a further increase at the freezing point, probably caused by the increase in the density. It is the main purpose of this section to discuss whether the changes in ε with temperature and pressure correspond exactly to changes in the density. An eventual deviation may be caused by the interaction of the individual molecular dipoles which is often described in terms of the Kirkwood–Fröhlich–Onsager theory using the well-known Kirkwood g-factor:[34]

$$g = \frac{(\varepsilon - \varepsilon_\infty)(2\varepsilon + \varepsilon_\infty)9kTM}{\varepsilon(\varepsilon_\infty + 2)^2 4\pi N_0 \rho \mu^2}.$$

ε is the measured static dielectric constant and ε_∞ is calculated from the refractive index ($n_D = 1.451$ at 20 °C and 1 atm)[35] according to the relation[36]

$$\varepsilon_\infty = 1.05\, n^2.$$

The change in ε_∞ with temperature and pressure was calculated with the help of the Lorenz–Lorentz equation. The necessary density data, ρ, have been taken from Wisotzki.[11] μ is the permanent electric dipole moment of an individual molecule that is usually obtained from dielectric measurements of dilute gases. However, most of the reported literature data refer to dilute solutions of cyclohexanone, μ being 2.8,[37] 2.9,[38,39] 2.95[40] and 3.01 D.[41,42] Another value of 2.87 D has been obtained from microwave spectroscopy.[43] A mean value of 2.9 D is used for the calculation of the Kirkwood g-factor. Also Corfield and Davies[12] report 2.9 D calculated from Onsager's equation; but of course this cannot be taken into account, because the application of Onsager's equation would mean $g \equiv 1$. On the other hand the results of Corfield and Davies show that the g-factor of cyclohexanone must be about unity, provided that the mean value of $\mu = 2.9$ D[37–43] truly refers to an individual molecular dipole. The latter statement is supported by the close value of 2.87 D[43] obtained without using any of the assumptions of the Kirkwood–Fröhlich–Onsager theory. The other constants of the Kirkwood–Fröhlich equation are explained in ref. (34); very similar calculations have also been done by Franck.[14,15]

Fig. 6 shows the Kirkwood g-factor as a function of pressure at three different temperatures in comparison with three isotherms of acetonitrile.[8] Whereas the g-factor of acetonitrile is distinctly less than 1, indicating preferred antiparallel correlation, the g-factor of cyclohexanone is very close to unity, in accordance with the above remarks concerning the results of Corfield and Davies.[12] The pressure and temperature dependences of g are similar for liquid cyclohexanone and liquid acetonitrile. At the freezing point of cyclohexanone (2360 bar at 30 °C, 1080 bar at 0 °C, 360 bar at −20 °C), however, a small increase in the *g*-factor is observed which means that the increase in ε at the freezing temperature does not exactly correspond to the increase in the density on freezing. Although this small change in *g* ($\approx 1.5\%$) during the melting transition is comparable with the sum of the relative uncertainties of the measured quantities involved, the increase in *g* was observed in all cases, also at atmospheric pressure: $g = 0.99$ in solid cyclohexanone at −35 °C, compared with $g = 0.97$ in liquid cyclohexanone at −20 °C. The higher value of the *g*-factor in the solid phase I is perhaps caused by the greater ease of molecular rotation of the cyclohexanone molecules compared with the liquid state. This assumption is supported

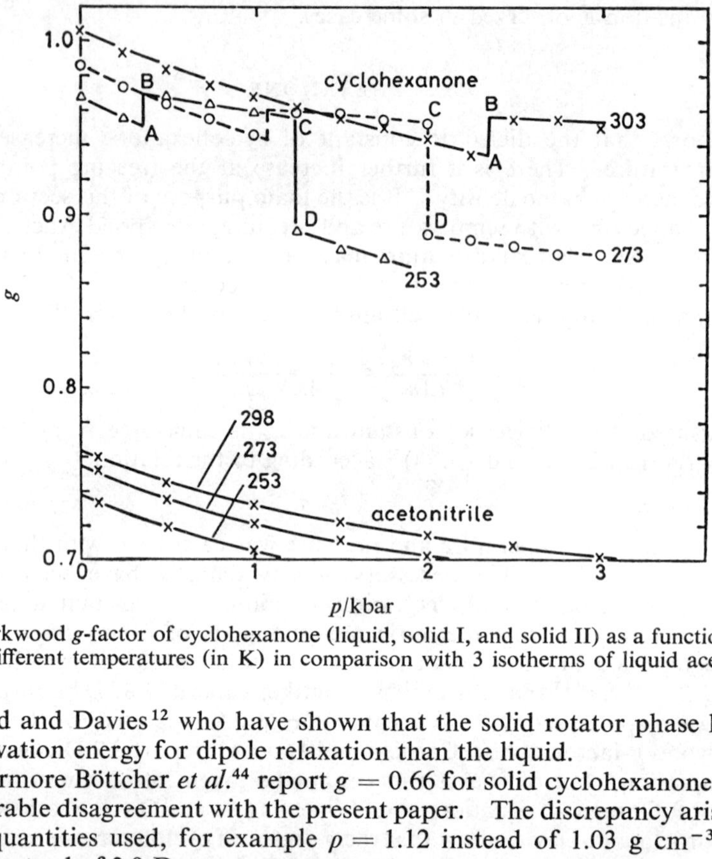

FIG. 6.—Kirkwood *g*-factor of cyclohexanone (liquid, solid I, and solid II) as a function of pressure for 3 different temperatures (in K) in comparison with 3 isotherms of liquid acetonitrile.

by Corfield and Davies[12] who have shown that the solid rotator phase has an even lower activation energy for dipole relaxation than the liquid.

Furthermore Böttcher *et al.*[44] report $g = 0.66$ for solid cyclohexanone at −40 °C, in considerable disagreement with the present paper. The discrepancy arises from the different quantities used, for example $\rho = 1.12$ instead of 1.03 g cm^{-3} [ref. (45)], $\mu = 3.08$ instead of 2.9 D, *etc.*

As far as the solid phase II which exists only at elevated pressures is concerned, no structural information is available. If solid II is also characterized by a cubic lattice the Kirkwood–Fröhlich equation may be applied to this phase as well. Reserving this limitation, the calculations have been extended to the solid phase II. Fig. 6 shows that the *g*-factor decreases considerably at the I–II solid–solid transition (2000 bar at 0 °C, 1240 bar at −20 °C), demonstrating that some part of the reorientational freedom must be frozen in. It does not seem reasonable to describe the lower *g*-value of the solid phase II in terms of antiparallel oriented dipoles or dimers, as has been done with certain success in the case of liquid acetonitrile and related compounds.[8,15,46,47]

Fig. 6 further shows that the steps of the Kirkwood g-factor are nearly independent of temperature and pressure. Obviously the change in g along a phase transition line is much less (almost negligible) than along an isotherm: g being about 0.95 (A), 0.965 (B), 0.955 (C), 0.89 (D), and 0.88 (E), respectively, the letters A–D referring to fig. 4 and 5. Similar behaviour was found with liquid acetonitrile, which has an approximately constant value of $g \approx 0.7$ at the melting curve.[8]

The author thanks the Deutsche Forschungsgemeinschaft for financial support.

[1] A. Würflinger and G. M. Schneider, *Ber. Bunsenges. phys. Chem.*, 1973, **77**, 121.
[2] B. Koppitz and A. Würflinger, *Colloid Polymer Sci.*, 1974, **252**, 999.
[3] C. Josefiak, A. Würflinger and G. M. Schneider, *Colloid Polymer Sci.*, 1977, **255**, 170.
[4] A. Würflinger, *Ber. Bunsenges. phys. Chem.*, 1975, **79**, 1195.
[5] R. Landau, A. Würflinger and G. M. Schneider, *Z. analyt. Chem.*, 1977, **286**, 216.
[6] A. Würflinger and J. Kreutzenbeck, *J. Phys. Chem. Solids*, 1978, **39**, 193.
[7] A. Würflinger, *Ber. Bunsenges. phys. Chem.*, 1978, **82**, 1080.
[8] A. Würflinger, *Ber. Bunsenges. phys. Chem.*, 1980, submitted.
[9] R. Landau, *Doctoral Thesis* (University of Bochum, 1980).
[10] R. Landau and A. Würflinger, *Rev. Sci. Instr.*, submitted.
[11] K. D. Wisotzki, *Diplom. Thesis* (University of Bochum, 1980).
[12] G. Corfield and M. Davies, *Trans. Faraday Soc.*, 1964, **60**, 10.
[13] D. N. Glidden, *Thesis* (University of New Mexico, 1972).
[14] E. U. Franck and R. Deul, *Faraday Disc. Chem. Soc.*, 1978, **66**, 191.
[15] E. U. Franck, in *Organic Liquids*, ed. A. D. Buckingham, E. Lippert and S. Bratos (John Wiley, London, 1978), p. 181.
[16] C. J. F. Böttcher and P. Bordewijk, *Theory of Electric Polarization* (Elsevier, Amsterdam, 1978), vol. II, p. 445.
[17] O. Hassel and A. M. Sommerfeldt, *Z. phys. Chem. (Leipzig)*, 1938, **40**, 391.
[18] R. R. Nelson, W. Webb and J. A. Dixon, *J. Chem. Phys.*, 1960, **33**, 1756.
[19] J. Timmermans and M. Kasanin, *Bull. Soc. Chim. Belg.*, 1959, **68**, 527.
[20] G. M. Schneider, *Z. phys. Chem. (Leipzig)*, 1964, **41**, 327.
[21] R. J. Jakobsen and Y. Mikawa, *Appl. Optics*, 1970, **9**, 17.
[22] M. Kamphausen and G. M. Schneider, *Thermochim. Acta*, 1978, **22**, 371.
[23] H. Arntz and G. M. Schneider, *Faraday Disc. Chem. Soc.*, 1980, **69**, 139.
[24] H. A. Chew and R. K. Chan, *Canad. J. Chem.*, 1973, **51**, 2141.
[25] G. Strobl, B. Ewen, E. W. Fischer, and W. Piesczek, *J. Chem. Phys.*, 1974, **61**, 5257 and 5265.
[26] Ishinabe, Takao, *J. Phys. Soc. Japan*, 1974, **37**, 315.
[27] W. Pechhold, W. Dollhopf, and A. Engel, *Acoustica*, 1966, **17**, 61; S. Blasenbrey and W. Pechhold, *Rheologica Acta*, 1967, **6**, 174.
[28] D. H. Bonsor and D. Bloor, *J. Mater. Sci.*, 1977, **12**, 1552.
[29] B. Ewen, G. R. Strobl and D. Richter, *Faraday Disc. Chem. Soc.*, 1980, **69**, 19.
[30] J. A. Pople and F. E. Karasz, *J. Phys. Chem. Solids*, 1961, **18**, 28 and **20**, 294.
[31] G. H. Findenegg in *The Liquid State*, ed. F. Kohler (Verlag Chemie, Weinheim, 1972), p. 194.
[32] F. Kohler, in *Organic Liquids*, ed. A. D. Buckingham, E. Lippert and S. Bratos (John Wiley, London, 1978), p. 260.
[33] A. Müller, *Proc. Roy. Soc.*, 1941, **178**, 227, and earlier papers.
[34] C. J. F. Böttcher, *Theory of Electric Polarization* (Elsevier, Amsterdam, 1973), vol. I, p. 258.
[35] J. A. Riddick and W. B. Bunger, *Techniques of Chemistry*, vol. II, *Organic Solvents*, ed. A. Weissberger, (Wiley-Interscience, New York, 3rd edn, 1970), p. 246.
[36] C. J. F. Böttcher, *Theory of Electric Polarization* (Elsevier, Amsterdam, 1972), vol. I, pp. 180 and 272.
[37] S. O. Morgan and W. A. Yager, *Ind. Eng. Chem.*, 1940, **32**, 1519.
[38] J. W. Williams, *J. Amer. Chem. Soc.*, 1930, **52**, 1831.
[39] I. F. Halverstadt and W. D. Kumler, *J. Amer. Chem. Soc.*, 1942, **64**, 1982.
[40] W. Holzmüller, *Z. phys. Chem. (Leipzig)*, 1937, **38**, 574.
[41] J. B. Bentley, K. B. Everard, R. J. B. Marsden and L. E. Sutton, *J. Chem. Soc.*, 1949, 2957 and 2963.
[42] K. Houssou, A. Proutière, A. Caristan, and H. Bodot, *J. Chim. phys., Phys. biol.*, 1977, **74**, 499.

[43] Y. Ohnishi and K. Kozima, *Bull. Chem. Soc. Japan*, 1968, **41**, 1323.
[44] C. J. F. Böttcher and P. Bordewijk, *Theory of Electric Polarization* (Elsevier, Amsterdam, 1978), vol. II, p. 451.
[45] R. J. Meakins, *Trans. Faraday Soc.*, 1962, **58**, 1962.
[46] W. Dannhauser and A. F. Flückinger, *J. Phys. Chem.*, 1964, **68**, 1814; G. P. Johari and W. Dannhauser, *J. Chem. Phys.*, 1968, **48**, 5114.
[47] K. R. Srinivasan and R. L. Kay, *J. Solution Chem.*, 1977, **6**, 357.

The Use of Powder Diffraction in the Study of Molecular Phase Transitions

By G. Stuart Pawley

Physics Department, University of Edinburgh,
Kings Buildings, Edinburgh EH9 3JZ

Received 7th January, 1980

An outline is given of the experimental techniques available for studying the structure of molecular systems through powder diffraction. A wide range of physical environments is possible with this technique making it appropriate for the study of phase transitions by means of structure. The difficulties encountered in the data analysis are enumerated and the limitations imposed by resolution are discussed. The state of the art is not yet fine enough for us to be able to put a high probability of success on neutron diffraction studies, though this is possible in practice. A new program to help in the search for unknown unit cells is described. The advantages of the fully operational program EDINP for the study of molecular phase transitions through the use of constrained refinements are stressed. A typical example of the type of system which can be studied in this way is presented, namely $C_6D_6.C_6F_6$, where the powder diffraction scans for the two phases are given for comparison.

The purpose of the present paper is to focus attention on a technique which is very rapidly developing and is outstripping the expertise needed for its analysis. Powder diffraction has been understood from the earliest days of crystallography, but has remained neglected until recently because of the unquestionable success of single-crystal diffraction work. Considerable stimulus was given to neutron powder diffraction by the work of Rietveld[1] and this field has flowered with the study of magnetic systems and with phase transitions in such systems as the perovskites, where small distortions of the unit cell usually occur. Success in these fields has depended in large measure on the approximate foreknowledge of the structure of the unknown phase. It is a very big step therefore to move to systems where phase changes occur between structures which are so unrelated that there can be no assistance in determining the structure of one phase simply by knowing the structure of the other. Such phase transitions are common in the molecular solid state, and it behoves us to develop the techniques to be able to attempt the analysis of such unknown phases.

EXPERIMENTAL TECHNIQUES

We should convince ourselves at the outset that the prize is worth the effort. In neutron scattering the technique of high resolution multicounter diffractometry is well developed, but there is still a possibility of further progress, increasing the resolution on even the best instruments such as D1A at I.L.L. Grenoble. Soller systems can now be made so that the ultimate possible resolution is achievable, this being governed only by the crystallite particle size. It appears, after considerable effort, that the resolution of D1A at present is not quite enough for us to put a high probability of success on a structure determination – we had hopes of presenting a success at this meeting, but will have to be satisfied in outlining the difficulties faced in such tasks.

With increased resolution there comes a loss in intensity, but as few powder diffraction analyses are statistics-limited, this should not unduly concern us. Clearly structures with supercells and therefore weak superlattice reflections will need higher count rates, but in these cases it often happens that a knowledge of the pseudocell is sufficient to enable the structure determination without the higher resolution.

One difficulty with neutron powder diffraction which has been almost totally neglected is the correct procedure for accounting for thermal scattering. It is not inconceivable that this problem be circumvented by the inclusion on a high resolution single counter diffractometer of a spin-echo device which would reject all those neutrons which were scattered with an energy change. This is possible to some extent by using a conventional analyser, but even then most of the troublesome scattering from the acoustic phonons is still not eliminated. A spin-echo powder diffractometer is technically possible, and simply needs the demand of the scientific community for it to be constructed.

But we should not restrict ourselves to considering only conventional neutron diffractometry. Pulsed source diffractometry is now possible on instruments such as the back-scattering apparatus on the Harwell LINAC, and in the early 1980s this will be possible on the higher flux spallation neutron source, SNS. The back-scattering technique is not limited in its resolution by sample particle size, and could well prove to be the superior technique. The difficulties with this technique lie in the diffraction peak shape which is very asymmetric and varies widely over one measurement set. However, it appears to the writer that a greater problem exists, and that concerns the range of measurements permitted. If a structure is known to some extent, then the improvement in knowledge of the structure lies in the measurement of the higher order reflections, at which this technique is undeniably successful. But for an unknown structure it is essential to measure the first-order lines, and these are produced by neutrons of, *e.g.*, 40 Å wavelength if back-scattered from a unit cell of 20 Å. Such cell edges are not rare. Although the neutron flux at 40 Å will be very weak, the problem of measurement is exacerbated by the fact that the long-wavelength neutrons travel so slowly that they are overtaken by faster neutrons from later pulses. It will be essential in our work to be able to reduce the pulse rate by a large factor, otherwise we will be denied these all-important reflections.

Another experimental source which should not be overlooked is the synchrotron X-ray source. Here again there are problems with the diffraction peak shape, but these are general problems which will be surmounted. These X-ray sources promise to be so powerful, as does the SNS, that full diffraction data from one sample can be collected in minutes or even in seconds. As a comparison, the conventional neutron diffractometer D1B at I.L.L. gives a respectable scan in about five minutes, though the resolution is not high.

Neutrons have the advantage over X-rays that they are not highly absorbed by the sample or its container. Thus it is possible to have quite a large specimen with a carefully controlled temperature environment and also with controlled pressures up to 35 kbar. The specimen size required restricts the pressure range possible, but most molecular transitions occur within this range.

Thus the techniques are possible for studying molecular systems over a wide range of conditions, but the analysis of the results now presents the most difficult problems.

STRUCTURE REFINEMENT

Logically it would be better to discuss the solving of structures before their refinement, but as the solution stage is the main stumbling block it will be left to last. One aspect of such refinements must be made quite clear, and that is that the powder technique cannot rival accurate single-crystal work in the precision of the final result. A number of papers have appeared (uncited) in which unreasonably small standard deviations are quoted without qualification. The fallacy which leads to these results has recently been analysed.[2] The information content of a powder scan is necessarily very limited, and therefore our scientific objectives will be limited. A very effective way of working within such limits is to employ constrained refinements[3] whereby use can be made of additional information. This is particularly advantageous for molecular systems.

If a molecular system undergoes a phase transition, it is to be expected that there is no change in the basic molecular structure, but that a rearrangement of the molecules takes place. The molecules retain their integrity. This is considerable information, and if it can be used in the data refinement it will improve the reliability of the other parameters being determined. Thus we are not concerned with the relative positions of the atoms in the molecules, these we know, but do want to know how the molecules rearrange themselves. The orientation is most usually determined by three parameters, such as Euler angles, and the structure can be described by a molecule of predetermined shape positioned through the use of these variable parameters.

The first molecular system to be studied by powder diffraction in this way was perfluorobiphenyl,[4] using Rietveld's programme.[1] In this example a molecule of 22 atoms was oriented by one Euler angle, but two other parameters were needed: the angle between the two ring systems and the length of the bond connecting them. The constraints necessary to retain the shape of the ring systems taxed the programme to the full, and were only possible because of the symmetry of the particular problem. However, the analysis showed that the technique is feasible and that it was worthwhile developing a programme that could cope with constraints of any complexity.

Such a programme has now been written and tested[4,5] and is now available for general distribution. The first tests of the programme were done on two unknown structures,[5] p-$C_6F_4Br_2$ and p-$C_6F_4I_2$, though in these cases the unit cells were determined by single-crystal studies. The systems are isostructural, except that the latter has a transition to a higher-temperature phase at about 80 °C. D.t.a. studies show this transition only on warming, and therefore a sample frozen from the melt was ground up and studied by powder diffraction, but it would seem that the grinding process precipitated the transition as the resulting diffraction pattern was consistent with the lower phase. A feeling for the capacity of the method can be gained from these examples. The unit cells have monoclinic angles very close to 90°, and such occurrences always lead to ambiguities. These ambiguities were, however, easily resolved by this technique. A number of false structural minima occur in these structures because the neutron scattering cross-sections for F, Br and I do not differ greatly. The false minima occur when the molecules are oriented wrongly by 60 and 120° in the molecular plane, but the true minimum was clearly indicated by the method.

EXAMPLES SHOWING THE SCOPE OF THE METHOD

The examples mentioned so far in this paper could have been studied just as successfully with single crystals, but there are many cases where single-crystal work is impossible and the powder technique comes into its own.

This is true of perbromobenzene, where the original attempts at single-crystal study were frustrated by the lack of good crystals, as this substance forms overgrowths even in the most favourable growing conditions.[7] The interest in this system is to know the molecular orientation in order to understand the cause for the overgrowing, and we therefore need to determine only the three Euler angles. A refinement of atomic coordinates is counter-productive as the information in a powder diffraction scan is insufficient to be able to improve on a molecule of predicted shape, and therefore only two molecule structure parameters were investigated, the C—C and C—Br bond lengths. The resulting structure was well-determined, but unfortunately no clear reason for the overgrowing emerged.[8]

There are many transitions of the order–disorder type, and the study of these will be greatly facilitated by the ability to collect full structural data at a given temperature in a small time. This was attempted on the system C_2I_4 using the instrument D1B at I.L.L. Earlier results examined by unconstrained refinement[9] did not give a clear picture of disorder, but when reexamined with EDINP (Edinburgh Powder Programme[5,6]) the disordering was significantly determined.[10] We hoped to be able to follow the change in disorder as a function of temperature, but now think that for such a study the information content of a low-resolution diffraction scan is insufficient.

In octafluoronaphthalene there is at least one pressure-induced phase transition at ≈ 0.8 kbar and probably a further transition at higher pressures.[11] The first transition has been elucidated through neutron powder diffractometry, but the problems outlined in the following section were avoided by the fact that the high-pressure phase was identical to the low-temperature phase, which had already been solved. One problem with the study of pressure-induced phase transitions is that a given sample may not experience an even pressure throughout its bulk. We have an example (pyrene, unpublished) where the diffraction scan appears to be the superposition of two diffraction patterns from two phases. As each pattern consists of many overlapping lines it is impossible to discard some data leaving a scan for just one phase. Consequently both phases must be refined together, and for such cases a special version of EDINP has been produced which will simultaneously refine any number of structures. This programme is not quite ready for general distribution.

Certain transitions may be caused by an instability against a mode of vibration. When this occurs, the mode frequency approaches zero and the amplitude increases, and it is possible that this manifests itself through the Debye–Waller factor. In molecular systems the type of the motion concerned is most likely to be of the rigid-body type, and the rigid-body Debye–Waller factors are therefore of interest. We have examined whether it is possible to determine these by constrained powder refinement,[12] and it is clear that the very best data are necessary if any reliance is to be put on such a result. Bearing in mind that the new techniques may well improve on those existing now, such studies are worth undertaking.

Many molecular systems undergo phase transitions to plastic phases, phases in which there is some degree of dynamic disorder. It is often not possible to grow single crystals in the crystalline phase, once more making powder diffraction the obvious tool. An example is given by pivalonitrile, as presented by one of the papers at this Discussion.[13] The tertiary butyl group consists of three methyl groups, each of which has considerable orientational freedom about the C—C bond. A special subroutine for EDINP was written for this case, in which each methyl group is oriented independently about its C—C bond. Many problems require the programming of a special constraint, but this is usually a simple procedure for anyone acquainted with FORTRAN.

STRUCTURE SOLUTION

In most of the examples quoted above, the unit cell or the structure was approximately known before the powder analysis was done. However, many cases exist where this is not so, and it is essential for us to develop the methods necessary to enable the solving of a structure without any crystallographic information. There are essentially three steps in this process, that of finding the unit cell, then the space group and then the full structure.

The first step is undoubtedly the most difficult, and has received some attention over the last decade or so. For a review of the present state of the art the reader is referred to the work of Shirley.[14] Procedures do exist for suggesting possible unit cells, and these have to be examined with the knowledge of the probable volume occupied by a molecule in mind. It is essential for any hopes of success in such a venture to have a very accurate knowledge of the zero scattering angle for a scan, for if this is just slightly wrong the correct solution may slip through the computational net, whereupon any subsequent work is entirely wasted. Next in importance is the resolution of the scan, as poor resolution results in close reflections being thought of as just one reflection, and when this happens the position of the average reflection does not correspond to any actual reflection, and false suggestions are supported.

Finding the appropriate unit cell becomes increasingly difficult if the low-order reflections are missing, and for this reason it becomes important to insist on having the ability to measure these reflections when pulsed-source diffraction is used. As all unit-cell solution techniques rest heavily on the lowest-order reflections, it is very important that these be resolved. This is probably the most limiting factor with conventional neutron techniques. A typical example of a neutron powder diffraction scan is given in fig. 1, from which one can see that the peak half-width at 30° is 0.3°.

It is unusual for the indexing programme to result in a clear correct solution for the unit cell. We have to be prepared to investigate a number of different possible solutions before choosing one in which to invest the effort of structure solution. To aid in this step we have recently produced a version of EDINP which does not require a trial structure before it can be run. The programme aims to fit the whole scan profile while refining the unit cell. Other parameters refinable if desired are the scan zero angle and the peak shape parameters. All this is exactly as in EDINP which, in this respect, is the same as Rietveld's programme.

The alteration in the functioning of the new programme concerns the crystallographic structure factors, which are replaced by independent intensities. By allowing these intensities to refine at the same time as the unit cell is refined, we are able to change the indexing of the scan in a smooth fashion, making allowances for the frequent occurrences of more than one reflection in one resolved peak. It is planned that each highly probable unit cell suggested by the indexing programme will receive some refinement by the new programme before a final assessment is made. Because this is a total profile fitting the result has to be very good before a solution is accepted for further study. Only one scan has been examined by the programme to date, cyclohexene-F_{10}, and only one unit cell, which may be incorrect, has been investigated. This cell fits the first 100 reflections except for two very small shoulders, but these discrepancies may well prove to be crucial and the result may have to be rejected.

One problem with this new programme has not been mentioned, and that is that two very close reflections cannot take on independent variable reflection intensities without enormous correlation. This correlation can be partially suppressed by constraining close reflections to have intensities which are nearly the same. The amount of correlation suppression can be controlled by the user. At the end of a refinement a set of reflection intensities is produced, with indices and correlations, and this data set may well be the best input for the next two stages of analysis.

If the unit cell is correct, the second stage, that of determining the actual space group, must be tackled. Either the list of reflection indices has to be scrutinised for possible systematic absences, or possible space group symmetries should be applied to the unit cell used with the programme just described. By either of these methods, or even by intuition, a space group must be proposed before the third stage can be attempted.

This third stage is a search for a possible structure. Possibilities can be postulated through packing arguments or calculations or through more sophisticated potential function lattice statics calculations. Probably the quickest way to test any proposed structure is to compare calculated intensities with the intensity set just mentioned, taking due care of the correlations. When a good trial structure is thus obtained, refinement with EDINP should be straightforward.

AN EXAMPLE

In conclusion an example will be described which has not yet been solved successfully. Two of the diffraction scans, obtained recently on D1A at I.L.L. are shown in fig. 1. The system under investigation is $C_6D_6.C_6F_6$, being a 1:1 complex between

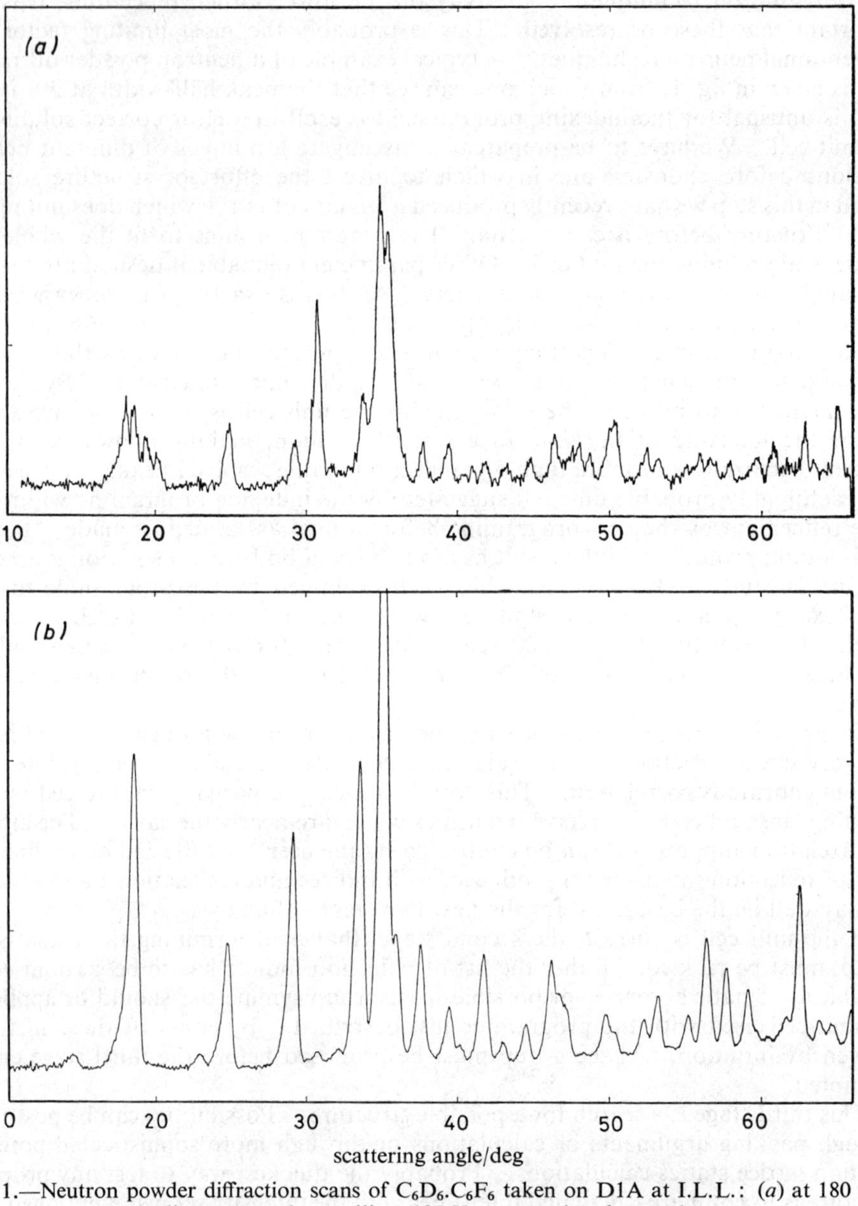

FIG. 1.—Neutron powder diffraction scans of $C_6D_6.C_6F_6$ taken on D1A at I.L.L.: (*a*) at 180 K on cooling, upper phase; (*b*) at 195 K on warming, lower phase.

benzene and perfluorobenzene. This has been studied most recently by Raman scattering,[15] and a transition was observed on cooling at 170 K which was not reversed on warming until 210 K. The first scan shows the result on cooling below 210 K, but still in the upper phase, while the second scan shows the result at a higher temperature in a warming sequence and therefore in the lower phase. These scans are consistent with others taken well away from the transition temperature. These scans are clearly distinct, but have a number of features in common. The common features probably indicate a relationship between the structure in the two phases. We surmise that the phase change in akin to a crystal-to-plastic phase, but the orientational disorder in the plastic phase could take on many forms, reorientation of individual molecules or reorientation of close pairs of dissimilar molecules which are rather more tightly bound together than the overall intermolecular binding. We hope to make progress on this system by searching for the unit cell of the upper phase structure by single-crystal X-ray methods, but our long-term objective is to be able to solve such structural probems entirely through the use of powder diffraction methods.

[1] H. M. Rietveld, *J. Appl. Cryst.*, 1969, **2**, 65.
[2] M. Sakata and M. J. Cooper, *J. Appl. Cryst.*, 1979, **12**, 554.
[3] G. S. Pawley, *Advances in Structure Research by Diffraction Methods*, ed. W. Hoppe and R. Mason (Pergamon Press, Oxford, 1972), vol. 4, pp. 1–64.
[4] G. A. Mackenzie, O. W. Dietrich and G. S. Pawley, *Acta Cryst. A*, 1975, **31**, 851.
[5] G. S. Pawley, G. A. Mackenzie and O. W. Dietrich, *Acta Cryst. A*, 1977, **33**, 142.
[6] G. S. Pawley, *J. Appl. Cryst.*, submitted.
[7] E. G. Boonstra and F. H. Herbstein, *Acta Cryst.*, 1963, **16**, 252.
[8] E. Baharie and G. S. Pawley, *Acta Cryst. A*, 1979, **35**, 233.
[9] B. C. Haywood and R. Shirley, *Acta Cryst. B*, 1977, **33**, 1765.
[10] G. S. Pawley, *Acta Cryst. B*, 1978, **34**, 523.
[11] G. A. Mackenzie, B. Buras and G. S. Pawley, *Acta Cryst. B*, 1978, **34**, 1918.
[12] G. E. Bacon, E. J. Lisher and G. S. Pawley, *Acta Cryst. B*, 1979, **35**, 1400.
[13] J. C. Frost, A. J. Leadbetter and R. M. Richardson, *Faraday Disc. Chem. Soc.*, 1980, **69**, 32.
[14] R. Shirley, Crystallographic Computing, *Proc.* 1978 *Summer School* (Delft University Press and Oosthoeks, 1978).
[15] G. A. Mackenzie, J. S. Overell and G. S. Pawley, *Solid State Comm.*, 1979, **31**, 431.

Thermoluminescence in Solid Methane

By Mary Anne White,† Thomas G. Ryan* and James A. Morrison

Department of Chemistry and Institute for Materials Research,
McMaster University, Hamilton, Canada

Received 29th November, 1979

Thermoluminescence from both CH_4 and CD_4 doped with benzene or toluene ($\leqslant$0.5 mol %) has been measured in the temperature region $10 < T/K < 40$. The methane was deposited from the vapour on a rotatable copper target, usually held at $T \approx 10$ K, and irradiated with ultraviolet light (predominantly of wavelength 2537 Å). Before the glow curves (emitted light intensity as a function of temperature) were recorded, phosphorescence and recombination luminescence from the specimens were allowed to decay isothermally. Certain sharp peaks in the thermoluminescence spectrum can be correlated with the known phase transitions in CH_4 and CD_4.

Under appropriate conditions, thermally-activated recombination of radiation-induced charged species in solids can give rise to luminescence. This thermoluminescence has been studied in both inorganic and organic solids and some correlation has been claimed to exist between peaks in the luminescence spectrum and solid–solid phase transitions.[1] More recently, the results of an investigation of radiothermoluminescence in eleven doped solid hydrocarbons, including methane, have been reported.[2] Some experimental factors such as sample size, glass-transition temperature and prior thermal treatment appeared to be important.

Experimental and theoretical study of solid isotopic methanes (CH_nD_{4-n}) has become rather extensive during the past few years.[3] In part, this is because of a range of interesting phenomena that the solids exhibit: orientational ordering of the molecules; quantum crystal behaviour; conversion between nuclear spin symmetry species; negative thermal expansion and other lattice dynamical oddities. The current state of understanding of solid CH_4 is well-summarized in a recent article.[4] Under its own vapour pressure, CH_4 forms two phases, the structures of which are, known:[5,6] phase I, $T > 20.4$ K, space group *Fm3m*, 4 molecules per unit cell; phase II, $T < 20.4$ K, space group *Fm3c*, 32 molecules per unit cell. The deuterated methanes (and CH_4 under pressure) form a third solid phase, phase III, but its structure has not been fully established as yet. Theoretical calculations suggest[7] the space group $P4_2/mbc$ with 16 molecules per unit cell but this is not completely consistent with thermodynamic data obtained for phase III.[8]

The essential point is that, for the solid methanes, the changes in molecular arrangements that occur *via* the phase transitions are now reasonably well understood. In phase I, the molecules are orientationally disordered although there is some evidence[9] that some orientational correlations persist in the temperature region a few degrees above the phase II → I transition point. In phase II, molecules on six of eight sub-lattices become orientationally ordered. Further ordering must occur in

† Natural Sciences and Engineering Research Council (Canada) 1967 Science Scholar, 1975–79. Present address: Inorganic Chemistry Laboratory, South Parks Road, Oxford.

‡ Present address: Research Division, Sherritt-Gordon Mines, Fort Saskatchewan, Alberta, Canada.

the formation of phase III but, as has been indicated, its exact nature is not yet completely established. Nevertheless, enough structural information is available overall so that it seemed appropriate to try to use it at this time to connect changes in thermoluminescence spectra with phase transitions in solid methane.

EXPERIMENTAL

CRYOSTAT

The cryostat, shown schematically in fig. 1 was a modified MD4A model made by Oxford Instruments. The flow of liquid helium from the main reservoir to a smaller pot beneath it could be controlled through a needle valve operated from the top of the cryostat. A copper plate of dimensions 20 × 20 × 2 mm, on which the samples were vapour-deposited, was connected to the lower helium reservoir by a copper rod which provided good thermal contact.

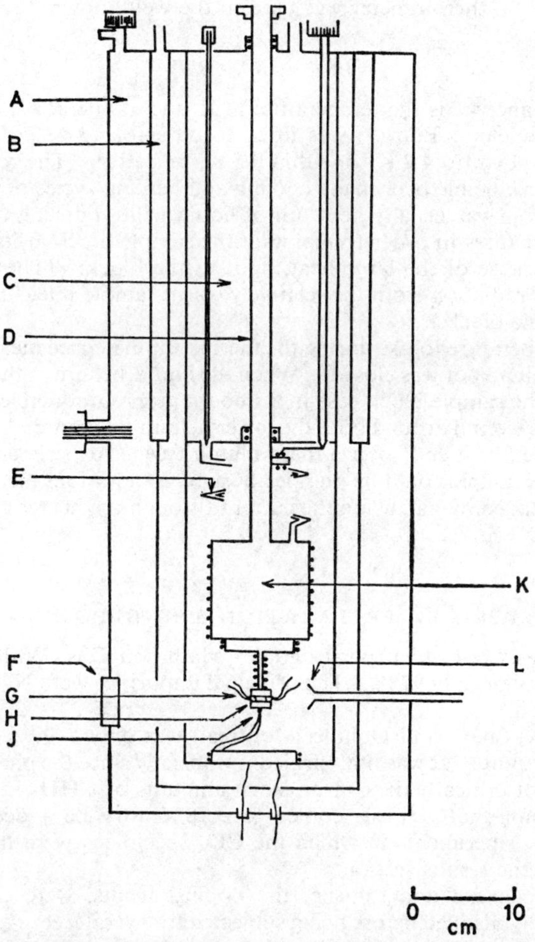

FIG. 1.—Schematic representation of the cryostat in cross-section. A, vacuum space; B, liquid nitrogen; C, liquid helium; D, rotatable shaft; E, helium flow tube; F, optical window; G, Speer resistance thermometer; H, sample block; J, copper–constantan thermocouple; K, liquid helium; L, sample inlet tube.

The principal modification to the basic MD4A cryostat was the introduction of a rotatable shaft through the main helium reservoir. This insert carried the lower helium pot and the sample block, and was installed to allow the rotation of the sample block to face either the sample deposition port or the irradiation/light detection window without loss of vacuum or refrigerant. (O-ring seals at the top and bottom of the shaft made this possible.) The sample block was rotated with a screw mechanism at the top of the cryostat, and its position, as determined on a goniometer, was reproducible within 1° of arc.

A carbon resistance thermometer (Speer Carbon Co, St. Marys, Pennsylvania) was attached to the back side of the sample block, and was used to determine the sample temperature during the thermoluminescence measurements. The thermometer was first compared with a copper–constantan thermocouple that was attached to the bottom of the block. The thermocouple had previously been calibrated at 4.2 and 77 K, which was sufficient to fix the deviation of the e.m.f., characteristic of the particular constantan wire used, from a standard table. The thermocouple and the thermometer were intercompared over the region $13 < T/\mathrm{K} < 38$, and gave absolute values of the temperature to within 0.5 K or better. At lower temperatures, copper–constantan thermocouples are rather insensitive and the resistance of the carbon thermometer was used in the well-known[10] resistance–temperature relationship:

$$\ln R = A + B/T \tag{1}$$

where R is the resistance, T is the temperature in K and A and B are constants of the particular Speer thermometer. Eqn (1) was fitted to data for $13 \leqslant T/\mathrm{K} \leqslant 20$, and gave the correct temperature down to 4.2 K to within 0.5 K, or better. The resistance–temperature relationship was reproducible over many cooling and heating cycles of the thermometer.

The lower reservoir was usually kept full of liquid helium during the sample deposition and irradiation procedures in order to maintain the sample block at its lowest temperature, which was 9.5 K because of the large heat input to the block. (The major source of this energy seemed to be radiation from the relatively warm sample inlet tube which was <2 cm away from the sample block.)

To allow the temperature to rise during the thermoluminescence measurements, the needle valve to the lower helium pot was closed. When the liquid helium in the pot had evaporated, the temperature of the sample block rose in a smooth and reproducible fashion, as shown in fig. 2. A heater was wound around both the lower helium pot and the copper rod connecting the pot to the sample block and, after a thermoluminescence experiment, it was used to heat the block so that the sample could be pumped away. In addition, the heating rate for some experiments was changed by balancing the current through the heater and the flow of helium into the smaller pot.

SPECIMEN PREPARATION AND DEPOSITION

The CH_4 used was ultra-high purity grade (Matheson Gas, Whitby, Ontario), with a stated chemical purity of $\geqslant$99.97%. The principal impurities were N_2 and C_2H_6; no purification was performed.

The CD_4 (Merck, Sharp and Dohme, Montreal) was only >99% chemically pure, and *ca.* 97% isotopically pure. It was not purified isotopically since the phase transition temperature in CD_4 does not critically depend on small amounts of CHD_3. In some experiments, the cylinder containing the CD_4 was cooled to −78 °C (dry ice + acetone) while a sample was removed. The experiments in which the CD_4 specimens were treated in this manner will be identified in the results section.

Reagent grade benzene and toluene, the doping agents, were dried over anhydrous $CaCl_2$ and fractionally distilled before being subjected to several freeze–pump–melt cycles on a high vacuum line to remove traces of dissolved gases.

The methane and the aromatic impurity, at pressures P_1 and P_2, respectively, where $P_1 \gg P_2$, were allowed to mix for a few minutes in the gas phase, before the homogeneous sample was deposited onto the sample block. The spray-deposition of the sample took place through the tube that is shown in fig. 1. The wide-mouthed stainless steel tube was connected *via* copper tubing to the sample reservoir on the outside of the cryostat and the flow of vapour into the cryostat was controlled by a needle valve. As the sample was deposited, the temperature of the sample block rose to ≈ 10 K or more, depending on the deposition rate (usually 2-20 mg min^{-1}), due to the energy released on solidification. The specimens contained up to 0.5 mol % of aromatic impurity, and ranged in size from ≈ 5 to 300 mg.

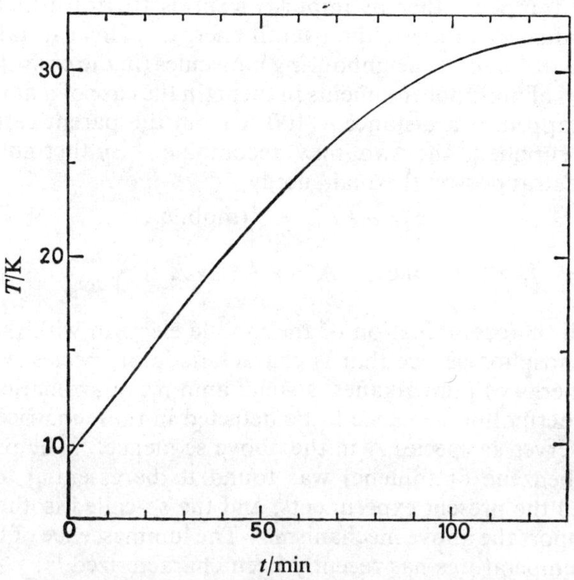

FIG. 2.—Typical heating curve for a thermoluminescence sample.

A freshly-deposited sample at its lowest temperature ($T \approx 9.5$ K) was a white translucent film but, as the temperature was increased during a thermoluminescence experiment, the sample took on a layered or flaky appearance.

IRRADIATION AND MEASUREMENT OF LIGHT EMISSION

A mercury pencil lamp (Spectronics Corp., Westbury, New York) of principal wavelength 2537 Å, and with an intensity of 2×10^{-3} W cm^{-2} at a distance of 2.5 cm, was placed at the same optical window which was used for light detection, approximately 6 cm from the sample block. When the sample had been irradiated, the light was removed and the light detector immediately placed in position. (The delay between the end of the u.v.-irradiation and the detection of the emitted light was usually <30 s.)

An RCA 6199 head-on photomultiplier tube was used to detect the light from the sample. The tube had its maximum sensitivity in the range $3000 \lesssim \lambda/\text{Å} \lesssim 6000$, with an absolute maximum at $\lambda = 4400$ Å. The tube output ($\approx 10^{-9}$ A) was amplified and recorded on a chart recorder.

To monitor the background response of the photomultiplier tube, a mechanical shutter mounted between the cryostat and the detector was closed and opened intermittently during the luminescence measurements.

RESULTS

GENERAL ASPECTS OF THE LUMINESCENCE

It has been postulated[11] that the mechanism responsible for photothermoluminescence is as follows. Photoionization of a species, A, in the solid takes place such that A is excited by a two-photon process:

$$A + h\nu \rightarrow A^* \rightarrow {}^3A, \quad (2)$$

$${}^3A + h\nu \rightarrow A^{**} \rightarrow A^+ + e^-, \quad (3)$$

and an electron is produced. (Similarly, in radiothermoluminescence, X-rays, electrons or γ-rays bombard A to produce A^+ and electrons.) The electron that is produced can be trapped either at impurity centres (to form anions) or at lattice defects in the solid so as to lower the overall energy. This can take place either by changing the orientation of the neighbouring molecules (in the case of a polar medium), or by the induction of multipole moments in them (in the case of a non-polar medium).[1] The electron is trapped at a distance $\leqslant$100 Å from the parent cation[1] and, if additional energy is supplied, the two may recombine. In thermoluminescence, the electron may be detrapped by thermal energy,

$$e^- + kT \rightarrow e^- \text{ (mobile)}, \quad (4)$$

$$e^- \text{ (mobile)} + A^+ \rightarrow A^* \rightarrow A + \begin{cases} h\nu_f \\ h\nu_p, \end{cases} \quad (5)$$

and the result will be recombination of the mobile electron with the cation, A^+, and fluorescence or phosphorescence that is characteristic of species A. Because of the high ionization energy of pure alkanes, a small amount of aromatic impurity must be added for photothermoluminescence to be detected in these matrices;[1] the aromatic compound then serves as species A in the above sequence. The presence of an aromatic impurity (benzene or toluene) was found to be essential to the observation of luminescence in the present experiments, and the so-called isothermoluminescence or ITL results support the above mechanism. The luminescence of benzene in various matrices at low temperatures has recently been characterized.[12]

Immediately following the deposition and irradiation of a methane + aromatic specimen, and before the temperature was allowed to increase, luminescence was observed. The light given off was of purple colour; a typical decay of the ITL is illustrated in fig. 3. Clearly, two processes are taking place, one short-lived and one long-lived.

Within $\approx$2 min after irradiation, the ITL decay is exponential with lifetimes consistent with those of solute phosphorescence (τ = 15 s for benzene and 12 s for toluene, in either CH_4 or CD_4). As has been demonstrated,[12,13] the phosphorescence lifetimes for benzene in CH_4, N_2 and Ar are very similar. Although the spectrum of the emitted light was not determined in the present experiments, similar studies on other systems have shown it to be consistent with phosphorescence of the solute molecules.[14]

After the solute phosphorescence has decayed sufficiently, the ITL decay follows the relationship

$$I = kt^{-m}, \quad (6)$$

where k is a scaling parameter, and $m \approx 1$. A similar decay of the long-lived isothermal luminescence has been observed in many other solute + solvent systems, and is generally accepted to be due to the slow recombination of trapped electrons with cations.[15–18] Whether the recombination takes place by diffusion or by electron tunnelling is a subject of much discussion; some other results aimed at elucidating the recombination mechanism will be discussed in detail elsewhere.[19]

THERMOLUMINESCENCE

Many factors, such as the sample size, heating rate and specimen preparation, have been reported to influence the glow curves of molecular solids, but they have not always been studied systematically. The reproducibility of many of the earlier results can therefore be questioned. An attempt was made here to investigate these factors thoroughly; the present results will illustrate the limitations placed on the absolute determinations of a glow curve for a particular matrix + solute system.

Sugawara and Tabata[2] reported grossly different thermoluminescence curves for alkane + toluene samples when the size was varied from 5 to 30 mg. In the present experiments, the shapes of the glow curves for methane + aromatic mixtures were found to be independent of the sample size from 2 to 200 mg. (Representative glow curves for CH_4 + toluene are shown in fig. 4.) In other experiments that are not illustrated, heating rates were varied from those given in fig. 2 to a constant 0.14 K min^{-1}; the shapes of the glow curves remained unchanged although the positions

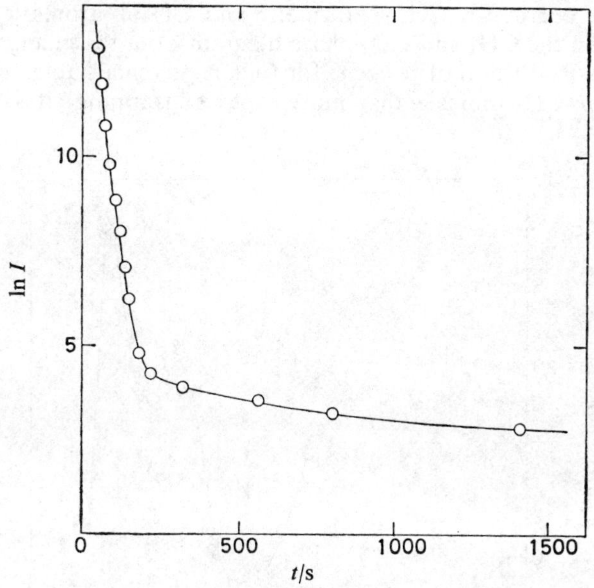

FIG. 3.—Decay of the isothermal luminescence. The results shown are for CH_4 + benzene at $T = 10$ K; the luminescence intensity is in arbitrary units.

of peaks shifted by as much as ± 1 K from one experiment to another. A possible source of the irreproducibility in the earlier work[2] may have been the extremely rapid heating rate used at low temperatures (as high as ≈ 20 K min^{-1} at the start of the determination of a glow curve).

As illustrated in fig. 4, the shape of the glow curve of CH_4 + toluene was also independent of the solute concentration. This was true for all methane + aromatic samples used, over a solute concentration range 0.001-0.5 mol %. In fact, it can be seen from fig. 4 that the luminescence intensities are proportional to the total number of moles of solute present.

Although the vapour deposition method has been used to prepare glassy or non-crystalline samples of CH_3OH and H_2O,[20] it is improbable that glassy methane was formed here because extremely slow deposition rates were not used. (If the deposition is carried out rapidly, the heat released on solidification should provide sufficient energy to crystallize the sample.) In the present experiments, the deposition rate was varied by more than a factor of 10 without any change in the main features of the glow curves. This result indicated that the samples used were probably crystalline and this was confirmed by the lack of dependence of the results on thermal treatment of the samples. (In solids which are known to form glasses, such as C_2H_5OH and $CH_3OH + H_2O$, thermoluminescence is strongly dependent on whether or not the samples were annealed.)[21,22]

Variation of the solute, although not often considered in thermoluminescence studies, has been shown to have little effect on the shape of the glow curves of squalane[23] and polystyrene,[24] as well as of many other organic matrices.[1] The prevailing view appears to be that the presence of the solute only serves to increase the luminescence, but some specific effects were found in the present experiments. For example, we may contrast fig. 4 which shows the thermoluminescence glow curve for CH_4 + toluene, with fig. 5, 6 and 7 which illustrate curves for CH_4 + benzene, CD_4 + toluene and CD_4 + benzene, respectively. It is not surprising to see some dissimilarities between the glow curves for CH_4 + aromatic and CD_4 + aromatic because of the differences between the CH_4 and CD_4 phase diagrams, but the change of the thermoluminescence on substitution of benzene for toluene is remarkable.

The results for CH_4 indicate that more types of trapping sites are available in

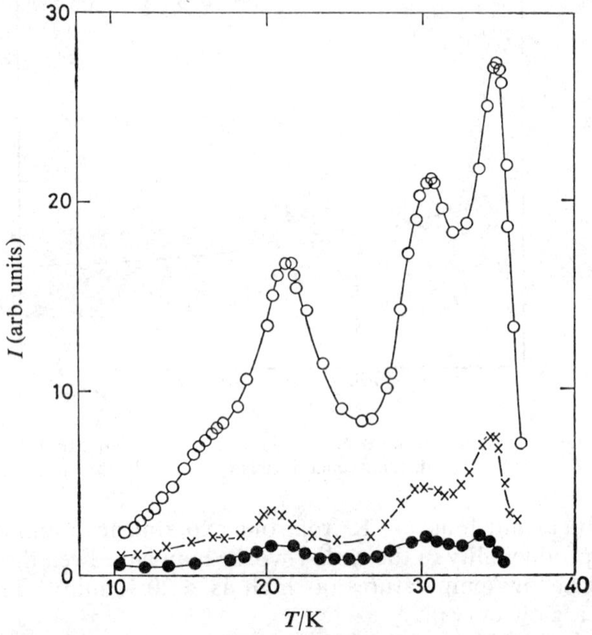

FIG. 4.—Effect of sample size and solute concentration on the glow curve. ○, 54 mg of CH_4 + 0.008 mol % toluene; ×, 12 mg of CH_4 + 0.008 mol % toluene; ●, 54 mg of CH_4 + 0.0009 mol % toluene.

CH_4 + toluene than in CH_4 + benzene. This can perhaps be reconciled with the possibility that the methyl group of toluene produces extra distortion of the matrix.

It can be seen from the results for CD_4 that extremely small amounts of extraneous impurities can also greatly influence the thermoluminescence profile. CD_4 untreated in any way and mixed with benzene always gave the glow curve displayed in fig. 7(*a*). On the other hand, when the cylinder containing the CD_4 was cooled in dry ice, the glow curve shown in fig. 7(*b*) was observed for 10 of 20 experiments; the other experiments yielded curves such as the one given in fig. 7(*a*). It is difficult to see that the cooling procedure could have had any effect other than to promote preferential adsorption of an impurity (or impurities) on the cylinder wall.

DISCUSSION

The presence of both sharp and broad peaks in the glow curves indicates that two processes give rise to electron detrapping. Each of the broad peaks are attributable to one or more Randall–Wilkins processes,[25] *i.e.*, thermally activated recombination of electrons and cations. Our results illustrate that the distribution of the electron trap depths and the distance from the trapped electron to the parent cation depend

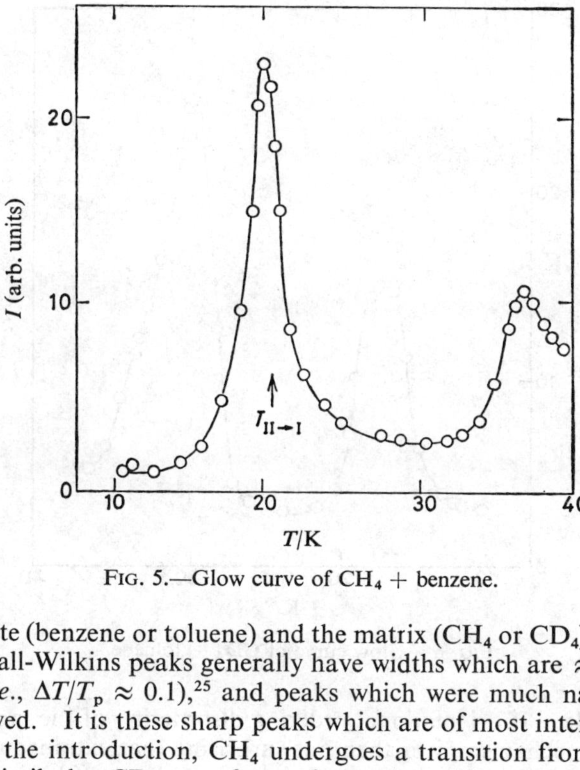

FIG. 5.—Glow curve of CH_4 + benzene.

on both the solute (benzene or toluene) and the matrix (CH_4 or CD_4) which were used. However, Randall-Wilkins peaks generally have widths which are ≈10% of the peak temperatures (*i.e.*, $\Delta T/T_p \approx 0.1$),[25] and peaks which were much narrower than that were also observed. It is these sharp peaks which are of most interest here.

As noted in the introduction, CH_4 undergoes a transition from phase II → I at $T = 20.4$ K. Similarly, CD_4 transforms from phase III → II at $T = 22.0$ K and from phase II → I at $T = 26.9$ K. From the results with benzene as additive, we see a clear correlation between thermoluminescence peaks and the solid–solid phase transitions: the sharp peak at $T = 20$ K in CH_4 + benzene and the shoulder and sharp peak at $T = 22$ K and $T = 27$ K, respectively, in CD_4 + benzene coincide with the transitions.

It is possible that the luminescence from accelerated recombination at the phase II → I transition in CH_4 + toluene is masked by processes that are associated with the thermally activated electron–cation recombination, *i.e.*, with the Randall–Wilkins process. Sharp peaks from the phase-transition-induced recombination appear to be also absent in CD_4 + toluene. This system was not investigated in more detail because of the occurrence of Randall–Wilkins peaks in the regions where effects of the phase transitions would be expected to be seen.

Since it seems probable that the methane samples were crystalline, it must be assumed that the phase transitions still take place in impure CD_4; perhaps the impurities bind tightly electrons that would otherwise be detrapped at the phase transitions.

At this time, we can only speculate about the microscopic processes that are responsible for the lowering of the barrier to electron–cation recombination at the phase transitions. Two obvious ones are changes in the arrangement of the centres of mass or in the orientational order of the molecules. Since the volume changes at the transitions are very small ($\approx 1\%$)[26] and since the basic crystal structure remains f.c.c. in the phase II $\rightarrow$ I transition, it seems improbable that displacements of the centres

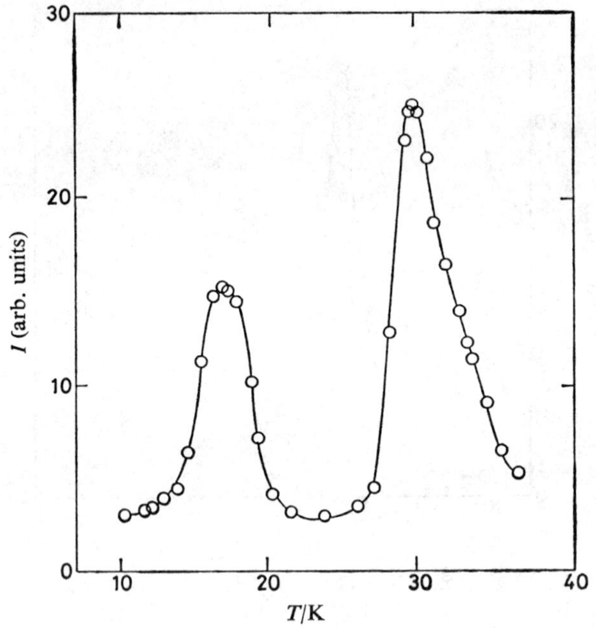

FIG. 6.—Glow curve of CD_4 + toluene.

of mass play a role. If orientational ordering/disordering of the molecules provides the basic release mechanism, it is then not surprising that extraneous impurity traps are so important. The energy changes corresponding to shifts of orientational order are very small (<100 J mol^{-1}).

In principle, the shapes and positions of the sharp peaks in the glow curves should provide information about the kinetics of the phase transitions. However, as noted earlier, the precise peak temperature varied by as much as 1 K from sample to sample. In addition, the peak width at half height ranged from 0.3 to 1 K for the phase II $\rightarrow$ I transition in both CH_4 and CD_4. These variations were probably caused by the existence of domain-type sub-structures in the samples with a resulting distribution of phase-transition temperatures. If the glow curve of a sample at a uniform temperature and in perfect thermal contact with the thermometer could be realized, a correlation between the peak widths and the two-phase coexistence region ought to be expected. (An X-ray diffraction study of CH_4 shows no appreciable temperature range of coexistence between phases I and II.[26] Similar experiments show a range of coexistence of 0.10 K for the phase III $\rightarrow$ II and II $\rightarrow$ I transitions in CD_4.)[6]

CONCLUSIONS

This experimental study of thermoluminescence in solid CH_4 and CD_4 doped with benzene or toluene has established that the phenomenon is independent of factors such as sample size, solute concentration and prior thermal treatment. In these respects, the results differ from those of a prior study[2] of thermoluminescence in CH_4 + toluene. On the other hand, glow curves are found to be strongly dependent on the nature of the solute and on small amounts of extraneous impurities.

The technique of specimen preparation by vapour deposition was satisfactory for an exploratory type of investigation. The absence of a dependence of the shapes of

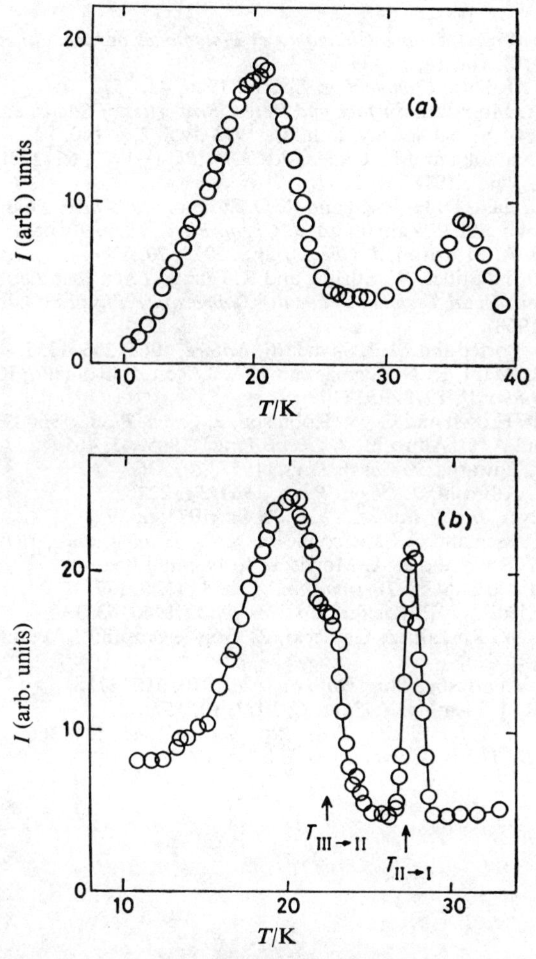

FIG. 7.—Glow curve of CD_4 + benzene: (*a*) untreated CD_4; (*b*) CD_4 cooled in dry ice.

glow curves on solute concentration indicates that the specimens had a uniform composition. To study peak shapes and positions more quantitatively, larger specimens whose temperatures could be controlled more accurately would be desirable. However, it would probably be very difficult to produce bulk specimens (even single crystals) of uniform solute distribution.

The results support the general proposition[1] that phase transitions can and do produce particular features such as sharp peaks in thermoluminescent spectra. But the recombination processes that lead to the abrupt light emission are strongly affected by traces of impurities. To extend the investigation, more attention would need to be paid to the purification of the materials.

We thank Drs. E. A. Ballik, M. L. Klein and Z. Racz for much helpful discussion and Mr. W. Scott for supplying liquid helium very efficiently. The financial support of the Natural Sciences and Engineering Research Council (Canada) is gratefully acknowledged.

[1] F. Kieffer and M. Magat, *Actions Chimiques et Biologiques des Radiations*, ed. M. Haissinsky (Masson, Paris, 1970), vol. 14, p. 135.
[2] I. Sugawara and Y. Tabata, *Chem. Phys. Letters*, 1976, **41**, 357.
[3] M. Bloom and J. A. Morrison, *Surface and Defect Properties of Solids*, ed. M. W. Roberts and J. M. Thomas (The Chemical Society, London, 1973), vol. 2, p. 140.
[4] T. Yamamoto, Y. Kataoka and K. Okada, *J. Chem. Phys.*, 1977, **66**, 2701.
[5] W. Press, *J. Chem. Phys.*, 1972, **56**, 2597.
[6] D. R. Baer, D. A. Fraass, D. H. Riehl and R. O. Simmons, *J. Chem. Phys.*, 1978, **68**, 1411.
[7] K. Maki, Y. Kataoka and T. Yamamoto, *J. Chem. Phys.*, 1979, **70**, 655.
[8] M. A. White and J. A. Morrison, *J. Chem. Phys.*, 1979, **70**, 5384.
[9] W. Press, A. Hüller, H. Stiller, W. Stirling and R. Currat, *Phys. Rev. Letters*, 1974, **32**, 1354.
[10] G. K. White, *Experimental Techniques in Low Temperature Physics* (Oxford University Press, Oxford, 2nd edn, 1968).
[11] W. A. Gibbons, G. Porter and M. I. Savadatti, *Nature*, 1965, **206**, 1355.
[12] E. P. Gibson, G. R. Mant, R. Narayanaswamy, A. J. Rest, S. Romano, K. Salisbury and J. R. Sodeau, *J.C.S. Faraday II*, 1979, **75**, 1179.
[13] M. R. Wright, R. P. Frosch and G. W. Robinson, *J. Chem. Phys.*, 1960, **33**, 934.
[14] W. M. McClain and A. C. Albrecht, *J. Chem. Phys.*, 1965, **43**, 465.
[15] P. Debye and J. O. Edwards, *J. Chem. Phys.*, 1952, **20**, 236.
[16] J. Bullot and A. C. Albrecht, *J. Chem. Phys.*, 1969, **51**, 2220.
[17] J. Kroh and J. Mayer, *Int. J. Radiat. Phys. Chem.*, 1973, **5**, 59.
[18] F. Kieffer, N. V. Klassen and C. Lapersonne-Meyer, *J. Luminescence*, 1979, **20**, 17.
[19] M. A. White, T. G. Ryan and J. A. Morrison, to be published.
[20] O. Haida, H. Suga and S. Seki, *Thermochim. Acta*, 1972, **3**, 177.
[21] J. Bullot A. Déroulède and F. Kieffer, *J. Chim. phys.*, 1966, **63**, 150.
[22] J. Kroh, J. Mayer, W. Roszak, Z. Galdecki, Z. Górkiewicz and B. Ptaszysńki, *Int. J. Radiat. Phys. Chem.*, 1974, **6**, 423.
[23] I. Boustead and A. Charlesby, *Proc. Roy. Soc. A*, 1970, **315**, 271.
[24] L. F. Pender and R. J. Fleming, *J. Phys. C*, 1977, **10**, 1571.
[25] J. T. Randall and M. H. F. Wilkins, *Proc. Roy. Soc. A*, 1945, **184**, 366.
[26] D. R. Aadsen, *Ph.D. Thesis* (University of Illinois, 1975).

Studies of Phase Transformations in Molecular Crystals Using the Positron Annihilation Technique

BY MORTEN ELDRUP

Chemistry Department,
Risø National Laboratory, 4000 Roskilde, Denmark

AND DAVID LIGHTBODY AND JOHN N. SHERWOOD

Department of Pure and Applied Chemistry,
University of Strathclyde, Glasgow G1 1XL

Received 10*th December*, 1979

An examination has been made of the *brittle/plastic* phase transformation in the molecular crystals cyclohexane, DL-camphene and succinonitrile using the positron annihilation technique. In each material, the transition is characterized by a distinct increase in *ortho*-positronium lifetime. The influence of impurities on the transition was examined for DL-camphene. Addition of the impurity tricylene in concentrations in the range 0.14-4.0 mol % resulted in a lowering of the transition temperature from 176 to 167 K and a broadening of the transition region.

Following emission from a nucleus a positron will annihilate with an electron of the environment with the emission of gamma quanta.[1] Annihilation may occur directly from the free state with the emission of two 0.51 MeV quanta or *via* the formation of the electron–positron bound state, positronium (Ps).

Positronium can exist in two forms: *para*-positronium (*p*-Ps) with the electron and positron having spins anti-parallel or *ortho*-positronium (*o*-Ps); spins parallel. In free space, *p*-Ps annihilates *via* a two-gamma decay with a mean lifetime $\tau_1 = 0.125$ ns whereas *o*-Ps undergoes a three-gamma annihilation, mean lifetime 140 ns. In condensed matter, there is a high probability of interaction between *o*-Ps and the electrons in the surrounding molecules. The positron may then annihilate with an electron of the opposite spin with the emission of two gamma quanta. This pick-off annihilation process dominates and reduces the *o*-Ps lifetime to the order of a few nanoseconds. Free annihilation of the positron in molecular solids is characterized by a lifetime $\tau_2 \approx 0.3$-0.5 ns.

The *o*-Ps annihilation process is sensitive to small variations in the electron density and hence molecular density of its environment. This renders it a potentially useful probe for the examination of small-scale structural processes such as vacancy[2] and void formation and phase transformations in molecular solids.

All of the materials studied belong to that state of matter usually described as *Plastic Crystals*.[3] At low temperatures they exist as brittle solid phases of low crystallographic symmetry, with little or no re-orientational molecular disorder. With increasing temperature, they undergo a lambda-type transition,[4–6] to yield a re-orientationally disordered form of higher symmetry (f.c.c. or b.c.c.). This latter form usually persists to the melting point. This change of state results in a significant change in molar volume[7] which should be reflected by a change in *o*-Ps lifetimes.

Although the phase behaviour of certain liquid-crystal systems have been investigated in detail using the positron annihilation technique,[8] the study of polymorphism in molecular solids has only been noted in a limited number of cases, *e.g.*, tripalmitin,[9] tristearin,[10] CH_3OH, CD_3OD and CG_2SH.[11] Only cyclohexane,[12]

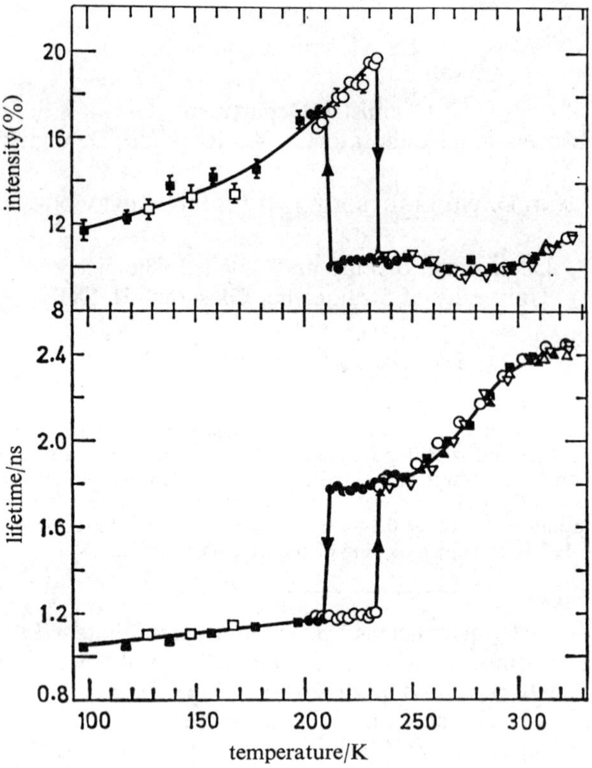

FIG. 1.—Succinonitrile. The temperature dependence of the average *o*-Ps lifetime, τ_3, and its intensity, I_3. The phase transition takes place at 234 ± 1 K. In all figures open symbols are for increasing, closed ones for decreasing, temperatures.

cyclo-octanone[13] and DL-camphor,[14] of the *plastic crystal* class have received previous mention. No detailed examinations have been reported.

The present study illustrates the degrees of sensitivity available to the technique in the region of the transition, where hysteresis and impurity effects are observed.

EXPERIMENTAL

All materials were purified by fractional sublimation, distillation or zone refining[15] and analysed using gas–liquid chromatography. Single crystals of succinonitrile (<1 p.p.m. total impurity content) prepared *in vacuo* by a Bridgman technique were sectioned into 1 cm diameter × 0.5 cm discs. The cyclohexane (<100 p.p.m. total impurity) and doped DL-camphene (0.14, 1.3 and 4 mol % tricylene) samples were prepared by freezing the degassed liquid and sectioning the resulting polycrystal as noted above.

The samples were mounted on either side of the positron source which consisted of 20 μCi ^{22}NaCl contained in an envelope of Kapton foil. This sample/source arrangement was mounted in a liquid nitrogen cryostat with a temperature control of ±0.5 K.

Lifetime measurements were made using a conventional fast–slow[1] or a novel fast–fast lifetime spectrometer[16] with time resolutions of 0.35-0.4 ns full width at half maximum. Zero time is marked by the prompt 1.28 MeV gamma which accompanies the positron emission and subsequent annihilations by the 0.51 MeV annihilation gammas. Accumulation

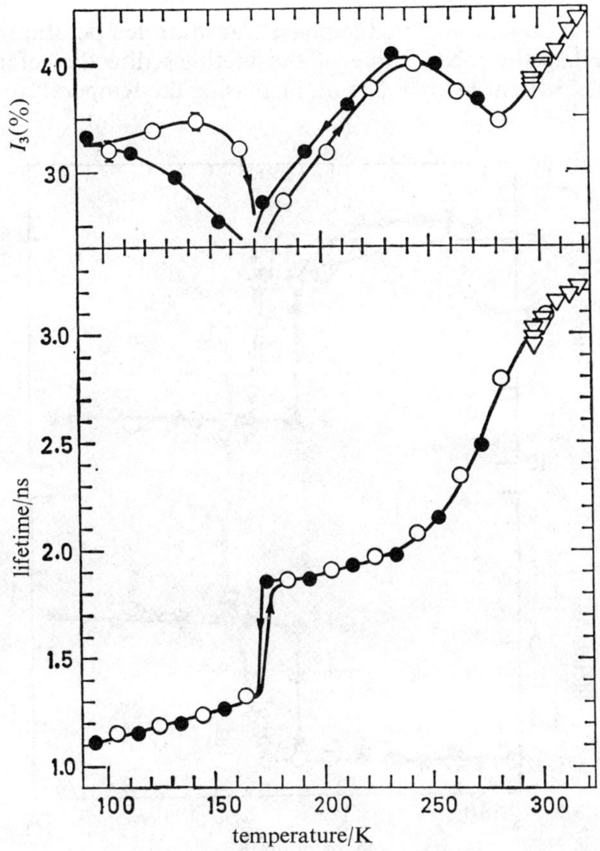

FIG. 2.—Camphene with 1.3% tricyclene. The temperature dependence of the average *o*-Ps lifetime, τ_3, and its intensity, I_3. The phase transition for material of this purity is defined at 173 ± 1 K (see also fig. 4).

of data was carried out over periods of 12-15 h to produce spectra containing ≈10^6 counts. All spectra were analysed using the computer programme POSITRONFIT.[17]

RESULTS

The spectra were resolved into three lifetimes, τ_1, τ_2 and τ_3, associated with *p*-Ps, free positrons and *o*-Ps, respectively. It is, however, the *o*-Ps lifetime which is of the greatest interest in this study since the others show little or no detectable change in value across the phase transition.

The shortlived *p*-Ps component is often experimentally difficult to resolve. The values of 0.05-0.25 ns determined in the present experiments are in satisfactory agreement with the theoretical value of 0.125 ns.

The variations of τ_3 and I_3 (the *o*-Ps yield) with temperature, for the materials studied, are illustrated in fig. 1-3. In all cases the lifetime increases almost linearly with temperature in the low-temperature phase. The phase transition is clearly identified by an abrupt change in τ_3. In succinonitrile and camphene a behaviour characteristic of *o*-Ps trapping at thermally generated defects is noted in the plastic phase. This sigmoidal variation of τ_3 and its interpretation has been discussed in detail elsewhere.[2]

It would seem rational to define the phase transition temperature as that temperature at which τ_3 has the mean value of the lifetimes directly before and after the transition; the assessment being made on increasing the temperature. On this basis,

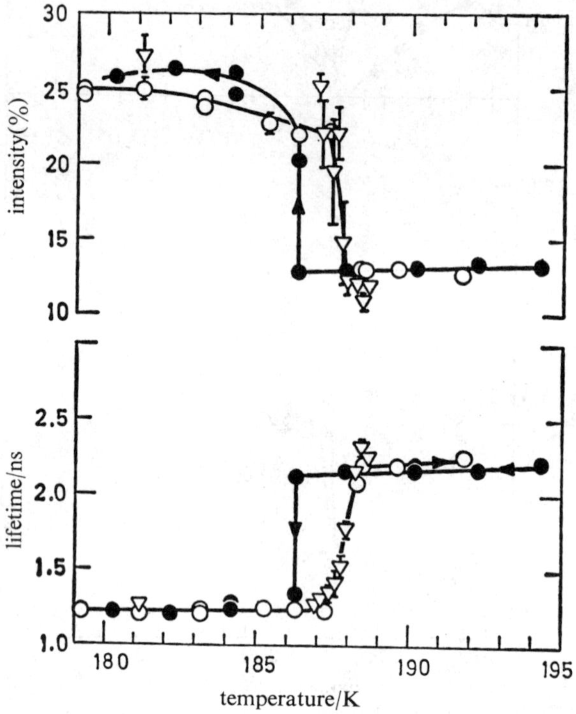

FIG. 3.—Cyclohexane with 100 p.p.m. impurities. The temperature dependence of the average *o*-Ps lifetime, τ_3, and its intensity, I_3, around the phase transition. The transition is found at 188 $\pm$ 0.5 K.

the phase transition from the low-temperature monoclinic phase to the high-temperature b.c.c. phase in succinonitrile occurs at 234 $\pm$ 1 K with an associated hysteresis of $\approx$25 K. Cyclohexane exhibits its transition from the low-temperature phase to the high-temperature f.c.c. phase at a temperature of 188 $\pm$ 0.5 K with $\approx$2 K hysteresis.

Fig. 4 and 5 illustrate the influence of the impurity, tricyclene on the characteristics of the phase transition in the camphene system. The transition from the low-temperature phase to the high-temperature b.c.c. phase at temperatures of 176, 173, 167 K (all with an uncertainty of 1 K) correspond to samples of 0.14, 1.3, 4.0% tricyclene concentration, respectively. The effect of dopant concentration seems also to be reflected in the width of the transition which increases with increasing concentration as illustrated in fig. 5(*b*).

DISCUSSION

In all three materials examined, the variation of *o*-Ps lifetime with temperature shows a significant change in the vicinity of the phase transformation. These changes can be roughly correlated with the change in free volume in the system.[18] The fractional increases in the densities of camphene and cyclohexane are 2.7 and 9.4%, respectively,[7] compared with lifetime increases of 0.52 and 1.0 ns. Within the limitations of the definition of purity of the specimens used in previous studies, the present estimates of the transition temperatures correlate well with the values previously quoted and probably provide a better estimate of these temperatures.

For succinonitrile, heat capacity[4] and n.m.r. studies[19] define the transition temperature as 233 and 236 K, respectively. The quoted purification schemes suggest a

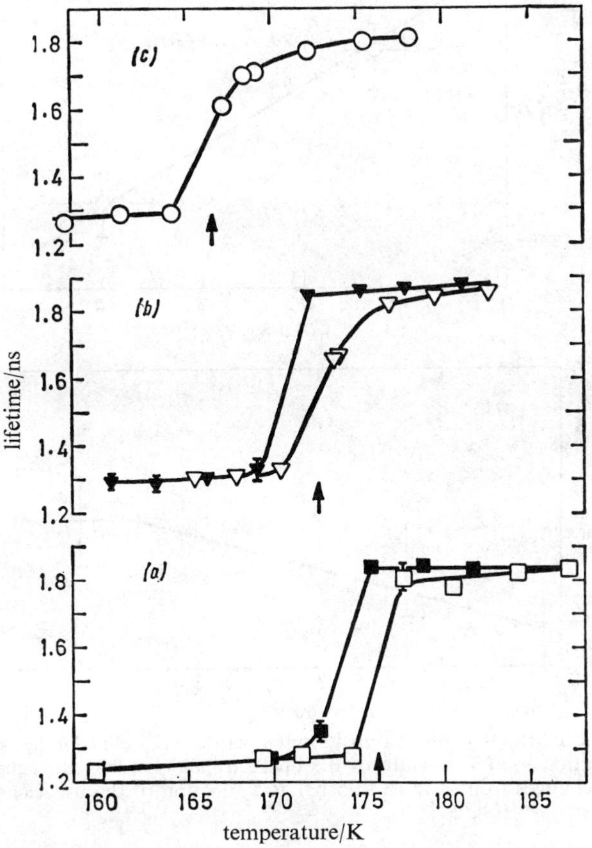

FIG. 4.—Camphene with (*a*) 0.14, (*b*) 1.3 and (*c*) 4.0% tricyclene. The shift in phase transition temperature as detected by the *o*-Ps lifetime, τ_3. The arrows show the temperature at which $\tau_3 = 1.55$ ns, taken to be the transition temperature.

similar purity of material in both cases. The present value of 234 K for potentially purer material is in adequate agreement with these values.

The purity of one previously studied sample of cyclohexane is better defined. Here assessments have been made using heat capacity measurements[6] (186.1 K) infrared absorption spectroscopy[20] (187 ± 0.5 K), n.m.r.[21] (186 K) and positron annihilation[12] (187 ± 4 K). A sample purity of 950 p.p.m. was noted for the material used in the heat capacity measurements. The present value is in good agreement with the above and is in the correct relationship to that obtained from the latter measurements allowing for the ten-fold improvement in purity (see below).

The values obtained for DL-camphene show the largest discrepancy with previous data. The previously published value of 153 K[5] is 24 K lower than that estimated for the present sample of highest purity. This difference probably arises from the presence of impurities in the original sample.

Reagent grade DL-camphene can be extremely impure (4-15% of tricyclene and

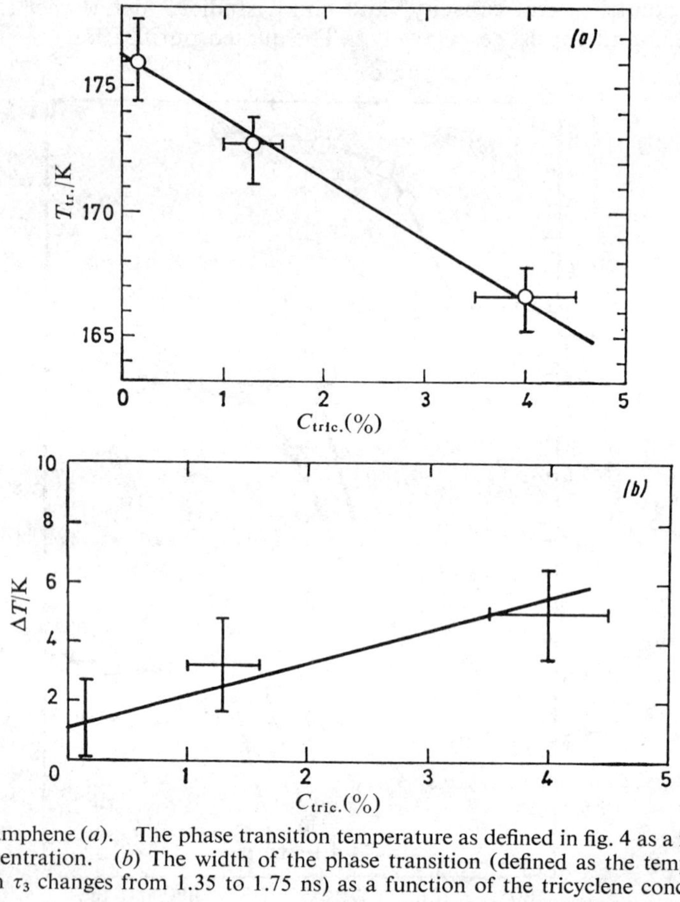

FIG. 5.—DL-Camphene (*a*). The phase transition temperature as defined in fig. 4 as a function of the tricyclene concentration. (*b*) The width of the phase transition (defined as the temperature range over which τ_3 changes from 1.35 to 1.75 ns) as a function of the tricyclene concentration.

lesser amounts of other impurities).[22] It is also notoriously difficult to purify. The influence of increasing tricyclene content on the phase transition temperature confirms this speculation (fig. 5).

Further confirmation of the present value has been obtained from n.m.r. studies[23] of DL-camphene of similar quality to that of the most pure sample (175 K).

This experiment, and to a lesser extent that on cyclohexane, demonstrates the considerable influence of included impurities on the transformation process. It confirms the view that experiments should be performed with samples of the highest purity obtainable.

The basic reason for the noted influence probably resides in the ease with which tricyclene forms solid solutions with DL-camphene.[22] Since the impurity molecule is more symmetrical than the host, its substitutional inclusion could lead to a lowering of the potential barriers opposing the orientational disordering. This in turn will lower the transition temperature. A similar effect has been noted previously in heat capacity studies of methane containing krypton as impurity.[24] A systematic lowering of the phase transformation temperature was found to accompany the progressive

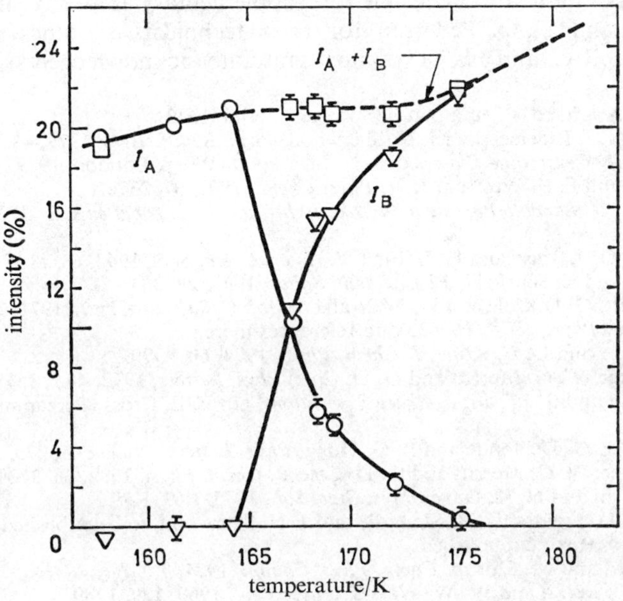

FIG. 6.—Camphene with 4% tricyclene. Intensities of *o*-Ps lifetime components with lifetimes characteristic of the low-temperature and the high-temperature phases, 1.3 ns (I_A) and 1.8 ns (I_B), respectively.

addition of the more symmetrical krypton atoms. A parallel broadening of the transition region was also noted in this case.

Above and below the phase transformations specific values of the *o*-Ps yields and lifetimes can be assigned to the separate phases. If we make the reasonable assumption that the gradual change observed in the *o*-Ps yield arises from the temporary coexistence of the two phases in the region of the transition, then we can attempt to estimate the relative concentrations of each. At the present time sufficient data for such an analysis only exist for the least pure camphene sample. The results are shown in fig. 6. With improved accuracy, it should be possible to extend this kind of analysis to the more pure systems.

The positron annihilation technique is presently suitable for the accurate definition of the phase transition temperature and potentially useful for the analysis of composition in the vicinity of the transition. More detailed structural interpretations derived from the nature of the annihilation process must await further work on this topic using a wider variety of materials. The degree with which the *o*-Ps lifetimes can be associated with density changes is noted above. A relationship exists here which can be developed. The variations in *o*-Ps yield (I_3) are more difficult to explain.

The formation of Ps takes place between the injected positron and one of the electrons created by the positron during the ionisation of the medium as it loses energy. The yield is thus dependent on the physical and chemical properties of the medium.[25] Similar processes are studied in radiation chemistry and, potentially, Ps yields could be correlated with those results. The current lack of such investigations in molecular crystals limits immediate progress. The potential of the positron annihilation technique for the study of solid–solid phase transformations is, however, clearly illustrated.

We thank O. E. Mogensen for discussions, S. J. Lund for his work on the lifetime spectrometry, H. Egsgaard Pedersen for supplementary impurity analyses and A. Bønke-Nielsen and N. J. Pedersen for their technical assistance. The financial support of the S.R.C. and NATO is most gratefully acknowledged.

1 *Positrons in Solids*, ed. P. Hautojarvi (Springer, Berlin, 1979).
2 M. Eldrup, N. J. Pedersen and J. N. Sherwood, *Phys. Rev. Letters*, 1979, **43**, 1407.
3 *The Plastically Crystalline State*, ed. J. N. Sherwood (Wiley, London, 1979).
4 C. A. Wulff and E. F. Westrum Jr, *J. Phys. Chem.*, 1963, **67**, 2376.
5 W. A. Roth, in *Landolt-Bornstein Physikalisch-Chemische Tabellen* EG IIIC, 2695 (Springer, Berlin, 1936).
6 J. G. Aston, G. J. Szasz and H. L. Fink, *J. Amer. Chem. Soc.*, 1943, **65**, 1135.
7 J. R. Green and C. Scheie, *J. Phys. Chem. Solids*, 1967, **28**, 383.
8 W. G. Merritt, G. D. Cole and W. W. Walker, *Mol. Cryst. Liq. Cryst.*, 1971, **15**, 105 and W. W. Walker, *Appl. Phys.*, 1978, **16**, 433 and references therein.
9 W. W. Walker and D. C. Kline, *J. Chem. Phys.*, 1974, **60**, 4990.
10 W. W. Walker, W. G. Merritt and G. D. Cole, *Phys. Letters*, 1972, **40A**, 157.
11 S. Y. Chuang and S. J. Tao, in *Phase Transitions*, ed. L. E. Cross (Pergamon, London, 1973), p. 363.
12 A. M. Cooper, G. Deblonde and B. G. Hogg, *Phys. Letters*, 1969, **29A**, 275.
13 W. W. Walker, W. G. Merritt and G. D. Cole, *J. Chem. Phys.*, 1972, **56**, 3729.
14 V. G. Kulkarni and N. K. Dave, *Phys. Stat. Sol.*, 1975, **B67**, K79.
15 J. M. Bruce, D. Lightbody, B. McArdle and J. N. Sherwood, to be published.
16 S. J. G. Lund, Risø, unpublished.
17 P. Kirkegaard and M. Eldrup, *Comp. Phys. Comm.*, 1974, **7**, 401.
18 W. Brandt, S. Berko and W. W. Walker, *Phys. Rev.*, 1960, **120**, 1289.
19 J. G. Powles, A. Begum and M. O. Norris, *Mol. Phys.*, 1969, **17**, 489.
20 G. N. Zhizhin, E. L. Terpugov, M. A. Moskaleva, N. I. Bagdanskis, E. I. Balabanov, and A. I. Vasil'ev, *Sov. Phys. Solid State*, 1973, **14**, 3028.
21 D. E. O'Reilly, E. M. Petersen and D. L. Hogenboom, *J. Chem. Phys.*, 1972, **57**, 3969.
22 N. T. Corke, N. C. Lockhart, R. S. Narang and J. N. Sherwood, *Mol. Cryst. Liq. Cryst.*, 1978, **44**, 45.
23 N. Boden, S. Hanlon, M. Mortimer and S. Ross, unpublished results referred to by N. Boden in *The Plastically Crystalline State*, ed. J. N. Sherwood (Wiley, London, 1979), p. 171.
24 A. Eucken and H. Veith, *Z. phys. Chem.*, 1936, **34**, 275.
25 O. E. Mogensen, *J. Chem. Phys.*, 1974, **60**, 998.

Diamagnetism and Phase Transformations in Molecular Crystals

By Józef W. Rohleder

Institute of Organic and Physical Chemistry,
Technical University of Wrocław, Wrocław, Poland

Received 28*th November*, 1979

The applicability of the Oriented Gas Model to describing the length and orientation of the principal axes of the susceptibility tensor in molecular crystals is discussed. The influence of temperature on diamagnetic anisotropy is interpreted as being a result of small but continuous changes in the orientation of molecular tensors. The elucidation of this effect is given by means of a " rigid box " model. The presentation of effects associated with a phase transformation is followed by the discussion of some damping effects occurring in high-temperature phases.

Unlike valence and ionic crystals, molecular crystals are composed of structural units which are bound together with very weak forces in comparison with the interactions between individual atoms in a given molecule. Because of this, molecules very frequently display features like those of molecules in the gaseous state and collective behaviour of the lattice appears only in rather rare circumstances (lattice vibrations, dynamical splitting of infrared frequencies, local electric field, *etc.*). In cases where the interaction effects can be neglected, the measurement of a macroscopic crystal property can lead to the knowledge of properties of the free molecule.

Among known physical properties only two seem to meet these conditions: diamagnetic susceptibility and intensity of infrared absorption, at least in certain spectral regions. The mean molar susceptibility of diamagnetic solids, $\langle\chi\rangle$, is of the order of -100×10^{-6} c.g.s. M units.* Therefore, molecular field effects can safely be neglected and we can assume that the field in the crystal is practically equal to the external field strength, $\boldsymbol{H}$.[1] This leads us to the Oriented Gas Model of diamagnetic susceptibility, presented in the next section. In following sections certain properties of real crystals are discussed in connection with anharmonic effects due to thermal motion.

DIAMAGNETIC SUSCEPTIBILITY OF AN IDEAL CRYSTAL PHASE

Measurements of diamagnetic anisotropy of crystals were introduced first by Krishnan[2] and a molecular interpretation of the macroscopic tensor $\boldsymbol{\chi}$ was given by Lonsdale and Krishnan.[3] Denoting by $\mathbf{K}$ the molecular tensor, given most frequently in *LMN* symmetry axes of the molecule, and by $\boldsymbol{c}^{(r)}$ the orientation matrix of *LMN* axes of the *r*th molecule with respect to the orthogonalized frame of the unit cell, abc^x, we can summarize the relationship between macroscopic and microscopic susceptibilities by means of the expression

$$\boldsymbol{\chi}\,(abc^x) = \frac{1}{Z}\sum_r \boldsymbol{c}^{(r)T}\,\mathbf{K}\boldsymbol{c}^{(r)} \tag{1}$$

* $cm^{-3}/4\pi$.

where c^T means c transposed. Matrix $c^{(1)}$ for the " first " molecule is to be taken from structural data, and for the other $Z - 1$ molecules

$$c^{(r)} = A_r c^{(1)} A_r^T. \tag{2}$$

A_r are matrices corresponding to the symmetry elements inherent in the symmetry factor group of the unit cell. Expression (1) may be called the Oriented Gas Model Approximation and in analogy with what is known from the infrared spectroscopy of crystals,[4] it is a special case of the additivity rule of physical properties. It may be expected that the model will hold only for well-ordered crystalline phases.

However, if molecular properties are known, some structural information can be obtained from crystal susceptibilities for disordered structures also. An example of such investigations may be found in ref. (5). Usually eqn (1) is used to calculate the molecular tensor **K** from measured crystal properties, and many examples of this application have been reviewed by Lasheen.[6,7] In principle, this problem can be solved uniquely only for triclinic crystals for which the number of experimentally available pieces of information exactly corresponds to the number of components of the tensor **K** (length of three principal axes and three orientation angles). In crystals with higher symmetry certain assumptions must be made about the orientation of the **K** axes which are difficult to verify. Recently, much progress has been made with the problem in two papers of Van den Bossche and Sobry.[8,9] Assuming that the additivity law holds for molecular susceptibility the authors have shown how to derive the particular tensors for the carbon skeleton and the substituents separately. Using the known increments we are able to obtain the orientation of the principal axes of **K** in an independent manner.[9,10]

The particular shape of $\boldsymbol{\chi}(abc^x)$ depends on the symmetry of the unit cell: the higher the symmetry, the more of the off-diagonal components are equal to zero.[1] For instance, assuming that the two-fold symmetry axis in a monoclinic crystal is parallel to the crystallographic axis b, we have

$$\boldsymbol{\chi}(abc^x) = \begin{pmatrix} c_{j1}^2 K_j & 0 & c_{j1}c_{j3}K_j \\ 0 & c_{j2}^2 K_j & 0 \\ c_{j1}c_{j3}K_j & 0 & c_{j3}^2 K_j \end{pmatrix} \tag{3}$$

where summation over j is inferred. The graphical picture of eqn (3) is, in all cases yet found, a triaxial ellipsoid with all axes negative.

Let us assume now that we are given a triaxial ellipsoid in a monoclinic crystal with the axes fulfilling the inequality

$$|\chi_1| > |\chi_3| > |\chi_2|. \tag{4}$$

The orientation of the χ_1 axes is assumed here, as usual, in such a manner that $\chi_3 \| b$, and χ_1 and χ_2 both lie in the (010) plane. The angle of orientation of χ_1 with respect to the crystallographic axis a, ψ, is meant to be positive if measured in positive monoclinic angle, β. With these assumptions we can always find a direction OP, fig. 1, such that the susceptibility along OP will be equal to χ_3. The section through OP and χ_3 is isotropic, *i.e.*, the crystal will show no anisotropy when hung in a magnetic field along its normal, OQ. Therefore, OQ is one of two magnetic axes of the crystal, and the axial angle, $2V$, is given by the expression

$$\sin V = \left(\frac{\chi_3 - \chi_1}{\chi_2 - \chi_1}\right). \tag{5}$$

Making use of eqn (3) we can see that the anisotropy of molecules and their distribution are factors determining the direction of macroscopic magnetic axes. Experi-

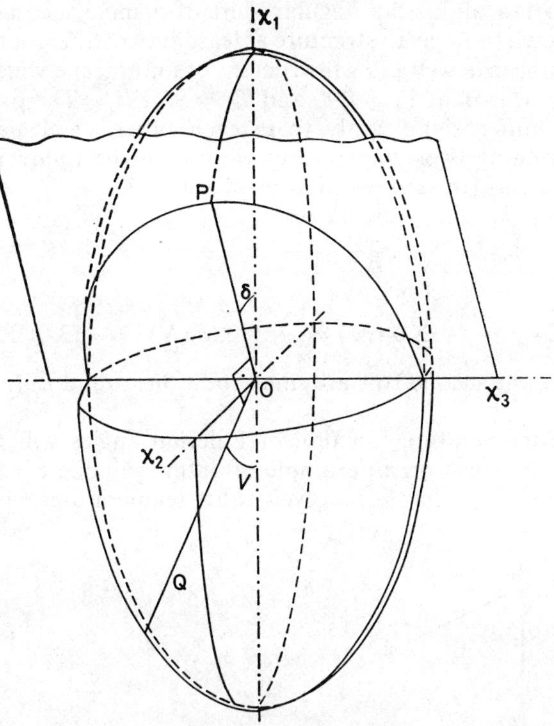

FIG. 1.—Circular section of an ellipsoid of magnetic susceptibility: susceptibility along OP is equal to χ_3. OQ is the direction of magnetic axis.

mental evidence of an isotropic section has been given in ref. (10) for the example of 1,5-dinitronaphthalene. Therefore, in analogy with the classification known in optics, we can divide dia- and para-magnetic crystals into biaxial, uniaxial and isotropic classes.

DIAMAGNETISM OF A REAL CRYSTAL PHASE

It is well known that the atoms execute thermal vibrations with an amplitude increasing as the temperature is raised. In the harmonic approximation the atomic vibrations can be presented in a compact form by means of tensors **T**, **L** and **S**, describing translational, librational and composed motion, respectively, of the molecule as a whole.[11–13] By a " real " crystalline phase we understand a periodic structure composed of molecules which execute thermal oscillatory motions. Other defects, such as stacking faults, dislocations or dissolved admixtures, will be ignored.

However, the vibrations cannot be harmonic, as is indicated by the anisotropy of thermal expansion, leading to a homogeneous deformation of the lattice. As a consequence, the actual position of the molecule will change slightly when the temperature is raised. The overall effect may be resolved into two components: a small translatory shift of the centre of mass and a small change of orientation of molecular axes with respect to the frame of the unit cell. Of primary importance here is the angular position of the molecule. The translatory shift leading to a small density change can be neglected because of the low accuracy of magnetic measurements. However, it can have a significant role in the much more accurate measurements of the temperature function of optical birefringence.[14–16]

Direct information about the angular shift of a molecule can be derived from structural data known for a given structure at least at two different temperatures. Let us take as an example the well known structure of anthracene which was investigated very accurately by Mason at $T_1 = 290$ and $T_2 = 95$ K.[17] The orientation matrices, $\boldsymbol{c}(290)$ and $\boldsymbol{c}(95)$, differ significantly from each other. Calculating the Eulerian angles χ, φ, θ, for these two temperatures we obtain the following values of mean temperature coefficients (in seconds of arc per degree K):

$$\frac{\Delta\chi}{\Delta T} = -11.5, \frac{\Delta\varphi}{\Delta T} = -3.6, \frac{\Delta\theta}{\Delta T} = +20.5 \ ''\mathrm{K}^{-1},$$

or

$$\delta_1(L) = -10.2, \delta_2(M) = -18.2, \delta_3(N) = -13.0 \ ''\mathrm{K}^{-1}.$$

The figures given in the second row are small rotations round *LMN* symmetry axes of the molecule.

Of course, the temperature function of Eulerian angles will not be, in general, linear. This is clearly seen on an example of anthraquinone crystal shown in fig. 2. The structure of this crystal has been solved at five temperatures[18] indicated by arrows on the abscissae.

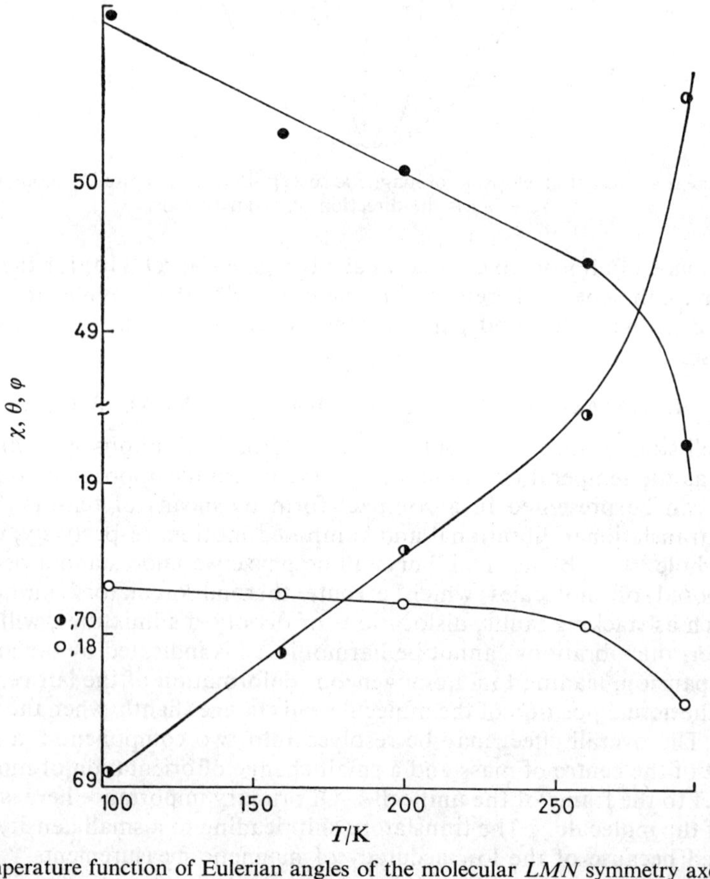

FIG. 2.—Temperature function of Eulerian angles of the molecular *LMN* symmetry axes in anthraquinone crystal. Points correspond to the values calculated from crystal structure data.[18] ●, χ; ○, θ and ◑, φ.

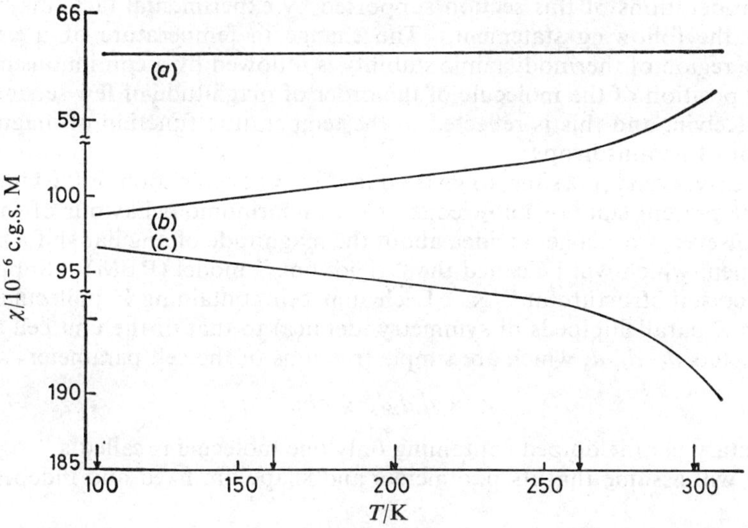

FIG. 3.—Temperature function of crystal susceptibilities in anthraquinone. Arrows indicate temperature points at which the structure was analysed.[18] (*a*) $-\chi_1$, (*b*) $-\chi_2$ and (*c*) $-\chi_3$.

We may now ask to what extent is the macroscopic susceptibility influenced by these small structural changes? The answer is illustrated in fig. 3 where the principal susceptibilities are shown for anthraquinone, calculated using eqn (1) and the temperature functions of Eulerian angles from fig. 2. Susceptibilities χ_2 and χ_3 change in a manner similar to the Eulerian angles but χ_1 is temperature independent. Unfortunately no experimental data are available for anthraquinone so that direct comparison is not possible.

Instead we may quote some results known for benzene. The crystal anisotropies (orthorhombic) have been measured by Hoarau *et al.*[19] in a wide temperature range. Two curves taken from this paper and shown in fig. 4 (solid lines) are similar in shape to those from fig. 3. A comment concerning the broken lines in fig. 4 will be made later in this section. Note that the mean crystal susceptibility, $\langle\chi\rangle$, is independent of temperature within the limits of experimental error. This has been experimentally proved for crystalline benzene by Hoarau and others in a temperature interval from 80 to 270 K.[19]

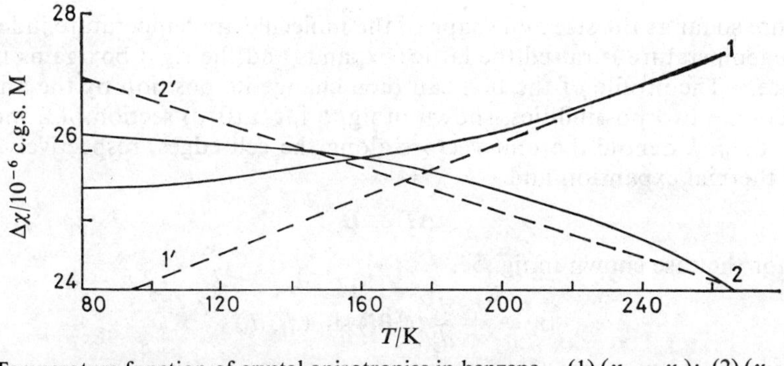

FIG. 4.—Temperature function of crystal anisotropies in benzene. (1) $(\chi_b - \chi_a)$; (2) $(\chi_b - \chi_c)$ taken from ref (19), broken lines (1′) and (2′) calculated according to RBM.

The considerations of this section supported by experimental facts may be summarized by the following statement. The change in temperature of a crystalline phase in the region of thermodynamic stability is followed by a continuous change in the angular position of the molecule of the order of magnitude of few seconds of arc per degree Kelvin, and this is reflected in the temperature function of magnetic susceptibility or of its anisotropy.

It is not easy, even if possible, to give a quantitative explanation of the temperature effects at the present state of knowledge of the anharmonic behaviour of the crystal lattice. However, we can get an idea about the magnitude of angular shifts by means of an argument which will be called the " rigid box " model (RBM). Suppose that we have a crystal structure at 0 K. Each unit cell containing Z molecules can be divided into Z parallelopipeds of symmetry identical to that of the unit cell and with edge parameters d_1, d_2, d_3 which are simple fractions of the cell parameters so that

$$Z \times d_1 d_2 d_3 = abc. \tag{6}$$

This elementary parallelopiped containing only one molecule is called a " rigid box " because we will assume that its parameters and shape are fixed and independent of

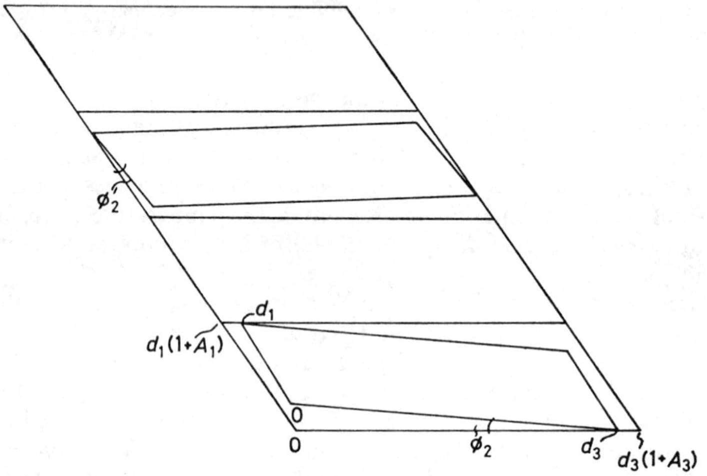

FIG. 5.—Two possibilities of rotation of (010) profile of a rigid box (d_1, d_3) within a cell $\{d_1(1 + A_1), d_3(1 + A_3)\}$, schematic.

temperature so far as the size and shape of the molecule are temperature independent. When the temperature is raised the lattice expands and the rigid box gains more and more space. The profile of the box can then change its position by the angle φ for which there are two possibilities, shown in fig. 5 for a (010) section of a monoclinic crystal. If $\boldsymbol{i}, \boldsymbol{j}, \boldsymbol{k}$ denote the unit vectors along the cell edges, respectively, α is the tensor of thermal expansion and

$$\alpha \Delta T \equiv \mathbf{B} \tag{7}$$

we have for the case shown in fig. 5

$$\varphi_2 = \pm \frac{d_1}{d_3} (\boldsymbol{i}^{\mathrm{T}} \mathbf{B} \boldsymbol{i}) \sin (d_1, d_3) \tag{8}$$

$$d_1(\boldsymbol{i}^{\mathrm{T}} \mathbf{B} \boldsymbol{i}) < d_3(\boldsymbol{k}^{\mathrm{T}} \mathbf{B} \boldsymbol{k}), \tag{9}$$

or

$$\varphi'_2 = \pm \frac{d_3}{d_1} (\boldsymbol{k}^T\mathbf{B}\boldsymbol{k}) \sin (d_1, d_3) \qquad (10)$$

if

$$d_1(\boldsymbol{i}^T\mathbf{B}\boldsymbol{i}) > d_3(\boldsymbol{k}^T\mathbf{B}\boldsymbol{k}). \qquad (11)$$

Analogous relationships will hold for another section of the unit cell. The choice between the two possibilities, eqn (8) or (10), can be made by means of inequalities (9) and (11), and the selection of sign must be made by correlation with the experimentally observed direction of changes of diamagnetic anisotropy or optical birefringence. Numerical values of the angle φ calculated from a, b, c, Z and $\boldsymbol{\alpha}$ for various crystals are summarized in table 1. The results given by RBM are very sensitive to $\boldsymbol{\alpha}$, as is

TABLE 1.—ROTATION ANGLES ROUND *LMN* AXES OF MOLECULES (seconds of arc per Kelvin)

	(*L*)	(*M*)	(*N*)	comments
			(A) anthracene	
1	−10.3 ± 0.1	−18.10 ± 0.07	−12.9 ± 0.09	directly from crystal structure: Mason[17]
2	−14.8	−14.0	−1.6	RBM, $\boldsymbol{\alpha}$: Mason[17]
3	−6.1	−3.5	0.5	RBM, $\boldsymbol{\alpha}$: Kozhin[20]
4	−18.2	−11.8	0.9	RBM, $\boldsymbol{\alpha}$: Ryzhenkov[21]
5	−10.2	−3.2	2.3	RBM, $\boldsymbol{\alpha}$: Jakubowski[22]
			(B) carbazole	
1	−9.8	0	0	from birefringence: Kusto[16]
2	−7.0	0	0	RBM, $\boldsymbol{\alpha}$: Świątkiewicz[23]
			(C) benzene	
1	73.7	−13.7	−12.8	from diamagnetic anisotropy: Hoarau[19]
2	60.3	−15.2	−37.2	RBM, $\boldsymbol{\alpha}$: Cox[24]

seen for anthracene. In certain crystals, *e.g.*, carbazole, only one degree of angular shift is allowed because of symmetry restrictions.

It may be interesting to see how far the angular shifts deduced from RBM are able to reproduce the temperature function of diamagnetic anisotropy. This is exemplified for benzene by the broken lines in fig. 4. If we assume, in accordance with molecular symmetry, that $K_L = K_M \neq K_N$, all principal susceptibilities, χ_a, χ_b and χ_c automatically become functions of only two Eulerian angles: φ and θ. Using mean temperature coefficients given by the RB model we cannot reproduce the nonlinear behaviour of $\chi_i(T)$. The position and mean slope of experimental curves in fig. 4 agree only qualitatively with predictions given by the model.

DIAMAGNETISM AND PHASE TRANSITIONS

So far we have dealt with the temperature behaviour of crystalline phases which are thermodynamically stable in some temperature region; in particular their symmetry remains unaltered within this region. During a phase transformation a sharp symmetry change occurs at the transition temperature, at least in the case of the first-order transition. This is usually accompanied by far greater changes in the orientation matrix, **c**, and, consequently, by more dramatic behaviour of magnetic susceptibility. This need not be true for crystals composed of molecules which have small or zero magnetic anisotropy.

Experiments performed on pentachlorophenol crystals (monoclinic) have shown[25] that, on heating, the diamagnetic anisotropy rapidly diminishes in the vicinity of 63 °C, corresponding to the transition temperature of this compound. Similar effects observed in 1,8-dinitronaphthalene[26] (orthorhombic) are illustrated in fig. 6.

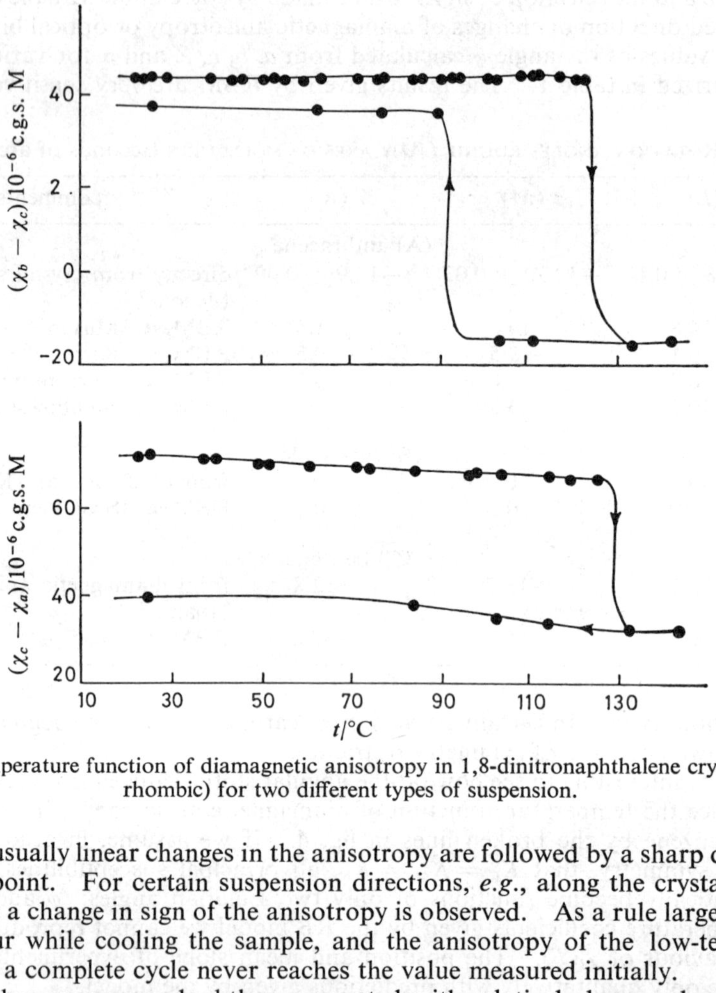

FIG. 6.—Temperature function of diamagnetic anisotropy in 1,8-dinitronaphthalene crystals (orthorhombic) for two different types of suspension.

Small and usually linear changes in the anisotropy are followed by a sharp drop at the transition point. For certain suspension directions, *e.g.*, along the crystallographic axis *a*, even a change in sign of the anisotropy is observed. As a rule large hysteresis effects occur while cooling the sample, and the anisotropy of the low-temperature phase after a complete cycle never reaches the value measured initially.

These phenomena are doubtless connected with relatively large volume changes accompanying these two phase transformations, which thus lead to a destruction of long-range order. Therefore, in spite of large effects, these and similar cases may not be very interesting from the structural point of view as they escape microscopic interpretation. More profitable experiments could be performed with crystals exhibiting a crystal-to-crystal transition, but according to our knowledge they have not yet been done.

However, the difference in density between low- and high-temperature phases can also lead to new and interesting conclusions concerning some aspects of the lattice dynamics of the high-temperature phase. A crystal sample, when hung on a thin thread in a homogeneous magnetic field, can execute torsional vibrations whose period is governed as much by the anisotropy of the crystal as by the elastic constant of the thread. If the sample oscillates in air, the motion is damped. This is partially caused by the viscous properties of air but is also due to the inner, microscopic state of the sample. Depending on the particular substance and its temperature the reasons for the appearance of an " inner viscosity " may be different. Using the oscillation method for quartz samples, which can surely be considered as a " rigid phase ",

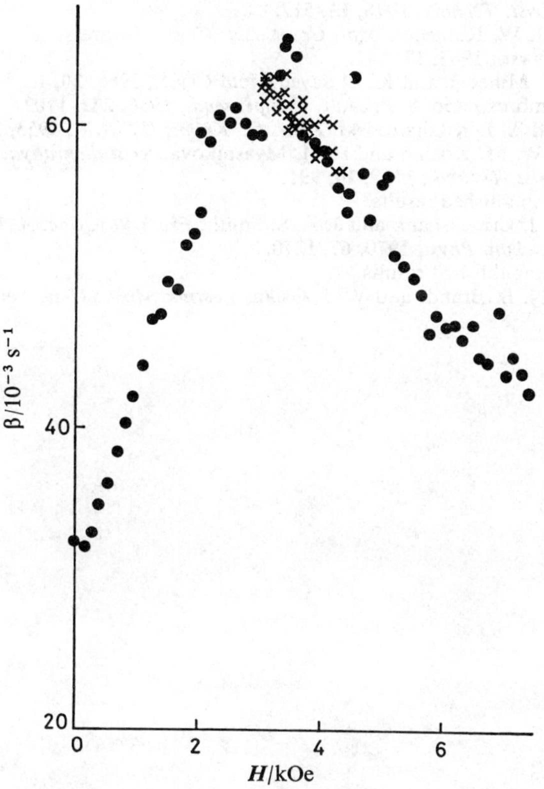

FIG. 7.—Damping constant, *β*, of torsional vibrations of a sample of potassium ferrocyanide trihydrate as a function of the field strength, *H*.

Braginski has shown[27] that it is an enormously sensitive tool for measuring the quality of the crystals. The high-temperature phase of a molecular crystal with its specific dynamics may be another example of enhanced damping. This is believed to be illustrated by fig. 7 which shows the damping constant measured in air for potassium ferrocyanide trihydrate, $K_3Fe(CN)_6$. The crystal suspended along the *b* axis shows very small anisotropy and in this case a sharp dependence of ***H*** is observed.

Further investigations are in progress.

[1] J. F. Nye, *Physical Properties of Crystals* (Clarendon Press, Oxford, 1957).
[2] K. S. Krishnan, B. C. Guha and S. Banerjee, *Phil. Trans. A*, 1933, **231**, 235.
[3] K. Lonsdale and K. S. Krishnan, *Proc. Roy. Soc. A*, 1936, **156**, 597.
[4] J. W. Rohleder and T. Luty, *Mol. Cryst. Liq. Cryst.*, 1968, **5**, 145.
[5] G. Fulińska-Wójcik and J. W. Rohleder, *Acta Phys. Polon.*, 1974, A**45**, 3.

[6] M. A. Lasheen, *Phil. Trans. A*, 1964, **256**, 357.
[7] M. A. Lasheen, *Acta Cryst. A*, 1968, **24**, 289.
[8] G. Van den Bossche and R. Sobry, *Acta Cryst. A*, 1974, **30**, 616.
[9] R. Sobry and G. Van den Bossche, *Acta Cryst. A*, 1974, **30**, 721.
[10] J. W. Rohleder and A. Mierzejewski, *Materials Sci.* (*Wrocław*), 1978, **4**, 97.
[11] D. W. J. Cruickshank, *Acta Cryst.*, 1956, **9**, 754.
[12] G. S. Pawley, *Acta Cryst.*, 1963, **16**, 1204.
[13] B. T. M. Willis and A. W. Pryor, *Thermal Vibrations in Crystallography* (Cambridge University Press, Cambridge, 1975).
[14] P. J. Bounds and R. W. Munn, *Mol. Cryst. Liq. Cryst.*, 1978, **44**, 301.
[15] J. W. Rohleder, *Krist. Technik*, 1978, **13**, 517.
[16] W. J. Kusto and J. W. Rohleder, *Mol. Cryst. Liq. Cryst.*, in press.
[17] R. Mason, *Acta Cryst.*, 1964, **17**, 547.
[18] K. Lonsdale, H. J. Milledge and K. El Sayed, *Acta Cryst.*, 1966, **20**, 1.
[19] J. Hoarau, N. Lumbroso and A. Pacault, *Compt. rend.*, 1956, **242**, 1702.
[20] W. M. Kozhin and A. I. Kitaigorodski, *Zhur. fiz. Khim.*, *U.S.S.R.*, 1953, **27**, 1676.
[21] A. J. Ryzhenkov, W. M. Kozhin and P. M. Myasnikova, *Kristallografiya*, 1967, **13**, 1028.
[22] B. Jakubowski, *Krist. Technik*, 1979, **14**, 991.
[23] J. Świątkiewicz, unpublished results.
[24] E. G. Cox, D. W. J. Cruickshank and J. A. S. Smith, *Proc. Roy. Soc. A*, 1958, **247**, 1.
[25] J. W. Rohleder, *J. chim. Phys.*, 1970, **67**, 1270.
[26] A. Mierzejewski, unpublished results.
[27] W. B. Braginski, N. B. Brandt and W. I. Osika, *Vestnik Mosk. Univ. Ser. III, Phys. Astron.*, 1970, **11**, 91.

GENERAL DISCUSSION

Prof. A. J. Leadbetter (*University of Exeter*) said: It is a great pleasure to see high-pressure techniques now being applied to the study of phase transitions in molecular solids. However, most of the techniques described are concerned essentially with static molecular properties, and dynamic properties provide a more sensitive probe of the intermolecular potential. What are the prospects for making such measurements as a function of pressure?

Dr. A. Würflinger (*University of Bochum*) said: The high-pressure equipment described in the paper allows dielectric measurements only below 10 MHz. There are only a few substances (*e.g.*, cyclopentanol) which reveal dielectric relaxation phenomena at such low frequencies. Cyclopentanol has already been studied dielectrically both at atmospheric[1] and elevated[2] pressures. A reinvestigation of the high-pressure behaviour of this substance is under way. In order to extend the measurements to substantially higher frequencies a different technique has to be used that requires a new high-pressure design.[3]

[1] J. R. Green, S. J. Dalich and W. T. Griffith, *Mol. Cryst. Liq. Cryst.*, 1972, **17**, 251.
[2] D. N. Glidden, *Thesis* (University of New Mexico, 1972).
[3] R. Pottel and E. Asselborn, *Ber. Bunsenges. Phys. Chem.*, 1979, **83**, 29.

Dr. B. Ewen (*University of Mainz*) said: Dr. Würflinger has shown us measurements on cyclohexanone which exhibit very clearly the existence of different phase transitions. Is anything known about the molecular origin?

Dr. A. Würflinger (*University of Bochum*) said: The solid I–solid III transition at atmospheric pressure has been discussed for example by Corfield and Davies.[1] It is generally assumed that the quasispherical cyclohexanone molecule rotates with little restriction in the high-temperature solid phase. As far as the high-pressure transition I–II is concerned no structural information is available. It should be mentioned that the dielectric decrease at the I–II transition is of the same order as the dielectric dispersion observed by Corfield and Davies at 50 MHz in the solid I state. They suggested the absorption to arise from a correlation of the neighbouring molecules. Furthermore, the possibility of the axial–equatorial ring inversion has been discussed, but according to their results this conformational change should not be detectable dielectrically.

It might be interesting to compare the phase behaviour of cyclohexanone with cyclohexane which also exhibits an additional high-pressure phase transition.[2] The structure of the high-pressure solid phase III is not yet known, and the question may arise whether this phase has something to do with the chair–boat inversion of cyclohexane. The large energy difference[3] between the chair and boat form does seem to rule out the possibility of the boat form to exist at low temperatures. However, Lüdemann *et al.*[4] found a *negative* activation volume for the ring inversion, using high-pressure n.m.r. spectroscopy. Possibly the intermediate twist-boat form (see figure below that has been taken from Squillacote)[3] has a smaller volume than the chair form, therefore the ring inversion could be favoured with increasing pressure.

If this were true, the solid phase III could be characterized by a steady conformational change between the two equivalent chair forms, whereas in the low-temperature phase II the ring inversion might be frozen in, thus reducing the possible orientations of the cyclohexane molecule to 1/2. This would correspond to an entropy change of *R*ln2 which indeed is observed for this transition.[2,5,6]

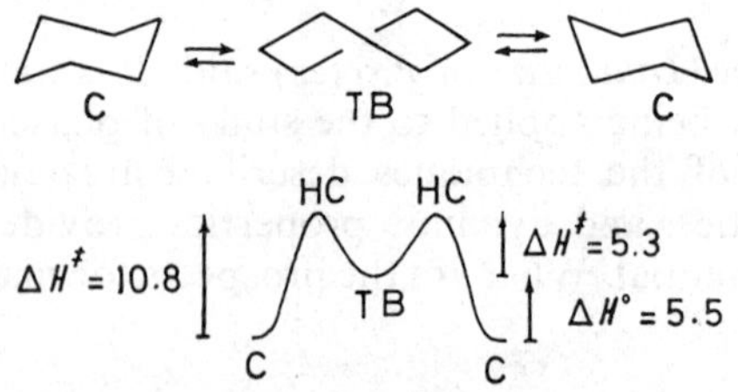

FIG. 1.—Energy diagram (units are kcal mol^{-1}) for ring inversion in the cyclohexane chair (C), showing the twist-boat (TB) and the half-chair (HC) *i.e.*, the transition state for the C to TB interconversion). Pseudorotation in the TB and HC is not shown.

In the case of cyclohexanone the entropy change of the solid I–solid II transition is much less[5,7] and it is open to question whether this transition is associated with the axial–equatorial interchange or not.

[1] G. Cornfield and M. Davies, *Trans. Faraday Soc.*, 1964, **60**, 10.
[2] A. Würflinger, *Ber. Bunsenges. Phys. Chem.*, 1975, **79** 1195.
[3] M. Squillacote, R. S. Sheridan, O. L. Chapman and F. A. L. Anet, *J. Amer. Chem. Soc.*, 1975, **97**, 3244.
[4] H. D. Lüdemann, R. Rauchschwalbe and E. Lang, *Angew. Chem.*, 1977, **89**, 340.
[5] K. D. Wisotzki, *Diplom. Thesis* (University of Bochum, 1980).
[6] H. Arntz and G. M. Schneider, *Faraday Disc. Chem. Soc.*, 1980, **69**, 139.
[7] A. Würflinger and J. Kreutzenbeck, *J. Phys. Chem. Solids*, 1978, **39**, 193.

Dr. J. A. Venables (*Sussex University*) said: It seems particularly difficult to grow large single crystals of the low-temperature forms of molecular solids, because of the volume change and defect production associated with phase transitions. These defects do not anneal out at all rapidly when the transitions take place far below the melting point. Even when single-crystal-like patterns are produced, troubles can arise from defects such as twins giving spurious reflections. Thus it seems that powder patterns in conjunction with packing calculations could be very helpful, as Dr. Pawley suggests.

An alternative approach which we have taken is to examine the necessarily small crystals produced, by high-energy electron diffraction in a transmission electron microscope. The crystals are produced by vapour deposition in a small enclosed cell, separated from the microscope vacuum by thin electron transparent windows: crystals in the 1-10 μm size range can give good single-crystal patterns after moderate annealing, and twinning and other defects can be identified in the corresponding electron micrographs. It may also be possible to observe the phase transitions as they occur.

Previous work of this type has enabled us to show that the α-N_2 structure is indeed *Pa*3, and twins were the almost certain cause of extra spots (in X-ray patterns) which had been interpreted as evidence for the *P*2, 3 structure.[1] Presently we are working on SF_6, whose low-temperature structure is unknown, and it is possible that the low-temperature structure of CD_4 could be tackled in this way. An elegant example of a similar approach is the work of Ramdas *et al.* on the various structures of anthracene, presented as a poster at this meeting.[2] Thus single-crystal electron diffraction to determine unit-cell dimensions may be a useful tool in conjunction with packing calculations, especially when X-ray or neutron single-crystal data are either not obtainable or are suspect.

[1] J. A. Venables and C. A. English, *Acta Cryst.*, 1974, **B30**, 929; G. J. Tatlock, R. Mevrel and J. A. Venables, *Phil. Mag.*, 1977, **35**, 641.
[2] S. Ramdas, G. M. Parkinson, C. M. Gramaccioli, G. Filippini, M. Simonetta and M. J. Goringe, *Nature*. 1980, **284**, 153.

Dr. G. S. Pawley (*University of Edinburgh*) said: It is good to know that there are other techniques which could be used in conjunction with the powder diffraction technique. Electron diffraction could well be used to suggest a unit cell for a structure, as this is the major stumbling block. With the unit cell established the technique of powder diffraction becomes much more powerful.

Dr. A. Hüller (*KFA Jülich*) said: Can Dr. Pawley's method also be applied to orientationally disordered molecular crystals where the individual molecules possess several equilibrium orientations? Phase II of the ammonium halides with two different orientations of the ammonium tetrahedron and a statistical distribution over the two possibilities may serve as a simple example.

Dr. G. S. Pawley (*University of Edinburgh*) said: It is quite possible to tackle structures with disorder in them, if constraints can be used to reduce the number of parameters significantly. This has been done[1] on tetraiodoethylene, where the disorder parameters could be found to a fair statistical reliability, whereas this was not possible in an unconstrained refinement.

[1] G. S. Pawley, *Acta Cryst. B*, 1978, **34**, 523.

Dr. S. Ramdas (*University of Cambridge*) said: This comment pertains to Dr. Pawley's work on structure refinement by powder diffraction. In the case of molecular crystals with usually high anisotropy, there is the problem of preferred orientations in the powder. In general we may encounter reflections mainly from the basal planes and those nearly parallel to them. There should be some way of finding the distribution of accessible planes and make the corresponding corrections on the measured intensities of these reflections.

Dr. G. S. Pawley (*University of Edinburgh*) said: The program does not contain any means of overcoming the problems of preferred orientation, but it would not be difficult to include some extra function to modify the intensities which is dependent on the Miller indices in any systematic fashion. However, prospective users should know that there is a device at I.L.L. for obtaining powder diffraction scans from a spherical sample which is set spinning on a cushion of air. This process soon averages out over all the crystallites in the sample, and a single crystal can even be used as the sample.

Dr. A. D. Taylor (*Rutherford and Appleton Laboratories, Didcot*) said: On pulsed neutron sources such as SNS it is obligatory to exploit time-of-flight techniques for powder diffraction, *i.e.*, the Bragg angle is fixed and the scan obtained from the polychromatic beam. Different detectors on the same Debye–Scherrer ring see identical patterns, in the absence of preferred orientation. The SNS powder diffractometers will have segmented annular detectors (2°) with each segment giving a simultaneous measurement of the powder pattern. This will allow preferred orientation effects to be routinely investigated.

In designing the SNS powder diffractometers, we have been conscious of the necessity of measuring the lowest-order reflections in any pattern. A low-angle detector is included specifically for this purpose and, on the High Resolution Powder Diffractometer, the provision exists to reduce the pulse repetition frequency of the diffractometer when required.

Dr. G. S. Pawley (*University of Edinburgh*) said: It is good to hear that the new instruments for the SNS are being designed so that we will be able to measure the lowest orders of diffraction to a sufficiently accurate level as to make unit cell determination almost routine. The program described has been modified for the LINAC already, but the form of the peak shape used is not ideal. Once this is known, further modifications are very easily made.

Dr. A. I. M. Rae (*University of Birmingham*) said: We have successfully used Dr. Pawley's powder profile refinement program including rigid-body constraints to refine X-ray data from the α-phase of malononitrile as reported in a comment on our paper.

Dr. G. S. Pawley (*University of Edinburgh*) said: The peak shape function which is necessary for X-ray work will not be as complex as that for the pulsed neutron machines (see answer to Dr. Taylor), and it is not surprising that the Gaussian form already programmed gives a satisfactory refinement.

Prof. G. Chapuis (*University of Lausanne*) said: I would like to remark that there is a class of phase transitions that can hardly be elucidated by powder diffraction, namely those associated with the appearance of satellites or diffuse scattering for which their orientation in reciprocal space is of great importance to derive the nature of the modulation or the disorder.

I would like to ask Dr. Pawley if the program EDINP has been used for analysing structures with weak intensities arising from superlattice reflections?

Dr. G. S. Pawley (*University of Edinburgh*) said: We have recently been able to solve a structure where the pseudocell is a cube of *ca.* 13 Å, but which forms a large cell which is doubled in all directions.[1] This would not have been possible without the use of constraints. This is probably about at the limit of complexity that can be tackled at the moment, but the high-resolution machines which are now being designed for the SNS and the synchrotron X-ray source may well change this.

The powder diffraction from a highly anharmonic molecular system shows very few peaks, and therefore this technique is not usable. It would appear that in some cases there is structure in the background of a powder pattern, as is evidenced by the work of Dolling and Powell on β–N_2 (work in progress). To extract information from this background a simulation calculation is required, which is what we intend to do.

[1] G. A. Mackenzie, R. W. Berg and G. S. Pawley, *Acta Cryst. B*, 1980, **36**, 1001.

Dr. M. Schott (*University of Paris*) said: Most conducting polymers do not form crystals. They are actually poorly organized and show only a few diffraction lines. But even the modest information extracted from these lines would be useful. Could your program be adapted to such problems?

Dr. G. S. Pawley (*University of Edinburgh*) said: The method can certainly be used for polymers, and already we are prepared for the refinement of isotactic polypropylene. In this example, one of the unit cell edges is an exact multiple of the polymer chain, so that, as the unit cell edge refines to its optimum value, the whole structure of the polymer must be altered. The constraint needed to do this was rather more complicated than the usual constraints applied, but now this is done it is hoped that other systems will need only a modification of this constraint.

Prof. A. J. Leadbetter (*University of Exeter*) said: I should like to describe briefly the results of a structural determination made by myself, J. C. Frost and R. M. Richardson using Pawley's EDINP profile refinement programme. This was for the material t-butyl cyanide discussed in our paper at this meeting, for which we have carried out a low-resolution structural study by profile refinement of the data for $(4\pi \sin\theta/\lambda) < 2.6$ Å^{-1} using results collected using the DIA powder diffractometer at I.L.L., Grenoble and the PANDA diffractometer at A.E.R.E., Harwell. The fully deuterated sample was supplied by Prof. G. W. Gray and Dr. J. W. Goodby, University of Hull.

Refinement was successfully carried out at several temperatures in the low-temperature phase ($T < 233$ K) from 5 K to just below the transition.

As pointed out by Dr. Pawley it is first necessary to establish the unit cell and this is often very difficult. However, by following the changes from the high-temperature tetragonal phase we could show the cell to the monoclinic with space group $P2$, $P2/m$ or $P2_1/m$. Subsequent refinement established the space group to be $P2_1/m$ with the following parameters at 5 K

$$a = 6.109 \text{ Å}, b = 6.863 \text{ Å}, c = 6.706 \text{ Å (all } \pm 0.007 \text{ Å,)} \ \beta = 95.6° \ z = 2.$$

The fit to the data is shown in fig. 2 and a projection of the structure along a is shown in fig. 3. In this constrained refinement the structure of the molecule was taken to be

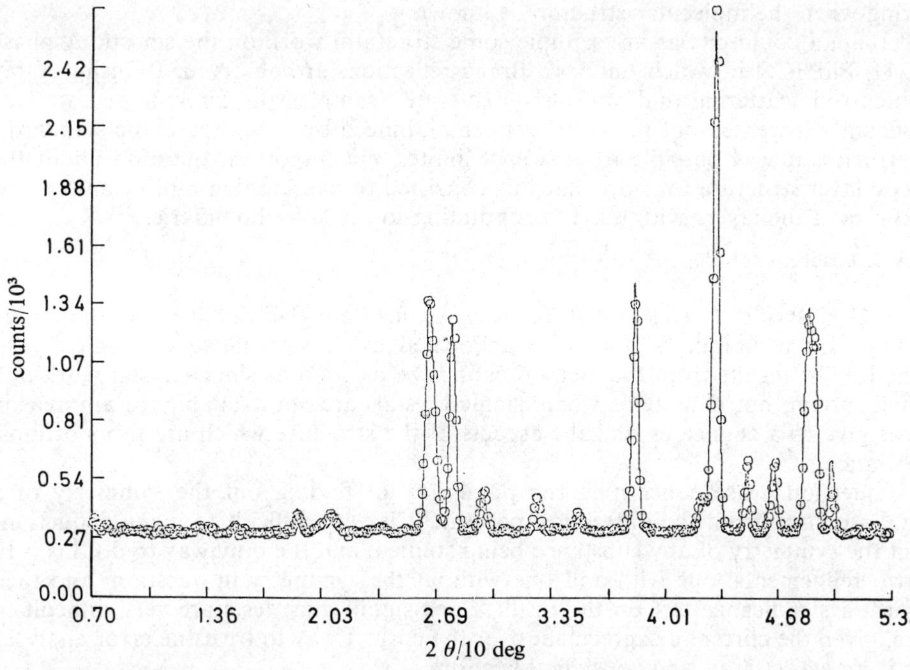

FIG. 2.—Calculated (full line) and experimental (circles) diffraction profile of t-butyl cyanide (D_9) at 5 K (Panda).

that deduced in the gas phase from electron diffraction and microwave experiments.

The structure of the tetragonal high-temperature phase has not yet been refined to the same extent but it is closely related to that of the low-temperature phase with the addition of considerable translational–rotational disorder (see our paper).

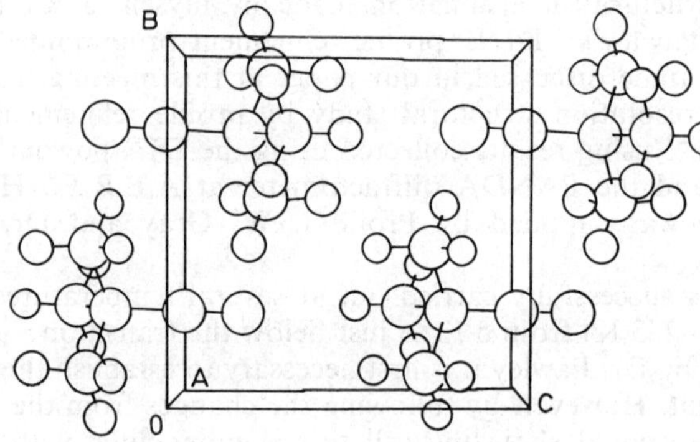

FIG. 3.—Structure of t-butyl cyanide at 5 K viewed down the *a* axis.

One might also remark that there is a danger of this discussion getting off track because Dr. Pawley is not advocating the use of powder in preference to single crystal methods but rather pointing out the possibilities when one is forced to use powders. Furthermore it may be that powder methods may provide a much more convenient route to the answer of specific limited questions, for example some aspect of molecular packing when the molecular structure is known.

Perhaps I could cite as an example some structural work on the smectic A phase of C_8H_{17}PhPhCN in which only *two* Bragg reflections are observed. By using X-ray and neutron scattering and variously deuterated samples the amount of data was significantly increased but remained very small indeed by crystallographic standards. Nevertheless it was possible to answer a limited but important question about the smectic layer structure to show that this consisted of overlapping molecular cores at the centre of the layers with alkyl tails pointing to the layer boundaries.[1]

[1] A. J. Leadbetter *et al.*, *J. Physique*, 1979, **40**. 375.

Dr. G. S. Pawley (*University of Edinburgh*) said: I am glad that it has been pointed out that this technique is not being advocated as an alternative to single-crystal methods. The results from the method cannot be as good as single-crystal work in a normal system, but in systems where single crystals are not available for any reason it does give us a chance to find the aspects of the structure which are to us of most importance.

A question arose concerning the possibility of finding out the symmetry of a crystal structure from the diffraction pattern. It is very difficult to make deductions about the symmetry of any substance being studied, and the only way to do this is to do two refinements, one with and one without the symmetry in question, and then perform a significance test on the result. Such significance tests are very difficult to make, given the current disagreement over the correct way to treat the error analysis. This is an aspect requiring considerable effort.

Dr. A. Hüller (*KFA Julich*) said: I would like to ask Dr. White two questions:

(1) What would the electron mobility be in an atomic crystal, say krypton, where the building blocks do not possess rotational degrees of freedom?

(2) What is the significance of the second maximum of luminescence around 37 K in CH_4 or 30 K in CD_4?

Dr. M. A. White (*Oxford University*) said: The electron mobility in an irradiated rare-gas solid is several orders of magnitude greater than that in most solid alkanes. This is primarily due to the molecular shape: as the molecules become more spherical (with the rare-gas molecules at the perfectly spherical limit), the anisotropy of the molecular polarizability decreases, and an increased electron mobility results. Therefore, the rotational degrees of freedom, *per se*, do not greatly influence the electron mobility; however, their effect on the determination of the molecular shape is important.

The polymorphism that is exhibited by many molecular solids can, of course, be associated with rotational degrees of freedom and at such solid–solid phase transitions the electron mobility would be greatly increased, resulting in thermoluminescence. However, it is not clear at this time whether this is to be associated primarily with increased rotational or translational motion, or both.

In answer to Dr. Hüller's second question, I remark that the origin of the broad thermoluminescence peaks (*i.e.*, those with widths of the order of several degrees) should be clearly distinguished from that of the sharper peaks which occur at the phase transitions. The latter are associated with recombination which is accelerated by the increased molecular motion at the solid–solid phase transitions; these peaks are found at $T \approx 20$ K in CH_4 + benzene (fig. 5 of our paper) and at $T \approx 22$ K and 27 K in CD_4 + benzene (fig. 7). All the remaining (broad) thermoluminescence peaks illustrated (fig. 4-7) can be ascribed to the thermal detrapping of electrons at temperatures that are characteristic of the activation energies necessary for their recombination with the parent cations. These are the so-called Randall–Wilkins peaks. The persistent peak at $T \approx 37$ K in the CH_4 + aromatic systems illustrated indicates the continued presence of electrons which are at a trap depth that can be thermally depopulated at $T \gtrsim 37$ K. Similarly, the results indicate the presence of an electron trap that will depopulate at $T \gtrsim 30$ K for CD_4 + aromatic.

Dr. M. Schott (*University of Paris*) (*partly communicated*): In connection with the question asked by Dr. Hüller, I believe that the transport properties of thermalized charge carriers in methane may be unimportant in Dr. White's experiments. Two reasons are (*a*) the low dielectric constant allows strong, long-range, coulombic electron–hole interactions implying very large capture cross-sections, and a Langevin recombination process, at least if the carriers' mean free paths are not more than a few hundred Å; (*b*) at low temperatures, recombination of carriers, both trapped, by tunnelling, may well be the dominant process outside the phase-transition temperatures. The isothermal luminescence of irradiated rigid organic glasses can be explained in that way.

It is puzzling that the emitted light is purple, at least when it can be seen: benzene and toluene are not expected to emit visible light. Could it be that reaction products are emitting as well? It is known that toluene for instance easily forms the benzyl radical C_6H_5—CH_2 and a hydrogen; I am not suggesting that benzyl emission is seen, since it should be green rather than purple, but other photochemical processes unknown to me are certainly possible. This remark raises the question of whether some of the light seen, especially at phase transitions, could be due to radical recombination, not to charge carrier recombination.

The thermoluminescence curves show that not all possible recombinations occur at the lowest phase-transition temperature (or below). Why should some charges become mobile, and others stay trapped? This could be related to the different microscopic environments of different traps. A more trivial explanation could be competition between recombination and retrapping, some charges liberated outside the capture radius of any charge of the opposite sign being retrapped before meeting any of these, in traps deep enough to hold them.

Dr. M. A. White (*Oxford University*) said: I would like to reply first to your comment concerning electron mobilities. Although tunnelling may be the dominant recombination mechanism in isothermal luminescence (and it is still not entirely clear that diffusion does not play a role in some systems), it is quite possible that at phase transitions, the lattice is sufficiently reorganized so that classical recombination processes may take place.

In reply to your second comment, perhaps I should qualify the statement of the colour of the emission. The glow curves that were reported were observed with a photomultiplier tube that was sensitive to wavelengths in the range 3000-6000 Å. The light intensities during thermoluminescence were very weak, and therefore spectral analysis was not carried out. The luminescence that was allowed to decay isothermally before heating the sample was, however, much more intense and, when observed by eye, it was seen to be violet in colour. This neither implies that the isothermal emission had a maximum in the violet range (since its maximum could have been in the u.v. region), nor that the thermoluminescence had a large violet component. It is, of course, possible that the violet emission that was observed was due to reaction products, as you suggest, or even due to small amounts of impurities.

Finally, I would add that we, too, were rather surprised to see the evidence of recombination at both phase transitions in CD_4 (rather than complete recombination at the lower transition). In addition to your suggested explanations, we could speculate that it is possible that this is due to intrinsic differences in the III→II and II→I phase transitions, with some electrons mobilized only at the upper transition.

Dr. M. A. White (*Oxford University*) (*partly communicated*). Some elaboration on the mechanism of energy absorption and luminescence in the methane solid might be useful. It is well known that it is difficult, if not impossible, to trap electrons in pure solid alkanes. In our experiments, we did not observe any luminescence when pure CH_4 or CD_4 was irradiated with 5 keV electrons. This was most likely due to the inability of these pure solids to trap electrons, with the result that the recombination process was too rapid for observation in our experiments. We also failed to see an emission when pure methane was irradiated by u.v. light. This is not the result of high electron mobility, but rather of the absence of ionic species when the pure sample is subjected to photolysis. As mentioned in our paper, a large population of triplet states is generated by u.v.-irradiation of the impure sample, so that the probability of second photon absorption is significant. This results in the generation of a cation–electron population, which leads to recombination luminescence. In summary then, the aromatic impurity is necessary for the observation of recombination luminescence in both photolysis and radiolysis experiments. In u.v. photolysis, it allows the production of an ion population and it creates electron traps. In radiolysis, the impurity generates electron traps but it is not necessary for the production of ionic species.

Prof. A. J. Leadbetter (*University of Exeter*) said: Could Prof. Rohleder tell us something more about his experimental method?

Prof. J. W. Rohleder (*Technical University, Wroclaw*) said: The method of measuring magnetic anisotropy of crystals is not new, and had been elaborated by Krishnan and co-workers by about 1935. In principle, the anisotropy is measured by means of the period of oscillation of the crystal when suspended on a thin and elastic thread in a homogeneous magnetic field. What seems to be new here is the fact that the amplitude of oscillations subsequently diminishes even in experiments carried out in a vacuum. This effect is due to the so-called " internal friction " within the solid and depends on the physical state of the sample.

Prof. G. J. Hills (*University of Southampton*) said: In his verbal introduction to his paper, Prof. Rohleder drew attention to the observation that the damping coefficients of the oscillations of the diamagnetic single crystals studied here were markedly dependent on the strength of the magnetic field. The reason for this is not clear although it seems not unlikely that mobile charges in the moving solid would respond to the magnetic field and thus enhance the damping. Any coupling of magnetic and electric phenomena in chemical systems is interesting and it would seem worthwhile to pursue the observation reported here for its own sake. I hope Prof. Rohleder will do this.

Prof. J. W. Rohleder (*Technical University, Wroclaw*) said: In a diamagnetic crystal we are concerned with closed electronic shells of structural units. In the most simple case of a NaCl or KCl crystal the electronic shells are more or less of spherical symmetry which presumedly cancels out the electromagnetic interactions. As a matter of fact, a very pure and nearly perfect NaCl crystal shows a low damping constant that is independent of the field strength. Experiments on these and other materials are still in progress and will be reported later.

Calorimetric Study of Transitions in Hexahydrated Fluorosilicates $MSiF_6.6H_2O$

By Ronald D. Weir,† Keith E. Halstead and Lionel A. K. Staveley

Inorganic Chemistry Laboratory,
University of Oxford, South Parks Road,
Oxford OX1 3QR

Received 5th December, 1979

Heat-capacity results are reported from 8 to 310 K for nickel fluorosilicate and from 7 to 320 K for the magnesium salt. These results, together with similar information on the manganese, iron, cobalt and zinc salts which will be published elsewhere, confirm that the nickel and zinc salts have no transition in the temperature region covered by the C_p measurements. Each of the other four salts has a transition, probably largely though not wholly isothermal. The entropies of transition are all rather small, being 2.35, 1.615, 3.76 and 5.29 J K^{-1} mol^{-1} for the magnesium, manganese, iron and cobalt salts, respectively. The trend of the molar entropies at 298.15 K of the salts from manganese to zinc is the same as that of the molar volumes, both quantities reaching a minimum at the nickel compound.

There seems no reason to doubt the conclusion previously reached largely on the basis of the diffraction evidence that the anions in these salts at room temperature are orientationally disordered, and possibly the cations as well. However, the available evidence suggests that the transitions are not primarily of an order–disorder kind, but are rather associated with changes in the hydrogen bonding between the water molecules and the fluorine atoms, such that the orientational disorder is relatively little affected by cooling through the transition.

In recent years, increasing interest has been shown in the hexahydrated silicofluorides of divalent metals of the general formula $M(H_2O)_6.SiF_6$, and especially in those for which M = Mg, Mn, Fe, Co, Ni or Zn. The investigations so far carried out have been primarily concerned with the structure of the crystals, the possibility of orientational disorder, and the dynamics of the torsional movement of the ions. The techniques employed include X-ray and neutron diffraction, proton and fluorine resonance experiments and optical examination.

At a glance, the formulae of these salts might suggest that they have relatively uncomplicated structures, with octahedral cations and anions packed in a simple way. In fact, the structures adopted at ordinary temperatures are essentially those of a slightly distorted caesium chloride lattice, but the ions are not necessarily exactly octahedral. The hydrated cation in particular may be distorted. Thus, in the cobalt salt it is elongated, in the nickel salt compressed and in the zinc salt almost regular.[1] Furthermore, the structures of the six salts at room temperature are not all identical, and while those with M = Mg, Mn, Fe and Co undergo transitions on cooling, the nickel and zinc salts do not. Such differences imply a subtle dependence of the properties of a particular salt on the nature or size of the metal atom. It appears that a key role must be played by the hydrogen bonding between the water molecules of the cations and the fluorine atoms of the anions. Just one piece of evidence for this, for

† Permanent address: Department of Chemistry and Chemical Engineering, Royal Military College of Canada, Kingston, Ontario, Canada.

example, is that the replacement of hydrogen by deuterium in the cobalt salt raises the transition temperature by 17 K.[2] It therefore seems that the explanation of the changes observed from one salt to another should be sought primarily in the influence of the metal on the hydrogen-bonding system.

An important early paper in this field was that in which Hamilton[3] reported the results of his neutron diffraction study of the iron(II) salt, and concluded that at room temperature this salt is orientationally disordered, in that each SiF_6^{2-} ion has two possible orientations, with an angular separation of 54°, while the cations likewise have two positions, separated by 18°. For each ion, the two positions are connected by rotation about a three-fold axis. Hamilton therefore considered this to be a case of static disorder. Since then, all the other five salts which we are considering have been stated on various grounds to be disordered, but as yet agreement has not been reached about the nature of this. Thus, Jehanno and Varret[4] expressed the view that in the iron, magnesium and manganese salts the disorder is dynamic rather than static.

These salts therefore present interesting problems which need further work for their solution. As far as thermodynamic measurements are concerned, these have so far been limited to d.s.c. and d.t.a. studies and to a few heat-capacity determinations of relatively low precision made over limited temperature ranges. It accordingly seemed worthwhile to measure the heat capacity of the six salts from ≈10 to 300 K in an adiabatic calorimeter, and this we have done. The primary purpose of the work was to confirm the presence or absence of transitions, and where these occurred to characterize them more accurately than hitherto. The detailed heat capacity results for the iron, manganese, cobalt and zinc salts will be reported in detail elsewhere. Here we shall only give the essential results for these four salts, together with the detailed results for the magnesium and nickel compounds. We shall then review some of the properties of the six salts in the light of our thermodynamic work.

EXPERIMENTAL

The heat capacities of the magnesium and nickel silicofluorides were measured in the smaller of the two calorimeters described by Waterfield and Staveley,[5] using samples of ≈35 g in each case.

RESULTS AND DISCUSSION

Values of the molar heat capacity C_p, and of derived thermodynamic functions, are presented in tables 1 and 2 for the magnesium and nickel salts, respectively. The experimental heat capacity results for the two salts are given in the Appendix. The C_p against T curve for the nickel salts shows no sign of any transition or thermal anomaly. The transition expected in the magnesium salt was duly observed. Equilibrium was established slowly in the transition region. Our estimate of the temperature at which the major part of the change occurs is 299.6 K. The heat capacity started to show an " abnormal " rise at ≈297 K. Two determinations were made of the total enthalpy of transition. In one, the sample was heated from 288.87 to 314.31 K, and in the second from 296.03 to 314.48 K. These gave 11 203.8 and 11 203.9 J mol^{-1}, respectively, for the total enthalpy gain in these intervals. Subtraction of the " normal " enthalpy intake, estimated from a freehand baseline heat-capacity curve, gave values of 703.5 and 703.6 J mol^{-1} for the enthalpy of transition, whence the entropy of transition, ΔS_t, is 2.35 J K mol^{-1}.

In table 3 we record for the six salts the values of the calorimetric entropy S° at 298.15 K and the temperature and entropy change for each transition. The chief results which emerge from our work are the following: (1) We confirm that whereas the magnesium, manganese, iron and cobalt salts all have a transition, the nickel and zinc salts do not, at least in the region ≈10-≈300 K. (2) The structural changes and the thermal hysteresis already known to be associated with these transitions strongly suggest that they are first order. Our results are consistent with the view that this is true of the major part of each transition, but generally this final stage is preceded by a rise in the heat capacity covering ≈2-5 K. (3) The entropies of

TABLE 1.—MOLAR THERMODYNAMIC PROPERTIES OF $MgSiF_6.6H_2O$

T/K	C_p/J K^{-1} mol^{-1}	$S°$/J K^{-1} mol^{-1}	$(H°-H°_0)$/J mol^{-1}	$[-(G°-H°_0)/T]$/ J K^{-1} mol^{-1}
10	1.95	0.63	4.8	0.14
15	7.17	2.3	26.8	0.51
20	15.0	5.3	80.8	1.30
25	24.2	9.6	178.8	2.49
30	34.0	14.9	324.3	4.08
40	53.3	27.2	760.8	8.20
50	71.5	41.1	1 386	13.4
60	88.75	55.6	2 187	19.2
70	105.6	70.6	3 159	25.4
80	122.3	85.8	4 298	32.0
90	138.5	101.1	5 602	38.8
100	154.5	116.5	7 067	45.8
110	170.7	132.0	8 693	53.0
120	186.2	147.5	10 480	60.2
130	202.0	163.0	12 420	67.5
140	217.6	178.6	14 520	74.9
150	232.8	194.1	16 770	82.3
160	248.2	209.6	19 170	89.8
170	263.0	225.1	21 730	97.3
180	276.9	240.6	24 430	104.8
190	290.4	255.9	27 270	112.4
200	303.5	271.1	30 240	119.9
210	316.2	286.2	33 330	127.5
220	328.6	301.2	36 560	135.1
230	340.5	316.1	39 900	142.6
240	352.8	330.8	43 370	150.1
250	365.6	345.5	46 960	157.7
260	382.1	360.1	50 690	165.2
270	388.3	374.8	54 570	172.7
280	400.2	389.1	58 510	180.1
290	413.2	403.4	62 580	187.7
298.15	431	415.0	66 000	193.6
310	444.9	434.2	71 810	202.6
320	459.5	448.5	76 330	210.0

transition are all relatively small, in no case reaching $\boldsymbol{R}$ ln 2 (=5.76 J K^{-1} mol^{-1}). (4) The value of $S°$ decreases from the manganese to the nickel compound, and then rises slightly on reaching zinc.

In transitions for which there are grounds for believing that an orientational order–disorder change is involved, and that a molecule or ion has n_2 distinguishable orientations in the low-temperature form and n_1 in the high-temperature form, then attempts to equate the observed entropy of transition ΔS_t with $\boldsymbol{R}$ ln (n_1/n_2) are more convincing, the greater the similarity in the structures of the low- and high-temperature forms and the smaller the volume change at the transition. Unfortunately, information on the volume and structural changes at the four transitions is not yet complete. That there may be significant differences in this respect is suggested by a comparison of the manganese and cobalt salts. For the former, Tsujikawa and Couture[6] reported that when cooled to liquid-helium temperatures (and therefore through the

TABLE 2.—MOLAR THERMODYNAMIC PROPERTIES OF $NiSiF_6.6H_2O$

T/K	C_p/J K^{-1} mol^{-1}	$S°$/J K^{-1} mol^{-1}	$(H-H_0^\circ)$/J mol^{-1}	$[-(G°-H_0^\circ)/T]$/ J K^{-1} mol^{-1}
10	1.62	0.51	4.0	0.11
15	6.1	1.9	22.5	0.4
20	13.1	4.5	69.0	1.05
25	21.0	8.3	154.2	2.1
30	29.2	12.8	279.7	3.5
35	38.5	18.0	449.0	5.1
40	47.7	23.7	664.5	7.1
50	65.3	36.2	1 229	11.6
60	83.5	49.7	1 972	16.8
70	101.5	63.9	2 897	22.5
80	119.8	78.6	4 004	28.6
90	137.5	93.8	5 291	35.0
100	154.8	109.1	6 753	41.6
110	172.3	124.7	8 387	48.5
120	189.8	140.4	10 190	55.5
130	207.0	156.3	12 180	62.6
140	223.5	172.3	14 340	69.9
150	240.5	188.3	16 660	77.2
160	256.8	204.3	19 140	84.7
170	272.2	220.4	21 790	92.2
180	286.7	236.3	24 580	99.8
190	300.2	252.2	27 520	107.4
200	313.8	267.9	30 590	115.0
210	326.9	283.6	33 790	122.7
220	339.4	299.1	37 120	130.3
230	351.6	314.4	40 580	138.0
240	364.1	329.6	44 160	145.6
250	375.0	344.7	47 850	153.3
260	388.2	359.7	51 660	161.0
270	395.6	374.5	55 590	168.6
280	405.9	389.0	59 590	176.2
290	418.3	403.5	63 710	183.8
298.15	428.1	415.2	67 150	190.0
310	438.5	432.1	72 290	198.9

transition), the crystals remained intact and transparent. By contrast, at the transition in the cobalt salt there is a colour change and cracks develop in the crystal. Indeed, Ghosh *et al.*[7] spoke of the crystal at the transition as undergoing " violent changes ". The temperature dependence of the volume of this salt as the transition is approached from below is interesting since the density *increases* by $\approx 4\%$ between 250 and 263 K,[8] above which temperature it starts to fall. However, in spite of these differences between the manganese and cobalt salts, Kodera *et al.*[9] concluded that their low-temperature forms are " quite similar ". For the magnesium salt, Syoyama and

Osaki[10] stated that the low- and high-temperature forms are " closely related ". As already mentioned, Hamilton[3] concluded that in the iron(II) salt both the anions and cations are disordered, each completely randomly between two positions, and Dereppe *et al.*[11] reported that disorder similar to this prevails in the magnesium and zinc salts. Ray *et al.*,[1] while forced to assume disorder in the cobalt, nickel and zinc salts, appeared to consider that this is confined to the fluorine atoms, to which they assigned two positions not related by any symmetry element and occupied to different degrees. Thus, for the cobalt salt they gave occupancy factors of 0.43 and 0.57. The distribution of the SiF_6^{2-} ions between two possible sites in these proportions would confer a configurational entropy of 5.68 J K^{-1} mol^{-1} on the salt. This is only slightly greater than the measured ΔS_t of 5.29 J K^{-1} mol^{-1} but, in view of the behaviour of this compound at the transition, this approximate agreement may be

TABLE 3.—SOME PROPERTIES OF HEXAHYDRATED FLUOROSILICATES $M(H_2O)_6.SiF_6$

M	Mg	Mn	Fe	Co	Ni	Zn
structure (high temp. form)	rhombohedral $R\bar{3}m$(?$P\bar{3}m1$)	hexagonal $R\bar{3}m$(?$P\bar{3}m1$)	rhombohedral $R\bar{3}m$(?$P\bar{3}m1$)	hexagonal $R\bar{3}$	hexagonal $R\bar{3}$	hexagonal $R\bar{3}$
structure (low temp. form)	monoclinic $P2_1/c$	monoclinic $P2_1/c$	monoclinic $P2_1/c$	probably monoclinic	(as above)	(as above)
molar volume[a]/cm^3	157.2	160.3	155.3	148.4	145.1	147.7
T_t/K	299.8	236.2	224.2	264.8	—	—
ΔS_t/J K^{-1} mol^{-1}	2.35	1.615	3.76	5.29	—	—
$S°$(298.15 K)/J K^{-1} mol^{-1}	415.0	455.2	447.5	446.6	415.2	437.0
$r(M^{2+})$/Å	0.72	0.82[b]	0.77[b]	0.74[b]	0.70	0.75

[a] Estimated for room temperature from unit cell parameters. [b] High-spin values.

fortuitous. For the nickel compound, Ray *et al.* gave the occupancy ratio as 0.33/0.67, but this substance has no transition, and indeed Kodera *et al.* found no difference in its structure between room temperature and 140 K. Frozen-in disorder with the proposed occupancy ratio would confer residual entropy of 5.3 J K^{-1} mol^{-1} on the crystal. For the zinc salt, the position with regard to the occupancy of the two anion sites seemed to be less clear.

In their n.m.r. study, Skjaeveland and Svare examined the linewidth changes in the Mn, Fe, Co, Ni and Zn salts, and found that for all of these compounds except the first the temperature of the linewidth narrowing due to SiF_6^{2-} reorientation was lower, and sometimes much lower, than that due to cation reorientation. Thus, for the nickel salt, the two temperatures were 150 and 375 K, respectively. For the manganese salt, the linewidth changes coincided at 225 K. Their results led them to conclude that the more strongly the water molecules are bound to the metal ion, the weaker are the H—F bonds and the easier the reorientation of the SiF_6^{2-} ions becomes. (The same sort of conclusion was reached by Jannek and Rager[13] from H and ^{19}F T_1 measurements on the magnesium and zinc salts.) It may therefore be significant that the two compounds for which the barrier to anion rotation is least (the nickel and zinc salts) are the two which have no transition.

The trend in the $S°$ values at 298.15 K (table 3) from manganese to zinc follows the trend in the molar volumes, and presumably reflects the increasing strength of the bonding between the metal ions and the six surrounding water molecules. With an increase in the strength of the M—H_2O bonding, one would expect some at least of the vibrational frequencies of the cation to rise, thereby lowering the entropy of the salt at any given temperature. As is well known, the stability of many six-coordinated complexes in the first transition metal series reaches a maximum at nickel, and in fact from the diffraction work it appears that the Ni—O bond is the shortest of the M—O bonds in the hydrated silicofluorides. Even so, it is perhaps surprising that the value of S° for the nickel salt is so much less than that of its neighbours. Frozen-in-disorder would make the value of S° less than the " true " entropy, but differences in this respect between the two salts without transitions, namely the nickel and zinc compounds, could scarcely account for much of the 22 J K^{-1} mol^{-1} difference between their S° values.

Some of the diffraction evidence indeed suggests that disorder in these compounds may be frozen-in on cooling. Hamilton's comparison[3] of the structure of the iron salt at room temperature and at 90 K led him to conclude that any transition falling between these two temperatures could not affect the nature of the disorder. Syoyama and Osaki[10] concluded that for the magnesium salt there are two different orientations for the anions in the low-temperature form, just as in the high-temperature modification. Moreover, two of the six compounds have no transition and for those that have ΔS_t is relatively small. It may well be, therefore, that the transitions reflect changes in the hydrogen bonding of a kind which have relatively little effect on the disorder. In this event, the interpretation of the ΔS_t values should be sought in changes in the frequencies of the numerous internal vibrational modes of the anions and cations and of their torsional oscillations. Any information bearing on such changes would be helpful, as would further diffraction work on the low-temperature modifications.

1 S. Ray, A. Zalkin and D. H. Templeton, *Acta Cryst. B*, 1973, **29**, 2741.
2 B. Ghosh, N. Chatterjee, A. N. Das, S. K. Dutta Roy and A. Pal, *J. Phys. C*, 1977, **10**, L527.
3 W. C. Hamilton, *Acta Cryst.*, 1962, **15**, 353.
4 G. Jehanno and F. Varret, *Acta Cryst. A*, 1975, **31**, 857.
5 C. G. Waterfield and L. A. K. Staveley, *Trans. Faraday Soc.*, 1967, **63**, 2349.
6 I. Tsujikawa and L. Couture, *J. Phys. Radium*, 1955, **16**, 430.
7 B. Ghosh, N. Chatterjee, A. N. Das and S. C. Nath, *J. Phys. C*, 1978, **11**, L243.
8 M. Majumdar and S. K. Datta, *J. Chem. Phys.*, 1965, **42**, 418.
9 E. Kodera, A. Torii, K. Osaki and T. Watanabe, *J. Phys. Soc. Japan*, 1972, **32**, 863.
10 S. Syoyama and K. Osaki, *Acta Cryst. B*, 1972, **28**, 2626.
11 J. M. Dereppe, P. W. Lobo and M. Van Meersche, *J. Chim. phys.*, 1964, **61**, 1076.
12 S. M. Skjaeveland and I. Svare, *Physica Scripta*, 1974, **10**, 273.
13 H.-D. Jannek and H. Rager, *Z. Naturforsch.*, 1975, **30A**, 1615.
14 K. Muthukrishnan and J. Ramakrishna, *J. Chem. Phys.*, 1973, **59**, 5571.

APPENDIX 1

EXPERIMENTAL MOLAR HEAT CAPACITY RESULTS FOR $MgSiF_6.6H_2O$

T/K	C_p/J K^{-1} mol^{-1}	T/K	C_p/J K^{-1} mol^{-1}
6.42	0.75	273.23	392.1
7.92	0.94	278.36	398.4
13.80	5.53	284.28	405.6
16.72	9.61	290.41	414.0
19.51	14.13	295.96	422.7
22.30	19.04	296.40	428.7
25.76	25.87	298.07	432.5
29.45	32.88	299.77	749.0
33.25	40.58	301.28	534.0
37.33	48.26	301.75	550.4
41.62	56.39	303.16	441.8
46.11	64.57	305.19	441.0
50.80	72.69	307.20	443.0
55.71	81.04	310.24	446.8
60.83	90.15	311.79	446.5
66.21	99.24	315.83	452.5
71.84	108.6	319.80	459.2
77.40	118.0		
81.12	124.2		
82.59	126.8		
86.69	133.2		
91.95	141.8		
97.36	150.2		
102.95	159.1		
108.29	167.7		
113.81	176.6		
119.50	185.6		
122.04	189.2		
127.45	197.9		
133.05	206.8		
138.83	215.9		
144.44	224.5		
149.88	232.8		
155.17	241.0		
160.68	249.4		
166.41	257.7		
172.00	265.7		
177.47	273.6		
183.16	281.2		
189.03	289.0		
194.81	296.8		
200.48	304.3		
206.04	311.4		
211.66	318.5		
217.51	325.6		
223.42	332.9		
228.23	338.5		
233.97	345.5		
239.62	352.4		
245.18	359.5		
250.35	366.5		
254.60	371.6		
259.37	380.9		
264.01	390.2		
268.62	386.8		

APPENDIX 2

EXPERIMENTAL MOLAR HEAT CAPACITY RESULTS FOR $NiSiF_6.6H_2O$

T/K	C_p/J K^{-1} mol^{-1}	T/K	C_p/J K^{-1} mol^{-1}
7.96	0.79	201.48	315.6
8.91	1.10	206.88	323.0
9.77	1.25	212.09	329.2
10.66	2.56	217.23	336.3
11.21	2.37	222.60	342.6
12.39	3.36	228.18	349.6
13.95	4.95	233.67	356.4
15.51	6.74	239.07	363.1
17.50	9.41	244.38	369.8
19.71	12.63	249.63	374.6
22.38	16.78	254.81	379.9
25.32	21.63	259.90	388.6
28.44	26.81	261.64	390.3
31.72	32.33	264.03	390.1
34.90	38.11	264.92	391.9
38.16	44.51	266.40	392.2
41.86	51.02	268.74	394.4
45.72	57.80	269.89	395.3
49.52	64.52	271.05	397.1
53.46	71.4	273.34	399.1
57.56	78.7	274.77	399.6
61.66	86.45	279.71	405.6
65.93	94.1	284.55	412.1
70.39	102.25	289.33	417.5
75.03	110.65	294.06	422.8
79.85	119.6	296.47	424.0
84.52	128.1	298.72	428.9
89.25	136.3	298.74	427.3
90.76	138.9	300.98	430.9
94.04	144.85	303.20	432.1
95.49	147.2	305.41	434.2
98.62	152.9	307.59	437.2
100.01	154.9	309.76	437.5
103.01	160.1		
104.76	163.2		
109.74	171.7		
114.53	180.0		
119.27	188.3		
123.98	196.4		
128.54	204.5		
132.96	212.1		
137.39	219.6		
142.06	227.4		
148.09	237.5		
152.72	245.0		
157.25	252.4		
161.68	259.4		
166.24	266.2		
170.95	273.1		
175.55	280.1		
180.62	287.5		
184.93	293.4		
190.56	300.8		
196.07	308.7		

Co-operative and Quadrupole Effects in Order–Disorder Transitions in Crystalline Ices

By P. Barnes, D. V. Bliss, J. L. Finney and J. E. Quinn

Department of Crystallography,
Birkbeck College, University of London,
Malet Street, London WC1E 7HX

Received 21*st December*, 1979

Using a previously developed *non-pair-additive* model of the water–water hydrogen bond, the relative stabilities of hydrogen ordered and disordered ice Ih, ices VII and VIII are examined. The results are consistent with both the observed hydrogen-ordering in the VII $\rightarrow$ VIII transition, and the lack of any similar ordering in hexagonal ice I. The symmetry relation between the two sub-lattices of VII and VIII, together with the non-linearity of the energy of the dipole–dipole interaction, suggest such ordering transitions may be favoured only when very special symmetry arrangements result in the cancellation of certain electropole energy terms.

1. INTRODUCTION

An adequate model of the water molecule and of the water–water hydrogen bond should, *inter alia*, be able to predict results consistent with the known properties of the various ices. In particular, such a model should ultimately be capable of explaining the complexities of the ice phase diagram, including the relative stabilities of the different structures, and the nature of the various phase transitions. In addition, the known large (40-60%) enhancement over the isolated molecule value of the average dipole moment in these ordered condensed phases provides a severe test of any model.[1] The origin of this enhancement in electron cloud distortion by the fields of neighbouring molecules implies that the known non-pair additivity of the water–water hydrogen bond[2] should be incorporated into the model.

We have developed a model of the water molecule which appears to fulfil these conditions, in particular giving a satisfactory account of the significant hydrogen-bond non-pair additivity, and the consequent cooperativity.[3-6] It has been tested extensively against the properties of dilute aqueous systems such as the isolated water dimer,[5] second virial coefficient[5] and small clusters of water molecules,[4] and has also been applied to studies of liquid water,[4,5] ion–water clusters ; and biopolymer–water interfaces.[8] We discuss here the application of the model to several ice phases, in particular focusing on the dipole-moment enhancement and the energetic and entropic contributions to phase transitions involving hydrogen ordering. The model is capable of explaining the different hydrogen-ordering behaviour of ice Ih and ice VII and throws some interesting light on the nature of and differences between the interactions involved in the different ice structures.

2. THE MODEL

Any charge distribution can be represented quite generally by a multipole moment expansion. For water, our preliminary calculations indicated that the expansion can be truncated after the quadrupole term without introduced serious errors in the electrostatic contribution to the energy of interaction of a molecular assembly, although higher terms can be incorporated later if necessary for particular cases. Non-pair additivity is incorporated by allowing the dipole moment to be perturbed under the influence of an electric field (either external, or internal from neighbouring molecules) through a polarisability tensor. In addition, a Lennard-Jones 6:12 term is used to handle the overlap repulsion and the van der Waals–London attractive component. Table 1 lists the values assigned to these various quantities: except for

TABLE 1.—VALUES USED FOR THE POLARISABLE ELECTROPOLE WATER MOLECULE[5] MODEL

dipole moment (μ_i)/D	0.0	0.0	1.855
dipole polarisability (α_{ij})/Å^3	1.444	0.0	0.0
	0.0	1.444	0.0
	0.0	0.0	1.444
quadrupole moment (Q_{ij})/D Å	−7.412	0.0	0.0
	0.0	−4.180	0.0
	0.0	0.0	−5.439
Lennard-Jones parameters:	$\sigma = 3.024$ Å		
	$\varepsilon = 0.764$ kcal mol^{-1}		

Axes are those defined by Glaeser and Coulson.[9] Lennard-Jones parameters refer to the form $E(r) = 4\varepsilon[(\sigma/r)^{12} - (\sigma/r)^6]$.

the Lennard-Jones parameters, which have been fitted with respect to the geometry and energy of the equilibrium dimer, the values used are obtained from experimental and/or theoretical work, and are *not* adjustable parameters. The full details are discussed elsewhere.[3–5,7]

In using the model to investigate a given assembly of n molecules, the electropole field from the other molecules acting on molecule i is calculated, and the resultant change in dipole moment calculated using the polarisability tensor α_{ij}. After all molecules have been cycled through, the fields are recalculated and further dipole moment increments made as necessary until a further pass results in insignificant changes in dipole moments. Repeating the procedure results in eventual convergence, 2-5 interactions being normally required to satisfy a 1% convergence condition. No averaging is used; the calculations are exact, the dipole moments of each molecule being augmented individually. Information is then extracted on the contributions to the total energy of the system from the various dipole–dipole ($\mu\mu$), dipole–quadrupole (μQ), quadrupole–quadrupole (QQ), polarisation, and Lennard-Jones terms. This whole procedure can be incorporated into standard Monte Carlo and molecular dynamics calculations.

3. ICE Ih

Ice Ih is particularly interesting in that it does *not* exhibit a disorder–order transition of the hydrogen atoms at low temperatures.

The open hexagonal structure, topologically equivalent to zinc blende or ideal tridymite, is illustrated in fig. 1. There is near-perfect tetrahedral bonding[10] and the two Bernal–Fowler[11] rules are fulfilled; these state quite simply that two hydrogen atoms are covalently attached to each oxygen, and only one hydrogen is associated with each hydrogen bond. Within these constraints, however, the hydrogens exist in a " quasi-disordered state ": there is no true crystallographic ordering of the hydrogens and thus no true unit cell once the hydrogen positions are considered. This quasi-disordered arrangement of hydrogens amongst an ordered oxygen lattice has been confirmed by neutron diffraction[12] and residual-entropy measurements.[13–17] It has also prompted the search for a transition to what would be expected to be an energetically more stable *hydrogen-ordered* phase of ice Ih. However, the weight of

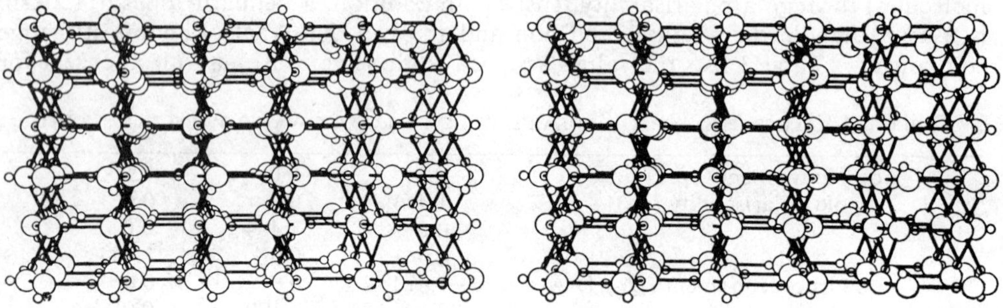

(*a*) (*b*)

FIG. 1.—Computer microfilm drawings of a 216-molecule repeating Ice I structure illustrating (*a*) a set of " hypothetical " fully-ordered positions, and (*b*) a set of " quasidisordered " hydrogen positions, the latter case being the one known to exist in real ice. The larger spheres represent oxygens, and the smaller spheres hydrogens, and the lines represent a set of purely notional (see text) hydrogen bonds each of length 2.76 Å. The PE model predicts that there should not be any low-temperature transition from state (*b*) to (*a*).

evidence currently is strongly against the existence of such a ferroelectric or piezoelectric low-temperature phase in pure ice.[18]

In order to examine theoretically the possibility of a disorder–order transition, structures were generated on the computer of (*a*) hypothetical fully-ordered ice Ih and (*b*) a quasidisordered ice Ih obeying the Bernal–Fowler rules as does the real structure. Periodic boundary conditions were used in all cases. A hypothetically ordered ice is obtained easily by indefinitely repeating a small unit cell of, say, 8 ice molecules (as in fig. 1). The system is then quasi-disordered using a method also used by Rahman and Stillinger[19] to randomly generate Bernal–Fowler configurations for evaluating the Kirkwood dielectric correlation " *g* " factor for ice I. The method is illustrated two-dimensionally in fig. 2. An initially hydrogen-ordered structure is hydrogen-disordered by moving the hydrogens around a randomly chosen *closed* circuit as shown, so that they read HO—HO—HO *etc*., in contrast to the initial OH—OH—OH *etc*. This new configuration, although no longer perfectly repeating, still obeys the Bernal–Fowler rules and has preserved its original (zero) electric polarity. These circuits may deviate from the basic lattice into the surrounding lattice of periodic images provided the structural mappings are maintained and the final closure of the circuit occurs in the real lattice. By generating large numbers of such " randomizing circuits " one can effect the disordering of an assembly of identical unit cells of ice I so that the final configuration is independent of its original state.*

In employing this procedure, up to 1000 randomizing loops were executed on a

* This is not strictly true as the original lattice has an inbuilt structural parity. The opposite parity, which is however energetically equivalent, can only be attained by introducing ionic and Bjerrum-bond defects into the lattice.

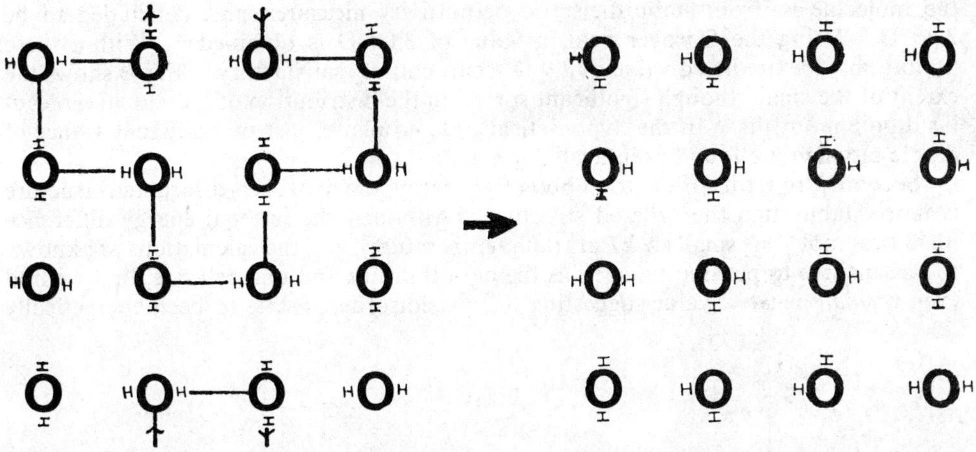

FIG. 2.—Schematic illustration of the disordering of a two-dimensional lattice using the technique of closed-loop generation. The left-hand perfectly-repeating lattice (repeats every alternate molecule) is made to undergo a disordering process around the closed circuit shown, by moving all the Hs on this loop to the opposite extremity of their respective bonds. A new non-repeating lattice (right-hand figure) is thus generated in which the two Bernal–Fowler rules (two Hs per O, one H per O—O bond) and the polar neutrality are maintained. As illustrated, the loop may extend out of the basic lattice, as long as closure finally occurs back in the basic lattice.

basic repeated cell of 216 ice molecules (equivalent to 14 disordering events per molecule). In order to improve the accuracy of the results, a larger 432 molecule orthorhombic (27.04 Å × 23.42 Å × 22.08 Å) basic cell was then used and an 11 Å distance cut-off was applied in the energy calculations. Moreover, seven polarisation iterations were used to achieve a stronger convergence condition of 0.2%. An Ewald summation was not used.

Table 2 summarises the results of this calculation and indicates the various contributions to the electrostatic part of the energy; the *total* energy of the disordered structure comes to -11.8 kcal mol^{-1}, in good (though likely fortuitous) agreement with an experimental value of -11.3 kcal mol^{-1} [ref. (19) and (20)].

Two major points stand out from the table. First, the average molecular dipole moment is increased to 2.90 D, some 50% greater than the initial isolated molecule value of 1.85 D. Comparison of this enhanced value with experiment is not straightforward. Assuming we require the average value of that dipole moment that reorients with the molecule, there is still uncertainty as to the value of the internal field in the crystal. This problem has been discussed in detail by Whalley;[21] using the Lorentz field, with a Kirkwood g factor of 3.00, the dipole moment that reorients with

TABLE 2.—ELECTROSTATIC CONTRIBUTIONS, EXPRESSED AS A PERCENTAGE OF THE TOTAL ELECTROSTATIC, TO THE ENERGIES OF ORDERED AND "QUASIDISORDERED" ICE Ih

	$\mu\mu$	μQ	QQ	polarisation	total[a] electrostatic	total[a] energy	average dipole moment, μ/D
hypothetical ordered	64	46	21	−31	−16.91	−11.4	2.88
quasidisordered	61	52	18	−31	−17.34	−11.8	2.90

[a] Energies are given in kcal mol^{-1}. The final total energy includes a Lennard-Jones term (see table 1) which is highly sensitive to the (unoptimised) parameters used.

the molecule is, from static dielectric permittivity measurements, concluded to be 2.45 D. Using the Onsager field, a value of 3.00 D is obtained.[21] Within these limitations, the predicted value of 2.9 D seems entirely satisfactory. Fig. 3 shows the extent of the small, though significant spread in the distribution of individual molecular dipole moments. In the hypothetical ordered phase, just two different values of dipole moment are found as indicated in fig. 3.

Secondly, in terms of the total potential energy, the hydrogen-disordered structure is more stable than the ordered structure. Although the internal energy difference (0.43 kcal mol^{-1}) is small ($\approx kT$ at room temperature), and the calculations are known to be sensitive to parameters such as the cut-off radius and unit cell size, that we find even a *small* negative value suggesting the quasidisordered state to been energetically

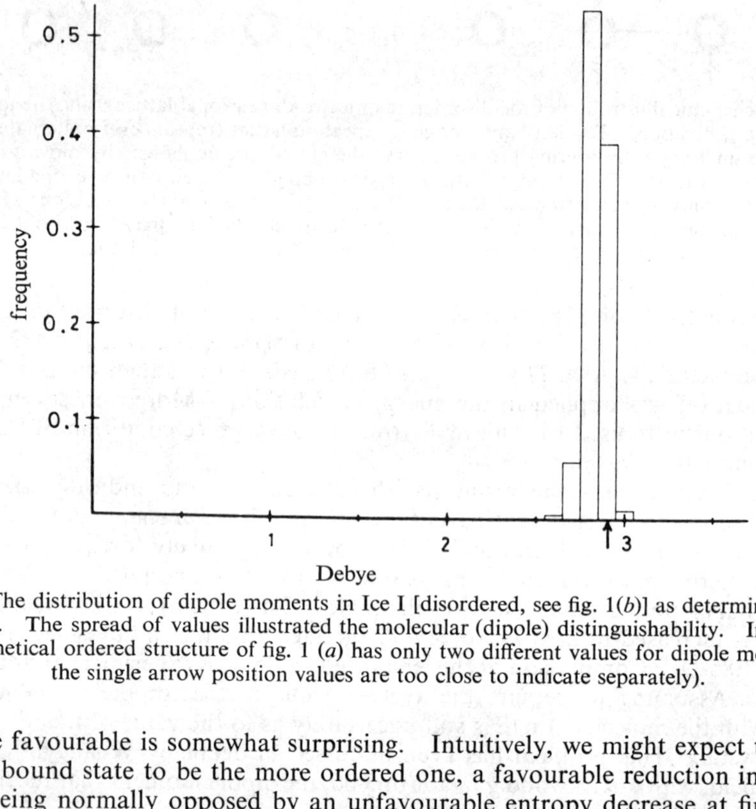

FIG. 3.—The distribution of dipole moments in Ice I [disordered, see fig. 1(*b*)] as determined by the PE model. The spread of values illustrated the molecular (dipole) distinguishability. In contrast, the hypothetical ordered structure of fig. 1 (*a*) has only two different values for dipole moments (at the single arrow position values are too close to indicate separately).

the more favourable is somewhat surprising. Intuitively, we might expect the more strongly bound state to be the more ordered one, a favourable reduction in internal energy being normally opposed by an unfavourable entropy decrease at finite temperatures. With ice Ih, however, the model suggests the opposite may occur. The ordered phase has the higher internal energy: thus any attempt to order the hydrogen positions will be opposed by both energetic and entropic driving forces. The non-existence of such a hydrogen-ordered ice Ih is thus consistent with this prediction of the model.

We can begin to conjecture a simple explanatory argument for this energy–disorder correlation in a polarisable dipolar system in the following way. For an assembly of N equal dipoles μ, the energy of interaction of any two will be proportional to $-\mu^2/r^3$. Assuming initially that only the n nearest neighbours around each molecule contribute significantly to the energy of the assembly, and that these near neighbours are all at the same distance r from their central molecule, the total dipole contribu-

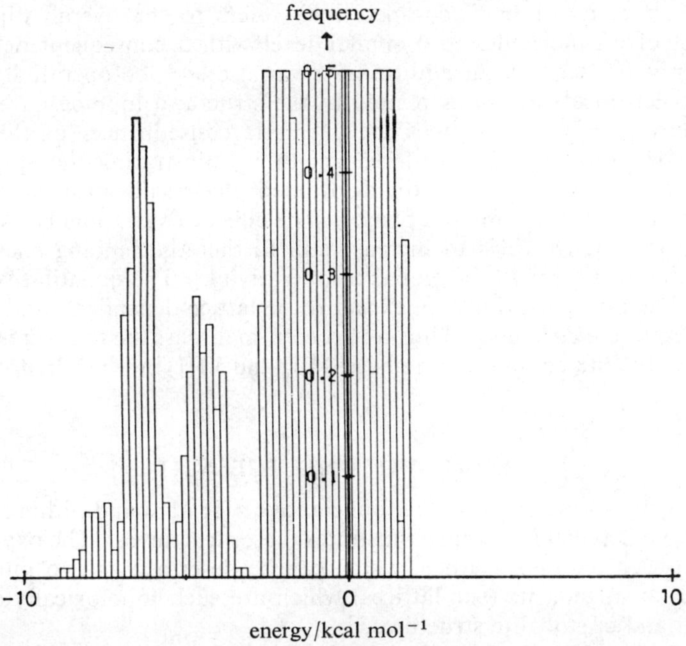

FIG. 4.—The pair-energy distribution (see text) obtained for Ice I using the PE model. The central bond arises from the larger number of weakly interacting pairs, whereas the low-energy peaks result mainly from first-, second- and third-nearest neighbour interactions. The lack of a complete (inverse) correlation between hydrogen-bond strength and separation distance, is a direct feature of non-additive models such as PE.

tion to the energy will be proportional to $-Nn\mu^2/r^3$. Now consider the same molecular assembly exhibiting a *spread* of dipole moments but with the same average value. Although the dipole energy of this system will depend critically upon the distribution in space of the various individual dipole moments, it is possible to set up trial assemblies in such a way that the total dipole energy is lower than for the ordered situation. For example, a domain within which each of m dipole moments has a similar value (say $\mu - \Delta$) will contribute a dipole–dipole energy of $-mn(\mu - \Delta)^2/r^3$, while another similarly sized domain will contribute $-mn(\mu + \Delta)^2/r^3$. Because of the non-linearity of the energy dependence on the dipole moments, the energy of these two domains taken together will be

$$-mn(\mu - \Delta)^2/r^3 - mn(\mu + \Delta)^2/r^3 \approx \frac{-2nm}{r^3}(\mu^2 + \Delta^2)$$

which is lower than the $-2nm\mu^2/r^3$ that would be obtained from two equally sized homogenous domains of identical dipole moments. The detailed energy balance of a general case is clearly more complex, depending critically on the distribution of the various dipole moments within the lattice, and also the interaction of each domain with its surroundings. Moreover, for the ices, the quadrupole contributions are significant (see table 2).

Nevertheless, in simple dipolar systems, we may expect certain lattice connectivities and " domain " structures to lead to the orientationally disordered state having the higher binding energy.

In the specific case of ice, we can consider two aspects of the co-operativity. First, what we might call " global " co-operativity refers to the overall dipole moment enhancement of all molecules to a similar level with a consequent decrease in the internal energy. Secondly, an additional " variational " co-operativity can be envisaged in which local variations related to the structural topology are favoured as discussed above. Fig. 4 quantifies the energetic consequences of this variational component for ordered and disordered ice Ih by illustrating the spread of near-neighbour interaction energies for the disordered phase as against the more discrete pattern of the hypothetical ordered phase. The effect of variational co-operativity in hydrogen-disordered ice Ih is to increase still further the binding energy over and above that obtained through the global co-operativity. The quantitative importance of the variational co-operativity we expect to be lattice-dependent, and we may well be led to different conclusions. Thus we proceed to discuss the structures and energies of a qualitatively different pair of ice phases (VII and VIII) in which hydrogen ordering does occur.

4. ICE VII AND ICE VIII

These two high-pressure forms are found above ≈ 18 kbar and within a temperature range of at least 200-360 K. Fig. 5 illustrates these structures. The oxygen positions define a body-centred cubic lattice, which can be described as two interpenetrating cubic ice (Ic) arrangements (sub-lattices) which are each topologically equivalent to the diamond and cristobalite structures.[22]

The two phases differ in the ordering of the hydrogens. Ice VII is the disordered phase, the hydrogens again obeying the Bernal–Fowler rules. There is, however, some evidence to suggest that some degree of hydrogen ordering is retained up to ≈ 350 K.[22,23] Cooling ice VII increases the degree of ordering until at ≈ 280 K the protons are almost fully ordered. This ordered phase is designated ice VIII and can be described within a tetragonal unit cell.[12]

We can find no direct evidence for the exact nature of the hydrogen ordering. Accordingly, model-building studies based upon a true crystallographic unit cell with twice the side of the oxygen unit cell revealed two possible modes of ordering as illustrated in fig. 5(*b*) and (*c*). In the first of these, designated VIIIa for the purposes of discussion, each sub-lattice is strongly polar, with all dipole moments parallel to a common direction; the other sub-lattice is similarly polar, but with dipoles arranged

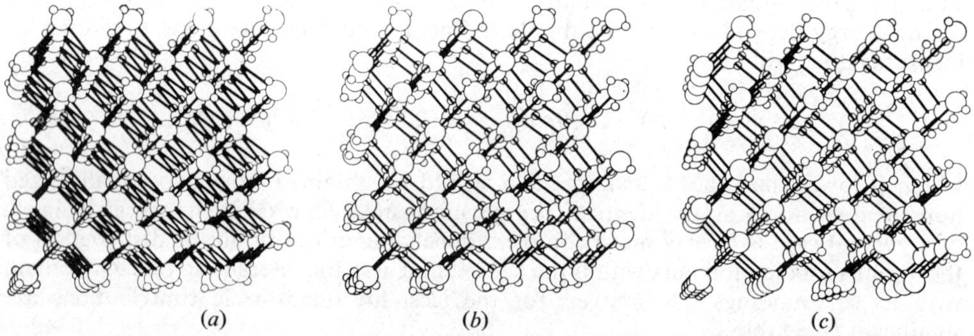

(*a*) (*b*) (*c*)

FIG. 5.—Computer microfilm drawings of 128 molecule repeating structures of (*a*) ice VII, (*b*) ice VIIIa, and (*c*) ice VIIIb (see fig. 1 for key). The two ice VIII structures have been determined from model-building studies though the antiparallel-dipole arrangement VIIIa (*b*) is found to be more stable than the alternative VIIIb (*c*). The polar ice VII (*a*) hydrogen-order has been derived from ice VIIIa using a variation on the disordering arrangement discussed in fig. 2.

anti-parallel to those in the first sub-lattice. The crystal is thus non-polar overall. A second ordering mode (VIIIb) is shown in fig. 5(*c*); each sub-lattice here is individually non-polar. Both VIIIa and VIIIb can be described by the same true crystallographic unit cell. We have been unable to identify other ordering modes for this cell, although other possibilities may well exist if larger unit cells are taken.

An elementary (repeating) cell of 128 oxygen positions consistent with the ice VIII positions were generated in the computer. A unit cell parameter of 3.4 Å was chosen to correspond to $\approx$273 K and 20 kbar; calculations for a smaller parameter corresponding to 25 kbar showed similar behaviour. Initial molecule orientations were fed in to describe the hydrogen-ordered structures of VIIIa and VIIIb. Using periodic boundaries, and a cut-off of 6.8 Å, the iterative energy calculation (section 2) was performed for both ordered structures. Preliminary Monte Carlo calculations were also performed using VIIIa as the starting configuration, constraining all oxygens to their starting positions, and allowing molecular rotations only. Short test runs at nominal temperatures of 263 and 313 K resulted in assemblies of slightly higher energy ($\approx$0.1 kcal mol^{-1}) than the starting (ordered) configuration, with no indication of a rapid co-operative disordering; in fact, no disordered configuration was generated with a lower energy than the starting ordered arrangement. At these nominal temperatures, our model ice VIIIa appears to be relatively stable. Sample models of hydrogen-disordered ice VII were produced from ice VIIIa using a procedure similar to that discussed in section 3 for ice Ih. However, as the initial polarity of each sub-lattice was non-zero, the polarity-conservation property of the method could not be retained and all structures were consequently polar over the small elementary cell used. Three sets of disordered hydrogen positions obeying the Bernal–Fowler rules were generated by simulating on a model in the laboratory the computer algorithm previously used.

Table 3 shows the results of energy calculations on these assemblies, together with

TABLE 3.—ELECTROSTATIC CONTRIBUTIONS, EXPRESSED AS A PERCENTAGE OF THE TOTAL ELECTROSTATIC, TO THE ENERGIES OF ICE VII AND ICE VIII

model		$\mu\mu$(%)	μQ(%)	QQ(%)	polarisation(%)	total[a] electrostatic	average dipole moment, μ/D
polarisable	VII	48	61	14	−23	−11.94	2.58
	VIIIa	0	132	−15	−17	−12.40	2.51
	VIIIb	73	0	47	−19	−11.74	2.53
" pseudo-ST2 "	VII	51	36	13	—	−10.63	
	VIIIa	0	110	−10	—	−10.47	2.35[b]
	VIIIb	70	0	30	—	−10.46	

[a] Energies are given in kcal mol^{-1}. [b] Dipole moment is as given by the ST2 potential; it is environment independent (polarisability $\equiv$ 0).

data obtained using point dipoles and quadrupoles equal to those of the popular pair-additive four-point-charge ST2 model of water.[18] For the polarisable model, the internal energy of the ordered phase VIIIa is clearly lower than that of either the disordered phase ice VII or the other ordered configuration VIIIb. The ordered configuration VIIIb is *less* strongly bound than the polar disordered phase. Thus the model predicts VIIIa should exhibit the preferred hydrogen arrangement of the two possibilities discussed. Moreover, the difference in internal energy between VII and VIIIa is of the same order of magnitude ($\approx$0.4 kcal mol^{-1}) as the contribution to free energy from the configurational entropy term at 273 K (0.24 kcal mol^{-1}). Thus, the predictions are qualitatively consistent with the experimental observation of a hydrogen-ordering transition, although these preliminary calculations using an unrefined potential are insufficient to allow us to quantify the transition itself. This can be contrasted with the " pseudo-ST2 " calculations (table 3) which predict a greater stability of the disordered phase.

The individual electrostatic contributions in table 3 are particularly interesting. Although the topologies of the oxygen positions are identical in all three structures, they show totally different major contributions. VIIIa has a zero dipole–dipole contribution, a consequence of the ordered arrangement ensuring that the net dipole field on any given water molecule is zero. The dipole–quadrupole contribution is almost totally dominant, the smaller quadrupole–quadrupole interaction being repulsive overall. This underlines the necessity for a realistic quadrupole moment, and suggests also that the quadrupole polarisability ignored presently in the polarisable model may well give rise to significant effects in special cases such as this. The VIIIb structure shows very different effects, in particular a large (73%) dipole–dipole contribution, with the dipole–quadrupole contribution cancelling to zero for reasons of symmetry. These relationships between energy terms and the symmetry of the lattice require further investigation; the less-extreme behaviour of hypothetical

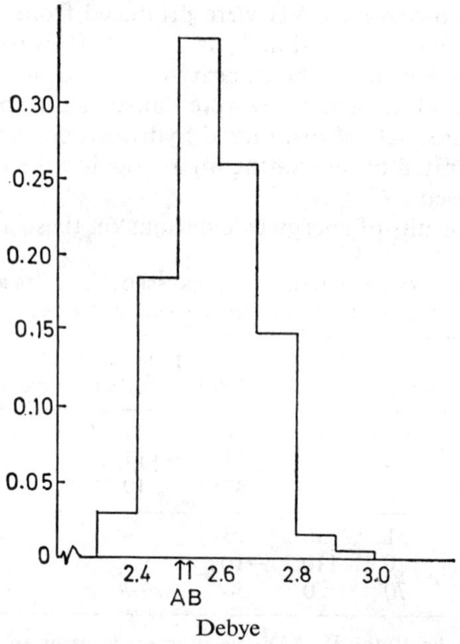

FIG. 6.—Combined distribution of dipole moments in ice VII simulations. Points A and B refer to the single values found in models of ice VIIIA and VIIIB.

ordered ice Ih (table 2) suggests the exceptional situation of lattice interpretation may be very significant for VIII.

The dipole–quadrupole term is the largest contribution to the energy of hydrogen-disordered ice VII, further underlining the necessity for a correct quadrupole moment. Again we can contrast this with ice Ih (table 2), where the dipole–dipole term is dominant. Even for the disordered ice VII, it appears that the interpenetrating lattice structure reduces the dipole–dipole contribution to stability, quadrupolar effects being augmented to compensate for this reduction.

The distribution of enhanced dipole moments in ice VII is shown in fig. 6; the average dipole moment is less than in ice Ih, perhaps reflecting the lower dipole–dipole interaction suggested by comparing that contribution for the two structures (table 3). Ice VIIIa shows a single value of μ at 2.51 D; this is $\approx 3\%$ lower than the average value for ice VII (2.58 D), perhaps consistent with the dipole field cancellation in the latter case.

5. SUMMARY

Assemblies of water molecules arranged into different structures are observed to show different behaviour with respect to ordering of the hydrogens. In pure hexagonal ice I, the hydrogens appear to be disordered at all temperatures, and no ferroelectric or piezoelectric phase has been found. In contrast, the interpenetrating cubic lattice structure of the high-pressure ice VII orders progressively as the temperature is reduced, becoming essentially fully ordered at ≈ 280 K and 22 kbar.

The ice Ih behaviour would seem at first sight to be anomalous, ordering of molecular assemblies normally being accompanied by a fall in potential energy which eventually overcomes the unfavourable $T\Delta S$ term as temperature is reduced. Examining the energies of both ice Ih and a hypothetically-ordered phase of identical topology predicts behaviour consistent with that observed experimentally. The energy of the (observed) hydrogen-disordered structure is lower than that of the hypothetical ordered (unobserved) structure, thus suggesting the ordered phase to be unstable in energy terms. The entropic effects will act in concert with the energy at all non-zero temperatures (neglecting effects due to lattice expansion) and the ordered phase will thus not be observed.

The average dipole moment of the water molecule in ice Ih is predicted to be enhanced by some 56% above the isolated-molecule value; this falls within the range of 32-62% estimated from static permittivity measurements. The significant dispersion of the enhanced dipole moment distribution (fig. 3) together with the non-linearity of a dipole–dipole energy term ($\alpha\mu^2$) may be one way of rationalising this unusual correlation of disorder with energetic stability. Whether or not a particular structure will show such behaviour will depend on a complex balancing of dipole moment distribution and the topology of their mutual interactions.

In contrast, the hydrogens in the two interpenetrating lattice structures of ice VII order progressively as the temperature is lowered. This observed behaviour is also consistent with the polarisable electropole model, which predicts that at least one mode of hydrogen ordering (designated VIIIa) is energetically more stable than the disordered phase by an amount which exceeds the opposing entropic contribution to the free energy by $\approx 60\%$. In contrast, a pair-additive potential modelled using point dipoles and quadrupoles obtained from the ST2 water model predicts ice VII to be more stable than either of the possible ordered phases considered here.

The very high symmetry of the interpenetrating lattice structure may be partly responsible for this difference of ordering behaviour. Whereas in ice Ih the major contribution to the energy arises from dipole–dipole interactions for both (observed) disordered and (unobserved) ordered structures, in VIIIa there is complete cancellation of the dipole–dipole term; there is a massive compensating contribution (≈ 16.4 kcal mol^{-1}) arising from the dipole–quadrupole interaction. We might speculate that in lattice-constrained dipolar systems, the absence of an order–disorder transition might be the general rule, with very special conditions of symmetry necessary before the ordering is consistent with a lower internal energy. Further work along these lines is in progress on ices Ic, II, III, VI and IX.

Part of this work arose out of a student project undertaken as part of the M.Sc. Crystallography course at Birkbeck College; D. V. Bliss thanks Northamptonshire Education Authority for financial assistance during the course. We are grateful to Dr. E. Whalley for helpful discussion and correspondence. J. E. Quinn is supported by an S.R.C. grant. Use of computer facilities at University of London Computer Centre and the S.R.C. Daresbury Laboratory is gratefully acknowledged.

[1] C. A. Coulson and D. Eisenberg, *Proc. Roy. Soc. A*, 1966, **291**, 445.
[2] D. Hankins, J. W. Moskowitz and F. H. Stillinger, *J. Chem. Phys.*, 1970, **53**, 4544.
[3] P. Barnes, *Report of Workshop on Molecular Dynamics of Water*, CECAM, Orsay, France, 1972, p. 77.
[4] P. Barnes, *Progress in Liquid Physics*, ed. C. A. Croxton (Wiley, Chichester, 1978), chap. 9.
[5] P. Barnes, J. L. Finney, J. D. Nicholas and J. E. Quinn, *Nature*, 1979, **282**, 459.
[6] J. L. Finney, *Faraday Disc. Chem. Soc.*, 1978, **66**, 80, 86.
[7] P. Barnes, J. L. Finney and J. E. Quinn, in preparation.
[8] P. Barnes and J. L. Finney, Abstr. 4th European Crystallographic Meeting, Oxford, 1977, A377.
[9] R. M. Glaeser and C. A. Coulson, *Trans. Faraday Soc.*, 1965, **61**, 389.
[10] D. Eisenberg and W. Kauzmann, *The Structure and Properties of Water* (Clarendon Press, Oxford, 1969), p. 72.
[11] J. D. Bernal and R. H. Fowler, *J. Chem. Phys.*, 1933, **1**, 515.
[12] B. Kamb, *Physics and Chemistry of Ice*, ed. E. Whalley, S. J. Jones and L. W. Gold (University of Toronto Press, Toronto, Canada, 1973), p. 28.
[13] L. Pauling, *J. Amer. Chem. Soc.*, 1935, **57**, 2680.
[14] J. F. Nagle, *J. Math. Phys.*, 1966, **7**, 1484.
[15] Y. Suzuki, *Cont. Inst. Low Temp. Sci.*, 1966, Series A21, 1.
[16] W. F. Giaque and J. W. Stout, *J. Amer. Chem. Soc.*, 1936, **58**, 1144.
[17] B. Kamb, *Acta Cryst.*, 1964, **17**, 1437.
[18] A. Rahman and F. H. Stillinger, *J. Chem. Phys.*, 1974, **60**, 1545.
[19] G. Dantl, *Phys. Kondens. Materie.*, 1968, **7**, 390.
[20] W. E. Sharp, *UCRL Tech. Rep.*, No. 7118, 1962.
[21] E. Whalley, *J. Glaciol.*, 1978, **21**, 13.
[22] N. G. Parsonage and L. A. K. Staveley, *Disorder in Crystals* (Clarendon Press, Oxford, 1978), chap. 8, p. 483.
[23] G. P. Johari and E. Whalley, *Physics and Chemistry of Ice*, ed. E. Whalley, S. J. Jones and L. W. Gold (University of Toronto Press, Toronto, Canada, 1973), p. 278.

Frozen-in States of Orientational and Positional Disorder in Molecular Solids*

BY HIROSHI SUGA

Chemical Thermodynamics Laboratory,
Faculty of Science, Osaka University,
Toyonaka, Osaka, 560 Japan

AND SYÛZÔ SEKI

Department of Chemistry,
School of Science, Kwansei Gakuin University,
Nishinomiya, 662 Japan

Received 3*rd December*, 1979

It has previously been shown that the ordinary liquid–glass transition is just one example of a class of " glass transitions " due to loss of equilibrium which may occur widely in metastable phases of condensed matter. In the present paper, glass-transition phenomena occurring in stable crystalline phases of $SnCl_2.2H_2O$, H_3BO_3, H_2O, pinacol hexahydrate, CO and $TlNO_2$ crystals are investigated. Deuteration effects on the kinetic properties of the freezing process are also reported. These glassy crystals differ from those realized in undercooled metastable phases in that they never encounter the Kauzmann paradox and are considered to belong to a new category of the state of aggregation of matter. Throughout the present studies, the usefulness of adiabatic calorimetry for the investigation of slow molecular motion in solids will be stressed, for which conventional spectroscopic techniques such as dielectric measurement are impractical because of the prolonged relaxation time.

The glass transition, at which a viscous liquid settles into the glassy state, has been recognized as a process of departure from thermodynamic equilibrium at some point in the cooling experiment. A decrease in heat capacity or thermal expansion coefficient of the liquid is the consequence of freezing of some degrees of freedom in the liquid. Thermodynamic properties such as enthalpy and volume become time-dependent around the glass-transition temperature T_g, the nature being governed by the kinetics of the process whereby equilibrium is approached. Thus, observation of a relaxational heat-capacity anomaly and verification of the existence of residual entropy have been used as the calorimetric characterization of a liquid in the glassy state.

Although a liquid is considered as representative of a disordered system, disorder prevails in various systems. Orientationally disordered crystals and liquid crystals may be looked upon as the embodiments of the dual nature of disorder existing in liquids. Orientational and positional disorder are mainly relevant to these respective phases. These systems usually change into ordered crystals on lowering the temperature through phase transitions. On rapid cooling, however, the disordered systems undercool and ultimately go into a " glassy crystalline state " or a " glassy liquid-crystalline state " *via* their " glass-transition " phenomena. The freezing phenomena observed calorimetrically in the undercooled metastable phases have

* Contribution no. 10 from the Chemical Thermodynamics Laboratory.

emarkable similarity to the ordinary glass transition in that the heat capacity jumps at the glass-transition temperature and the extrapolated excess entropy below T_g shows the Kauzmann paradox. The famous review article [1] by Kauzmann on the nature of the glassy state is transferrable to the description of the natures of the glassy crystal or the glassy liquid-crystal without any serious modification of concept. Thus, the ordinary glass transition has proved to be just one example of a class of glass transition which may occur quite frequently in metastable phases of condensed matter. This idea is supported by the experimental observations of multiple glass-transition phenomena for differently disordered phases of one and the same substance. It is interesting to note that the glass transitions for undercooled phases of liquid and orientationally disordered crystal of ethanol took place at essentially the same temperature. This observation suggests the importance of the role of reorientational motion in determining the relaxational nature of the configurational change in the vitrification process of the liquid. The residual entropy for glassy crystalline ethanol, 4.24 J K^{-1} mol^{-1}, smaller than that for the glassy liquid, 12.85 J K^{-1} mol^{-1}, is in conformity with the existence of positional order in the former state. These experimental findings have been summarized in a review article.[2]

In the present paper we would like to present many examples of " vitrification " or " freezing-in " phenomena in stable crystals. The only modification of the concept of the glassy state is to remove the limitation of metastability that has been observed without exception to be associated with that state. Again, the glass transition is caused by a relaxational effect, through which some process in a disordered stable crystal occurs too slowly at low temperature to permit thermodynamic equilibrium to be established during the laboratory experimental time-scale. The effects of deuteration on the kinetic parameters associated with the degree of freedom that was frozen are also reported.

EXPERIMENTAL

CALORIMETRIC SYSTEM

The heat capacity of solids can be measured accurately by a low-temperature adiabatic calorimeter with an intermittent heating mode. In this mode, a heating period of about 1.2 ks is generally used, during which the sample temperature is increased by 2-3 K. The heating period is followed by a temperature drift period of 10-20 min. The first 4 or 5 min of the drift period is required for the sample and the cell assembly to reach a uniform temperature distribution. Then the calorimetric temperature drift observed for the later part of the period reflects the combined effect from a small heat leakage due to incomplete adiabaticity and from any relaxational thermal anomaly that the sample might show.

The quasiadiabatic drift is relatively small and changes monotonically as the calorimetric temperature is raised. In order to detect sensibly any relaxational thermal anomaly of the sample, the adiabatic condition is required to be realized as perfectly as possible. In our system, a sample cell is surrounded by two (inner and outer) adiabatic shields and by a massive thermal station. Adiabatic conditions are achieved by use of high vacuum and four-channel temperature controllers with p.i.d. mode. A digital voltmeter is used for the energy measurement and the duration of energy input is determined by an electronic timer. Temperatures of the sample cell are measured by a 25 Ω platinum resistance thermometer and d.c. Müller bridge[3] or a.c. double bridge.[4] In the completely automated calorimetric system constructed[5] recently, readings of an automatic a.c. double bridge are fed into a mini-computer as a function of time. A program then proceeds to calculate a temperature drift rate from the latest six data. A decision is made whether the drift rate is smaller than a pre-set value, usually 7×10^{-8} Ω s^{-1} or 0.7 $\mu K\ s^{-1}$ approximately. If the answer is " yes ", the program judges that the thermal equilibrium is established and calculates the apparent heat capacity in terms of IPTS-68 on the basis of a conversion table stored in a data bank.

If the answer is " no ", the cycle of temperature measurements is repeated. Table 1 reproduces a part of the experimental results taken with unattended operation for the ordered phase of crystalline cyclohexanol.

The magnitude of the quasi-adiabatic drift is usually less than the preset value. Therefore, any persisting large temperature drift beyond this value is an indication of relaxational thermal effect originating from the sample. The nature of the temperature drift differs according to thermal history of the sample. Samples were either quenched or annealed in the calorimeter cell. Quenched samples were produced by fast cooling at a rate of the order of 1 K min^{-1}. Annealed samples were produced either by slow cooling or by keeping the sample tempera-

TABLE 1.—REPRODUCTION OF A PART OF THE EXPERIMENTAL HEAT CAPACITY DATA OBTAINED BY THE FULLY AUTOMATED ADIABATIC CALORIMETER

A, resistance (in Ω); B, real time (in s); E, equilibrium temperature (in K); C, heat capacity (in J K^{-1}); D, average temperature (in K); G, real time when the next cycle is started (in h, min, s); H, input energy (in J); Y, total input time (in s).

5.709 694	A	5.931 621	A	6.149 571	A	6.363 621	A
4.079	B	5.700	B	7.211	B	8.812	B
5.709 691	A	5.931 621	A	6.149 571	A	6.363 628	A
4.112	B	5.733	B	7.244	B	8.845	B
5.709 649	A	5.931 621	A	6.149 571	A	6.363 628	A
4.174	B	5.766	B	7.277	B	8.878	B
5.709 674	A	5 931 621	A	6.149 571	A	6.363 621	A
4.207	B	5.799	B	7.309	B	8.911	B
5.709 679	A	5.931 621	A	6.149 571	A	6.363 628	A
4.241	B	5.832	B	7.343	B	8.944	B
5.709 679	A	5.931 621	A	6.149 571	A	6.363 624	A
4.284	B	5.865	B	7.736	B	8.977	B
85.554 173	E	87.556 242	E	89.522 641	E	91.454 699	E
22.500 4	C	22.935 2	C	23.344 4	C	23.750 1	E
84.533 582	D	86.555 207	D	88.539 411	D	90.488 670	D
11 127	G	13 748	G	20 300	G	22 951	G
45.918 231	H	45.904 483	H	45.886 583	H	45.861 427	H
1 200	Y	1 200	Y	1 200	Y	1 200	Y

ture at 10-20 K below T_g of the quenched samples. Quenched samples showed exothermic temperature drifts which were then followed by endothermic drifts. Only endothermic drifts were observed for well annealed samples. The sign of the drift depends on whether the enthalpy of the actual sample lies below or above its equilibrium value. Fig. 1 illustrates schematically the enthalpy against temperature relation in the glass-transition region. Here, T_g is defined conveniently as the temperature where the temperature drifts change their sign. The ordinate is the enthalpy of sample cell loaded with a quenched sample. The bold curve shows the manner in which the enthalpy of a system increases during the course of heat-capacity measurement. In a heating period an energy input leads to an enthalpy increment. The slope of the bold curve corresponds to the heat capacity C of the system. In a drift period in the temperature range below T_g, the calorimetric temperature rises spontaneously because of the irreversible approach of the sample to an equilibrium state. In this process, the enthalpy of the whole system is kept constant since the system is in an adiabatic condition, so that the process is shown by the bold horizontal line. The rate of enthalpy relaxation of the sample may be expressed by the following equation

$$-\Delta\dot{H}_{conf} = \Delta H_{conf}/\tau,$$

where ΔH_{conf} is the configurational enthalpy still to be relaxed before thermal equilibrium is established, $\Delta\dot{H}_{conf}$ is the time derivative of this quantity, and τ the relaxation time. Here, the configurational enthalpy stands for the component of enthalpy that is associated with the frozen degrees of freedom. $\Delta\dot{H}_{conf}$ is measured as $C(dT/dt)$. Therefore, the equation enables us to determine the relaxation time from the observation of temperature drift rate.

Above T_g, the temperature drift becomes endothermic and the relaxation time becomes shorter as the temperature is raised. As a result, $\Delta\dot{H}_{conf}$ becomes dependent on time, which

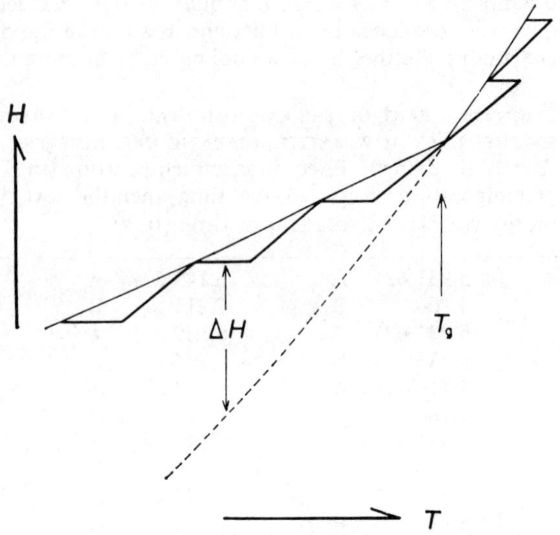

FIG. 1.—Schematic diagram illustrating the variation of enthalpy with temperature in a glass-transition region. Dashed curve corresponds to an ideal equilibrated value of the enthalpy.

means that the above equation can not be applied straightforwardly. However, this short relaxation time allows us to follow the whole enthalpy relaxation until thermal equilibrium is established. In this case the relaxation time can be determined by the equation,

$$\ln[T(t) - T(\infty)] = A - t/\tau,$$

where $T(\infty)$ is the equilibrium temperature after the whole enthalpy relaxation is completed, and A is a constant.

RESULTS AND DISCUSSION

FREEZING PHENOMENA IN SEVERAL STABLE CRYSTALS

$SnCl_2.2H_2O$

Stannous chloride dihydrate has a layer structure in which double layers of $SnCl_2$ molecules are sandwiched between a well-defined hydrogen-bonded network of the water molecules, composed of 4-membered and 8-membered oxygen rings alternately. A dielectric phase transition was found to occur at 235 K by Kiriyama *et al.*[6] by the use of dielectric, n.m.r. and neutron diffraction techniques and was attributed to the order–disorder change of protons in the network. This phase transition was accompanied by a symmetric divergence of heat capacity,[7] reflecting the two-dimensional nature of the interacting system. A relaxational heat-capacity anomaly[8] was found to occur at 150-160 K where the heat capacity depended on the rate at which the crystal had been cooled through the region. A step-like increment in the heat capacity appeared at higher temperatures for the more rapidly cooled crystal, as is shown in fig. 2.

Thermal equilibrium was quite difficult to attain in this temperature range. Fig. 3 shows the rate of spontaneous temperature rise of the calorimeter cell after a uniform temperature was realized in the cell. This relaxational thermal anomaly can be explained as follows. Long-range order with respect to proton position in the network develops as the temperature is lowered through the transition temperature T_t. However, as is evident from the extended region of the heat-capacity anomaly, the development of order is not completed at T_t but continues over a considerable temperature range. Finally, the crystal reaches a temperature where the relaxation

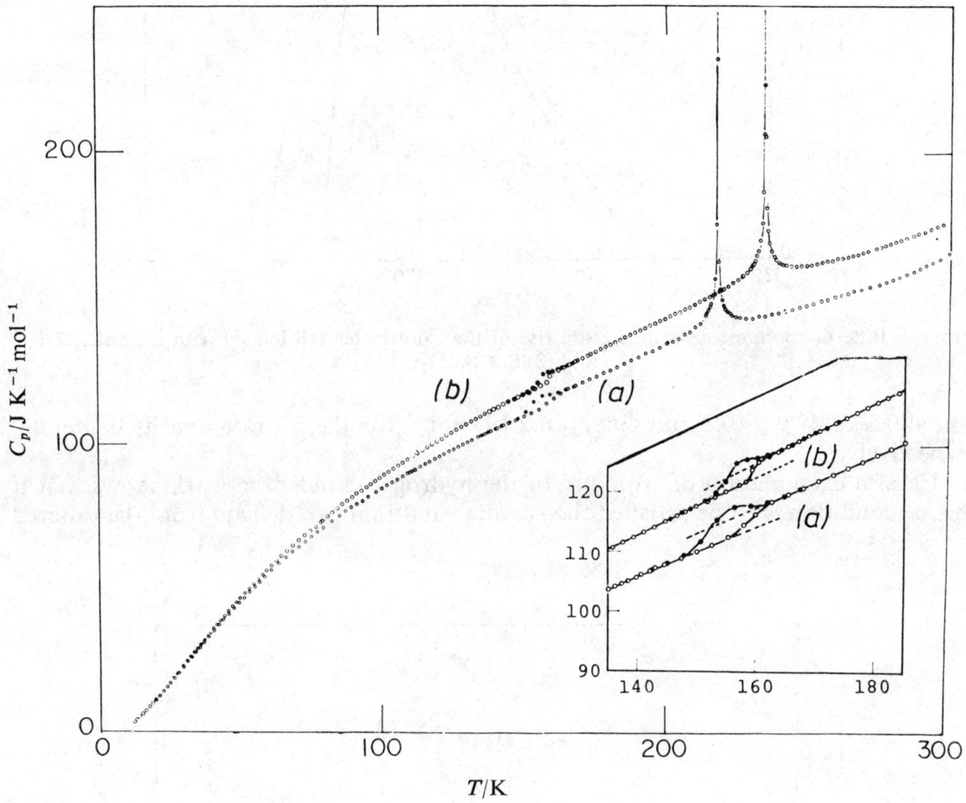

FIG. 2.—Heat capacity of (*a*) $SnCl_2.2H_2O$ and (*b*) $SnCl_2.2D_2O$ crystals, with an expanded view of the heat capacity around T_g. Open circles are for the sample cooled rapidly (-1.7 K min^{-1}), closed circles for the sample cooled slowly (-0.01 K min^{-1}).

time for proton redistribution responded to a temperature change exceeds the experimental time-scale, leading to the disappearance of the heat capacity expected for the configurational degree of freedom of proton system.

The heat evolution rate was integrated numerically with time, to give the enthalpy against temperature relation. The proton configurational enthalpy was separated from the rest by taking the low-temperature heat capacity as the background contribution. These treatments enabled us to calculate the relaxation time of the system. The results are plotted against reciprocal temperature in fig. 4. It is remarkable that the drift rate against temperature curves of fig. 2 could be reduced to the straight lines in fig. 4 in this manner. A linear least-squares fit to these data gives activation

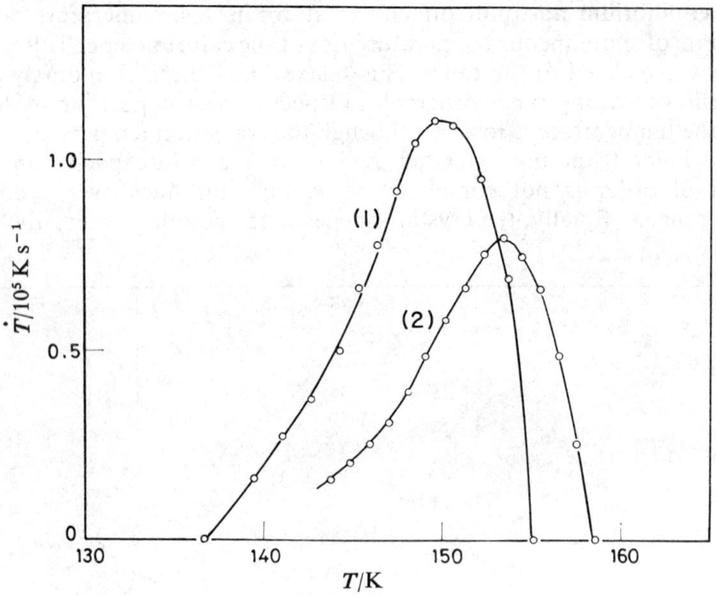

FIG. 3.—Rate of spontaneous temperature rise of the calorimeter cell loaded with (1) $SnCl_2.2H_2O$ and (2) $SnCl_2.2D_2O$.

enthalpies of 49.9 $\pm$ 0.3 and 49.8 $\pm$ 0.2 kJ mol^{-1} for the hydrate and its deuterate, respectively.

Careful examination of structure of the hydrogen-bonded network shows that if the ice condition is to be satisfied the smallest unit that can deviate from the ordered

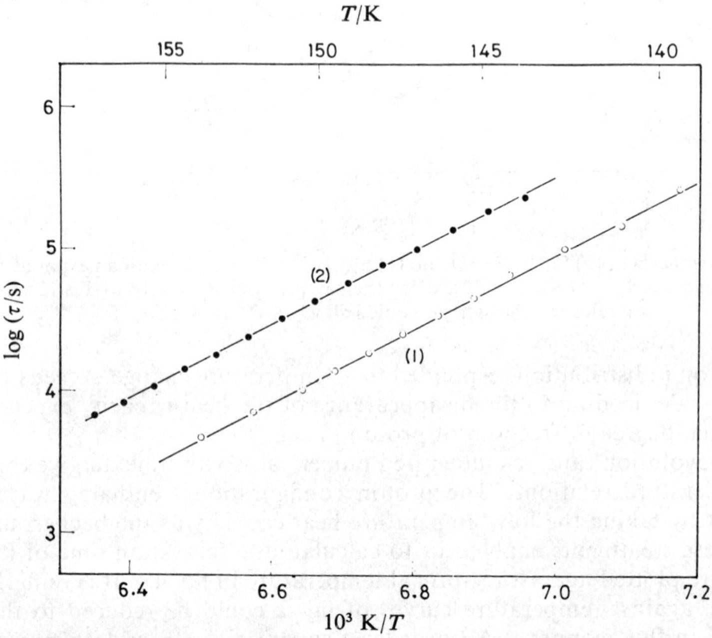

FIG. 4.—Arrhenius plot of the enthalpy relaxation time τ of (1) $SnCl_2.2H_2O$ and (2) $SnCl_2.2D_2O$.

proton configuration is the 8-membered ring. The activation enthalpy may be related to the motion in which the eight protons move in unison in this ring. Fig. 5 gives the heat capacity of the systems $SnCl_2.(H_2O)_x(D_2O)_{2-x}$ with x ranging from 2.00 to 0.03 around their glass-transition regions. The heat-capacity jump at T_g is considerably smaller than those of the glassy liquid. The glass transition in $SnCl_2.2D_2O$

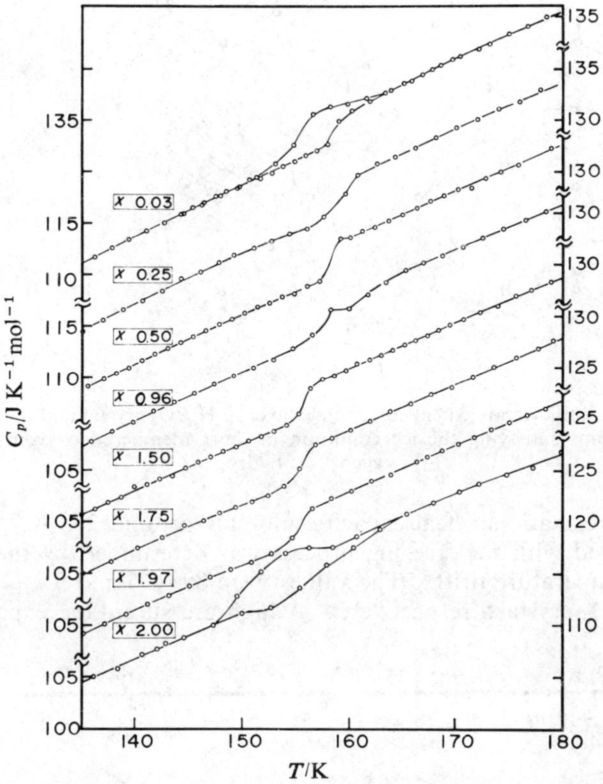

FIG. 5.—Heat-capacity curves of $SnCl_2.(H_2O)_x(D_2O)_{2-x}$ in the region of glass transitions.

occurs at ≈5 K higher than in $SnCl_2.2H_2O$. The glass-transition intervals of the solid solutions are not broader than those of the hydrate or the deuterate crystals. This means that correlation region of the atomic motion is so large that the local isotopic concentration does not much deviate from the average concentration. This supports the earlier model based on the structure of hydrogen-bonded network that the smallest unit of the atomic motion consists of eight protons.

H_3BO_3

Crystalline orthoboric acid is another example of a hydrogen-bonded system with two-dimensional structure, being essentially a molecular arrangement of planar $B(OH)_3$. Each hydroxyl group is linked by hydrogen bonding to two others belonging to different molecules and forms a 6-membered ring, which is arranged in a triangular lattice pattern as shown in fig. 6. Protons are situated on the O · · · O lines, in an ordered arrangement at room temperature.[9]

The heat-capacity curves[10] of H_3BO_3 and D_3BO_3 are given in fig. 7, and show the

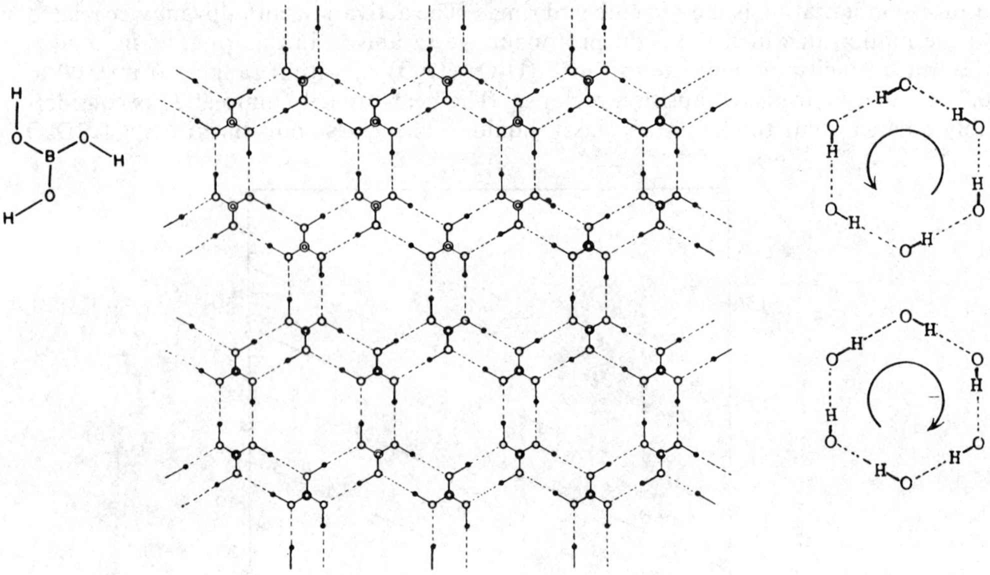

FIG. 6.—Schematic atomic arrangements of a single layer in H_3BO_3 crystal and the two arrangements of the hydrogen atoms satisfying the ice condition in the 6-membered oxygen ring. ◎, Boron; ○, oxygen; ●, hydrogen.

occurrence of a relaxational heat-capacity anomaly around 290 K. The activation enthalpy associated with the freezing process was determined by the same analysis of the rate of temperature drift. The values were 89 ± 5 and 97 ± 5 kJ mol^{-1} for H_3BO_3 and D_3BO_3 crystals, respectively. A high-pressure d.t.a. experiment showed

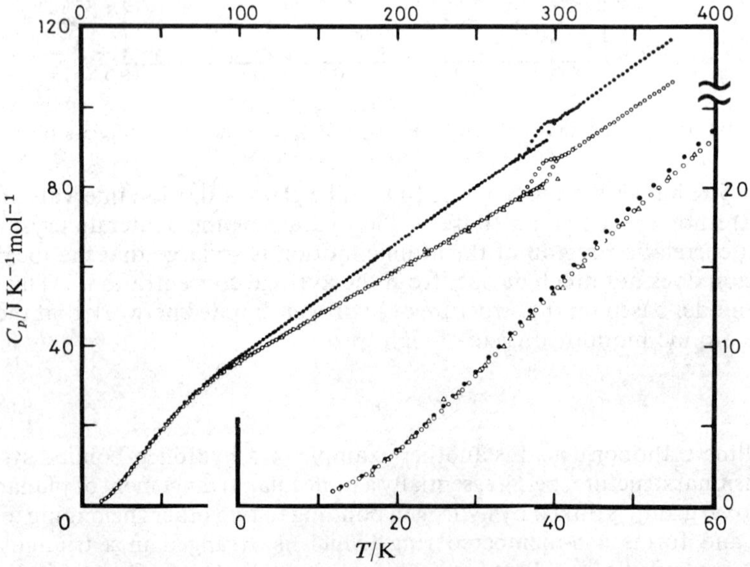

FIG. 7.—Heat capacity of H_3BO_3 (△, Johnston and Kerr) and D_3BO_3 (○, ●, present data) crystals. Two data points around T_g correspond to the samples cooled at the rate of —5.6 and —0.008 K min^{-1}, respectively.

that there existed no other phase transitions before their melting points are reached, 443.1 K for H_3BO_3 and 440.6 K for D_3BO_3.

This glass-transition phenomenon can be interpreted again in terms of a frozen disordered configuration of the proton system at the low-temperature stage of ordering process. Craven and Sabine[9] assigned uniquely all the atomic position at room temperature. However, in our interpretation, ordering of the protons (or deuterons is not completely attained. The 6-membered oxygen ring, which is assumed to be the individual ordering unit of the system, has two permissible arrangements; *i.e.*, clockwise and counter-clockwise under the extended ice condition shown in fig. 6. Interaction among the rings will favour a ferro-state in which all of the rings are the same arrangement (either all clockwise or all the reverse, similar to ferroelectricity). The entropy of this model is $(Nk/2)\ \ln 2$ in the completely disordered state, where N is the total number of rings. Mean-field theory was applied to obtain the order parameter of the ring system from the ΔC_p value at T_g. Here, the order parameter σ is defined as $\sigma = (2n - N)/N$, where n is the number of rings in the preferred arrangement. The order parameter estimated at 290 K was 0.94 for H_3BO_3 and 0.96 for D_3BO_3. This degree of estimated ordering is probably not ruled out by the neutron diffraction experiment. These results enabled us to calculate the hypothetical transition point to be 536 K for H_3BO_3 and 587 K for D_3BO_3. The values are well above their melting points. Theoretical calculation[11] on the basis of the triangular Ising lattice model gave 378 K for H_3BO_3 and 404 K for D_3BO_3 as the hypothetical transition temperature. This may imply that the interaction among the rings has a three-dimensional nature in spite of the structural two-dimensionality. In any event the present explanation is reasonable in the absence of any other degree of freedom responsible for the freezing phenomenon.

H_2O

The structural features of hexagonal ice are that, while the oxygen atoms form a regular lattice, the hydrogen atoms are arranged randomly in that structure. The residual entropy[12] observed for ice crystals was interpreted by Pauling[13] as a completely disordered pattern of the location of protons. The dynamical situation of this disorder has not been clarified. Since such a disordered structure contradicts the third law of thermodynamics, the search for a possible ordering of the proton system has been the main subject of recent investigations. In 1936, Giauque and Stout[12] noticed a slow thermal equilibration in ice crystals around 100 K. They measured the heat capacities of ice samples with different thermal histories, but they did not observe any significant difference of heat capacity among these samples.

Our reinvestigation[14,15] showed the existence of a relaxational heat capacity anomaly around 100 K for H_2O crystals and around 115 K for D_2O crystals, as shown in fig. 8. Exothermic temperature drifts were observed for quenched samples, while endothermic ones occurred with annealed samples. Spontaneous temperature rises during the annealing experiments were followed until they ceased completely and the data were analysed to yield relaxation time by assuming an exponential approach to the equilibrium. The results are summarized in table 2 and are plotted in fig. 9, along with the relaxation time data obtained by thermally stimulated depolarization[16] (t.s.d.) and dielectric step-response[17] experiments. The current peak due to t.s.d. observed at 110 and 124 K in H_2O and D_2O ices was considered to be due to relaxation of a frozen orientation polarization of the water molecules in both crystals. These data indicate essentially two facts. First, the degree of freedom associated with the enthalpy relaxation is the same as in the dielectric relaxation. Secondly,

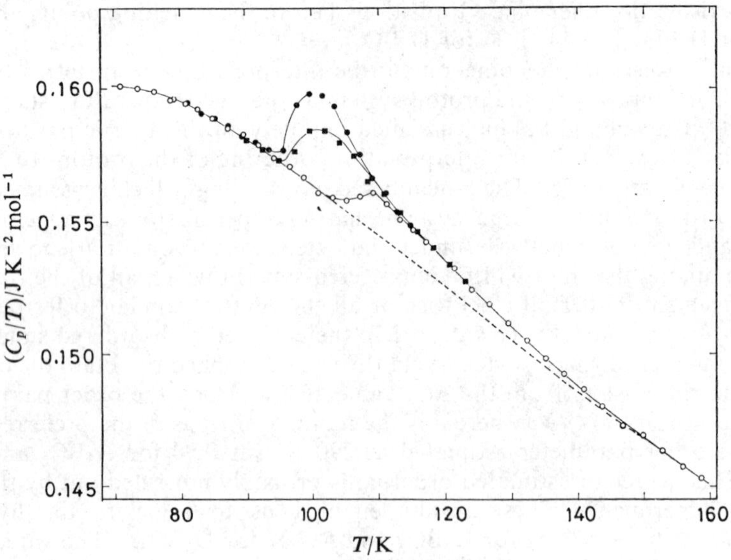

FIG. 8.—Plots of C_p/T against T for hexagonal ice subjected to different thermal treatments. ○, rapidly cooled (—1 K minB^1); □, annealed (94.4 K, 71 h); ●, annealed (89.4 K, 624 h); (- - -) assumed lattice heat capacity.

some configurations of the water molecules realized by that degree of freedom must differ energetically from others. This is the necessary condition for the relaxation to be observable calorimetrically.

Campbell and others[18] calculated the lattice energies of ice crystals in terms of multipole interactions and found significant differences among the energy of various configurations which must affect the thermodynamic properties of the crystal at a temperature around 100 K. All these facts lead to a picture of the ordering process of protons in ice crystals. The water molecules are in an orientationally disordered state at high temperatures. As the ice crystals are cooled, some of the proton configurations are favoured by the intermolecular interactions, thus realizing a partially ordered state (short-range order). The corresponding configurational heat capacity increases gradually as the temperature is lowered.

The configurational heat capacity would increase further and reach a maximum

TABLE 2.—RESULTS FOR THE ANNEALING EXPERIMENTS OF ICE CRYSTALS; T_a IS THE ANNEALING TEMPERATURE, t_a THE ANNEALING TIME, AND τ THE RELAXATION TIME.

ordinary ice			heavy ice		
T_a/K	t_a/h	τ/h	T_a/K	t_a/h	τ/h
89.4	624	145	110.0	166	44.8
94.4	71	20.0	112.9	92	17.2
97.4	55	13.6	113.5	68	12.2
99.4	45	10.4	117.9	31	6.50
102.5	9	3.82	120.9	28	4.63
104.5	7	1.76	122.1	9	1.97
107.6	2	1.01			

value at a hypothetical transition point where an (long-range) ordered state might be realized, if the crystal were cooled at an infinitely slow rate. The prolonged relaxation time, however, hinders the crystal from reaching the equilibrium state in an actual experiment and the configurational disorder is frozen at a certain temperature. The corresponding decrease in the configurational heat capacity occurs around 100 K

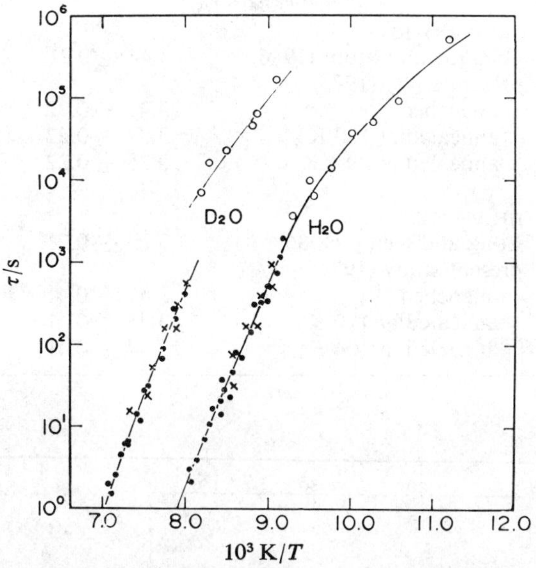

FIG. 9.—Relaxation against time data of ice crystals determined by dielectric and calorimetric methods. ○, calorimetric method; ×, TSD method; ●, step-response (or bridge) method.

for H_2O ice and around 115 K for D_2O ice, respectively. Consideration of the values of residual entropies given in table 3 shows that we are concerned with the very initial stage of the ordering process and the hypothetical transition point seems to be located much lower than 100 K. Nobody knows how nature would consume time exceeding far beyond a human life-span in removing all the entropy of ice crystals.

PINACOL HEXAHYDRATE

Pinacol(2,3-dimethyl-2,3-butane diol) hexahydrate gives an interesting example of a stable crystal in which two degrees of freedom may be frozen. The crystal has a three-dimensional hydrogen-bonded network[19] composed of layers of water molecules and hydroxyl groups of the pinacol molecules. The diol molecules are entrapped in the voids in this framework with two-fold orientational disorder. The behaviour[20] of dielectric dispersion and absorption of the crystal is closely related to that in hexagonal ice. The large value of dielectric constant is due to reorientational motions of the water molecules. Thus, both the diol and the water molecules are considered to be orientationally disordered at room temperature.

Fig. 10 shows the heat capacity curves[21] of the hydrate and its deuterium analogue, $[^2H_2]$pinacol hexadeuterate. Both the hydrate and the deuterate underwent a phase transition at 192.14 and at 191.17 K, respectively. The associated entropy change, 5.10 and 5.28 J K^{-1} mol^{-1}, suggested that the transition is due to orientational order–disorder change of the diol molecules ($\boldsymbol{R}\ln 2 = 5.76$ J K^{-1} mol^{-1}). The

TABLE 3.—RESIDUAL ENTROPY OF HEXAGONAL ICE

THEORETICAL	
	S_{res}/J K^{-1} mol^{-1}
Pauling (1935)	3.22
Nagle (1966)	3.408 ± 0.0008
EXPERIMENTAL	
ORDINARY ICE	
Giauque and Stout (1936)	3.44 ± 0.21
present study (1976)	
quenched	3.41 ± 0.22
annealed at 100 K	3.37 ± 0.22
annealed at 89.4 K	3.35 ± 0.22
HEAVY ICE	
Long and Kemp (1936)	3.22 ± 0.21
present study (1979)	
quenched	3.47 ± 0.41
annealed at 110 K	3.45 ± 0.41
annealed at 104 K	3.44 ± 0.41

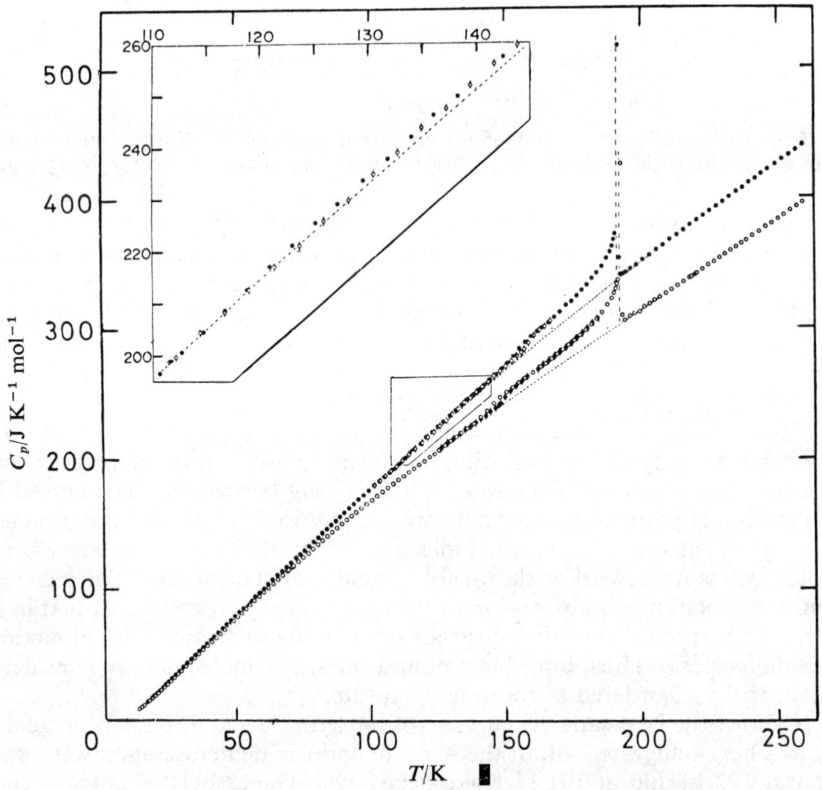

FIG. 10.—Heat capacity of pinacol hexahydrate (○, ○) and [$^2H_2$]pinacol hexadeuterate (●, ●) crystals. Open symbols: annealed, closed symbols: quenched.

mechanism was also supported by an n.m.r. observation[22] that both the crystals showed almost the same magnitude of discontinuity in $T_{1\rho}$ at the transition temperatures. The negative isotope-effect of deuteration on the transition temperature may be understood by the elongation of O· · ·O hydrogen-bond distance on deuteration in the relatively short hydrogen-bonded system,[23] leading to weakening of the interaction energy between diol molecules.

The heat-capacity curves have additional small anomalies around 150 K for the hydrate and around 130 and 160 K for the deuterate, respectively. The data in the anomalous regions are shown in fig. 11 as the form of C_p/T against T. All these small anomalies have a relaxational nature. Two candidates for the molecular origin of

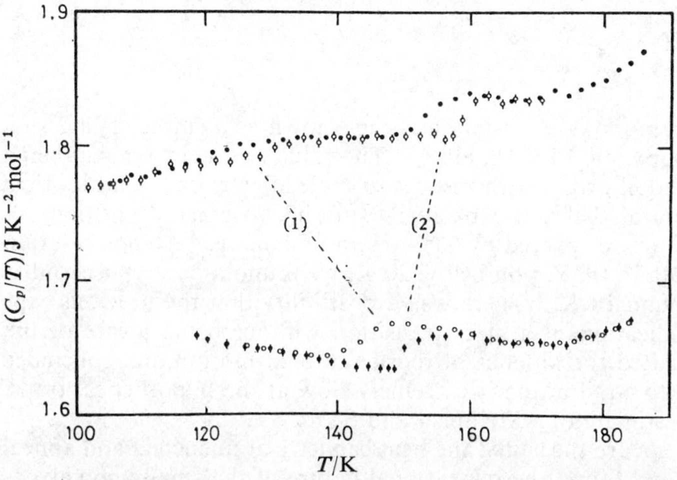

FIG. 11.—Plots of C_p/T against T for pinacol hexahydrate (lower curve) and [$^2H_2$]pinacol hexadeuterate (upper curve) around the glass transition regions. The barred circles (○ and ●) are for the rapidly cooled samples (-1.7 K min^{-1}), the plain circles (○ and ●) for the slowly cooled samples (-0.008 K min^{-1}). Two molecular mechanisms (1) and (2) of the freezing processes in the hexadeuterate are assumed to happen to be incorporated with the single freezing process in the hexahydrate.

the freezing processes are the reorientational motions of the diol and the water molecules. It had already been shown that some ordering of the orientation of the diol molecules occurs at the phase transition. However, the ordering may be incomplete at that temperature as in the case of $SnCl_2.2H_2O$. The residual disorder would cause a glass transition when it freezes into a static disorder. On the other hand, the water molecules are in a dynamically disordered state at room temperature which will also change into static disorder on cooling as in the case of hexagonal ice. If we take 1 ks as the demarcation between dynamic and static disorder, the Arrhenius parameters given by Cook and Meakins[20] predict that the glass transition will occur at 170 K.

For the deuterate, we may tentatively assign the upper glass transition to the freezing of the reorientational motion of water molecules and the lower to that of diol molecules. The n.m.r. experiment[22] revealed that the reorientation of water molecules occurs at a higher temperature than that of diol molecules. The activation enthalpy derived from the temperature drift rate around the upper glass transition of the deuterate 33 ± 4 kJ mol^{-1} is equal to the activation enthalpy (33 kJ mol^{-1}) of the water reorientation in the hydrate obtained by n.m.r.

On the other hand, both the mechanisms are assumed to be occurring during the glass transition of the hydrate. The glass transition occurs at an intermediate temperature between the two glass-transitions of the deuterate, and none occurs above or below. The glass-transition region of the hydrate is broader than either of the glass-transition regions of the deuterate. The increase in the freezing temperature of the water reorientation upon deuteration is in accord with the cases of $SnCl_2.2H_2O$ and ice crystals. The lowering of the freezing temperature of the diol reorientation can be correlated with the negative effect of deutration on the transition temperature. The same lattice expansion will enlarge the void in which the diol molecule reorients and thus decrease the height of potential barrier hindering the motion. This will result in faster reorientation of the diol molecule when compared at the same temperature and result in the lower T_g in the deuterate than in the hydrate.

CO

α-Carbon monoxide crystal has been known as a stable crystal which possesses a residual entropy,[24] 4.4 J K^{-1} mol^{-1}. The value was interpreted as being due to failure to establish equilibrium with respect to molecular end-for-end reorientation. Later measurements of the heat capacity[25–27] found no phase transition below 14 K, the lowest temperature covered by Clayton and Giauque.[24] Upon re-examination of the published data in the region below 20 K, we found a barely discernible heat capacity anomaly around 18 K. There was a possibility that the previous experimenters observed an indication of a glass transition without being aware of the effect. Curl *et al.*[28] calculated the kinetics of reorientational motion and concluded that the rate of approach to equilibrium is extremely slow at the hypothetical transition temperature of 5 K estimated by Melhuish and Scott.[29]

Atake *et al.*[30] re-measured the heat capacity of quenched and annealed samples of CO crystals and found the relaxational nature of the small anomaly. The nature of the thermal anomaly was essentially the same as that of ice crystals. The effect of annealing was such that the heat-capacity anomaly becomes larger and its maximum is displaced to lower temperature. All this evidence leads to a conclusion that some degrees of freedom of motion become frozen in; we assume, in the absence of direct experimental evidence, that it is the reorientational degree of freedom that is frozen. The present crystal may be a promising model substance for the glassy state of a simple Ising system, and a theoretical approach is highly desirable.

$TlNO_2$

Thallium(I) nitrite crystallizes in the CsCl structure[31] at room temperature. The V-shaped NO_2^- ion must be in an orientationally disordered state in the crystal in view of the octahedral symmetry of the anion site. On cooling the sample, a phase transition was observed[32] at 282.4 K, as shown in fig. 12. The associated entropy change is 23.6 J K^{-1} mol^{-1}, and the crystal seems to settle into an ordered phase. Dielectric dispersion and relaxation were found, however, between 80 and 180 K in the frequency range 200 Hz-100 kHz. The real and imaginary parts of the dielectric permittivity are drawn in fig. 13, and the dielectric relaxation time τ derived from them by assuming a single relaxation process is plotted in fig. 14. The magnitude of the dispersion suggests that reorientational motion of NO_2 – ion is responsible for the relaxation.

A relaxational heat-capacity anomaly was found around 60 K, at which the dielectric relaxation time becomes 10^{-4} Hz, approximately the inverse of the calorimetric time-scale. Also plotted in fig. 14 are the relaxation times as determined by enthalpy relaxation rate. Both sets of data lie well on a straight line whose slope gives the activation enthalpy, 19 kJ mol^{-1}, for the process. These experimental findings suggest that the low-temperature phase of the crystal still has some orientational disorder. The magnitude of the dielectric absorption increases with increasing temperature. With the assumption of a two-state model of the orientational disorder,

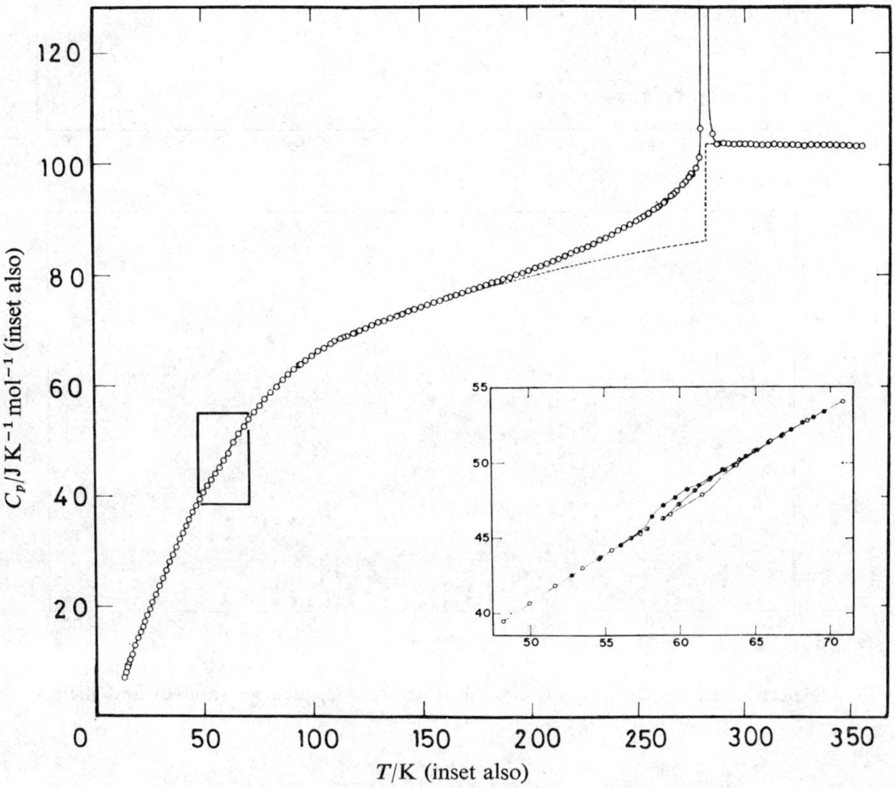

FIG. 12.—Heat capacity of $TlNO_2$ crystal, with an expanded view of the heat capacity around T_g. Series 1 (○), quenched sample ($\simeq -1$ K min^{-1}) measured with the average rate of 4 K h^{-1}; series 2 (◑), quenched sample measured with the rate of 1 K h^{-1}; series 3 (●), annealed sample at 57.3 K for 39 h.

this indicates that there is an energy difference larger than ***RT*** between the two minima, which are separated by the activation enthalpy 19 kJ mol^{-1}.

The simplest disorder compatible with the cubic symmetry of the high-temperature phase is one which allows $4 \times 3 = 12$ orientations for the nitrite ion. The associated entropy is equal to 20.7 J K^{-1} mol^{-1} and explains the most part of the observed transition entropy. However, the low-temperature phase retains orientational disorder. Thus, the disorder in the high-temperature phase appears more extensive than is expected from the 12-fold degeneracy. Interestingly, the present crystals exhibit high plasticity, as organic " plastic crystals " do.[33] High plasticity may thus be characteristic of highly disordered crystals[34] irrespective of their molecular or ionic nature.

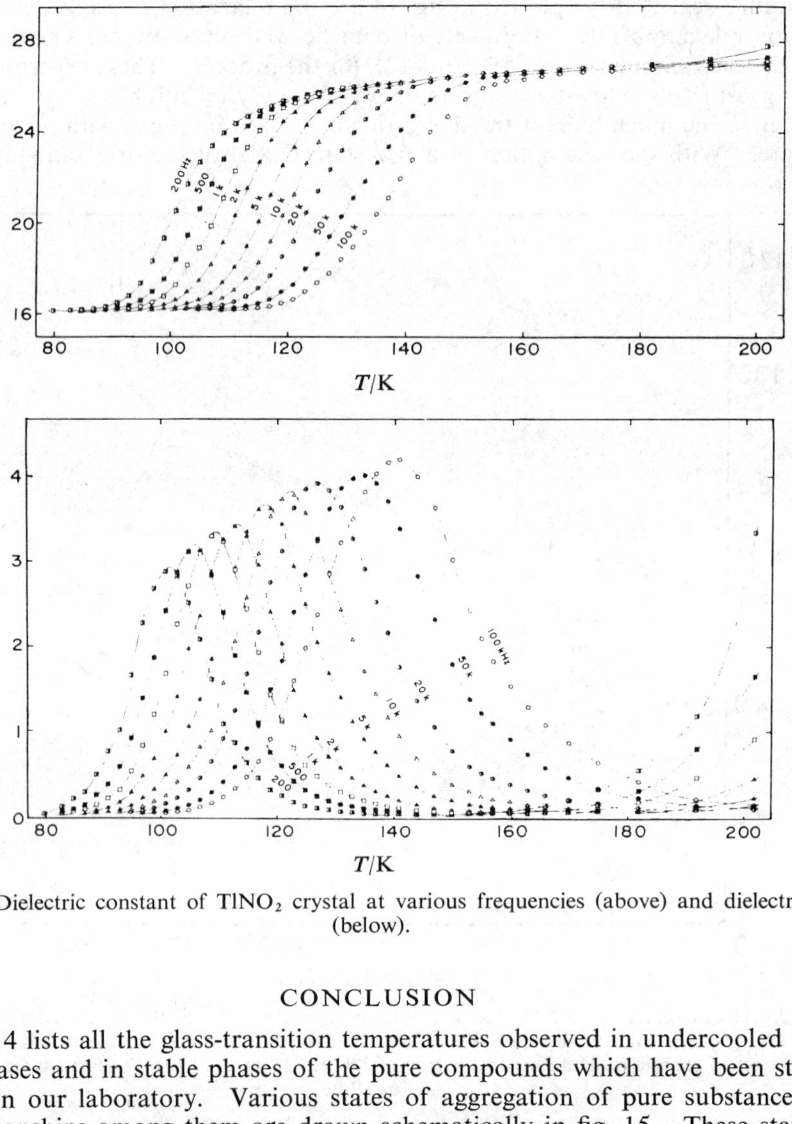

FIG. 13.—Dielectric constant of $TlNO_2$ crystal at various frequencies (above) and dielectric loss (below).

CONCLUSION

Table 4 lists all the glass-transition temperatures observed in undercooled metastable phases and in stable phases of the pure compounds which have been studied hitherto in our laboratory. Various states of aggregation of pure substances and interrelationships among them are drawn schematically in fig. 15. These states of aggregation are classified according to the mechanical, structural and thermodynamic characteristics. All the glass-transition phenomena reported in this paper are the ones observed in stable phases and differ from those occurring in metastable phases in that they never encounter the embarrassing situation of the Kauzmann paradox. In this sense, these crystals offer a new realm of the glassy crystal.

From the thermodynamic point of view, the glassy state is a kind of non-equilibrium state of a solid which changes into an equilibrium state *via* T_g on heating. Although previous workers have paid attention to glass transitions occurring in metastable phases of liquid owing to the drastic changes in various properties, we have no reason to exclude the same situation in stable phases. There seems no logical necessity to confine the existence of non-equilibrium state to the metastability of the phase. Only the existence of disorder at high temperature is of primary concern.

Stability is a relative concept based on limited observation. For example, the famous phase transitions[35] in ferrocene around 165 K have been believed to be the ones occurring in the stable crystalline phase of this compound. We have found[36] recently, however, that this was not true and that the more stable phase existed below 242 K. The stable low-temperature phase can be produced by holding the undercooled crystal in the temperature range 170-220 K for several days. However, we are not convinced that the newly found phase is the most stable one of crystalline ferrocene below 242 K. If we consider stability of phase with respect to chemical change[37]

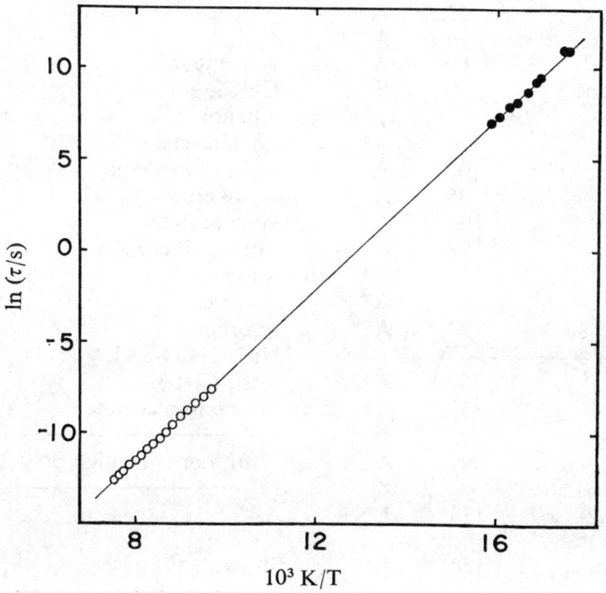

FIG. 14.—Arrhenius plot of the relaxation-time data determined dielectrically (○) and calorimetrically (●). $\Delta H\ddagger = 19.5$ kJ mol^{-1}.

or even nuclear reaction, the concept of stability in nature becomes quite limited. The successful observations of these physical properties of an organic compound, instead of a mixture of CO_2 and H_2O, is entirely due to frozen degrees of freedom with respect to chemical potential. The whole story concerning the glassy state is summed up in the phrase—physical freezing phenomena occurring in chemically frozen systems. We are merely passengers travelling in an eternal flow of time.

Two varieties of freezing phenomena were observed in stable crystals. One is exemplified by $SnCl_2.2H_2O$, H_3BO_3 and pinacol hexahydrate in which the frozen degree of freedom relates to a long-range order. The others are ice, pinacol hexahydrate, CO and $TlNO_2$ in which the freezing occurred before the crystals reached their hypothetical transition points. The verification of existence of the residual entropy in the former case may be difficult, owing to the smallness of the value and also to the principle of its determination, which is similar to the method of weighing a captain by the difference of the weight of a ship loaded and unloaded with said captain! However, the existence of a relaxational heat-capacity anomaly has been clearly demonstrated calorimetrically. The basic principle was that relaxational behaviour occurred whenever the relaxation time of a particular degree of freedom became comparable with the time required for a single heat-capacity measurement,

TABLE 4.—COMPLETE LIST OF THE GLASS-TRANSITION TEMPERATURES T_g OBSERVED IN VARIOUS PHASES OF PURE COMPOUND STUDIED IN OUR LABORATORY.

substance	T_g/K	method	substance	T_g/K	method
(I) glassy crystals			(II) glassy liquids		
			(*a*) low molecular-weight compounds		
cyclohexanol	150	A*	propene	55	B
2,3-dimethylbutane	76	A*	CCl_4	60	B
1,2-difluoro-1,1,2,2-tetrachloroethane	60	A*	isopentane	65	A
	90		$CHCl_3$	79	B
			ethanol	90	A
cycloheptane I	100	A*	cyclohexene	92	A
II	100	A*	dichloromethane	93	B
III	96	A*	methanol	103	B
cycloheptatriene	106	A*	vinyl acetate	123	A
cyclohexene I	92	A*	isopropylbenzene	126	A
III	93	A*	water	135	B
cycloheptanol	135	A*	*m*-cresol	200	A
ethanol	90	A*	resorcinol	250	A
— — — — — — —			Ni(3-CH_3O-SAL iso-$C_3H_7)_2$	297	A
$SnCl_2.2H_2O$	150	A*			
$SnCl_2.2D_2O$	155	A*	magnesium acetate	469	D
D_3BO_3	298	A*	(*b*) inorganic high polymers		
1,2-difluoro-1,1,2,2-tetrachloroethane	130	A*	As_2S_3	473	C, A
				495	C
methylisothiocyanate	150	A*	Sb_2S_3	489	B
H_3BO_3	290	A*	(*c*) organic high polymers		
heavy ice	115	A*	polytetrahydrofuran	185	A
ordinary ice	100	A*	poly(ethylene oxide)	206	A
pinacol.$6H_2O$	155	A*	polyoxacyclobutane	193	A
[$^2H_2$]pinacol.$6D_2O$	130, 165	A*	polyoctamethyleneoxide	193	A
$TlNO_2$	61	A*			
CO	18	A*,E			
(III) glassy liquid crystals					
cholesteryl-hydrogen-phthalate (cholesteric)	295	A**	PENTOBUTA (smectic)	181	A**
			HEXOBUTA (smectic)	180	A**
N-(*o*-hydroxy-*p*-methoxy-benzylidene-*p*-butyl aniline) (nematic)	204	A**	OCTABUTA (smectic)	197	A**
			HEXOETHA (smectic)	187	A**
BOBUTA (smectic)	112	A**			

(A) Undercooling of liquid. (B) Vapour condensation. (C) Rapid precipitation. (D) Thermal decomposition. (A*) Undercooling of crystal (metastable and stable phases). (A**) Undercooling of liquid crystal. (E) Collaboration with Chihara laboratory.

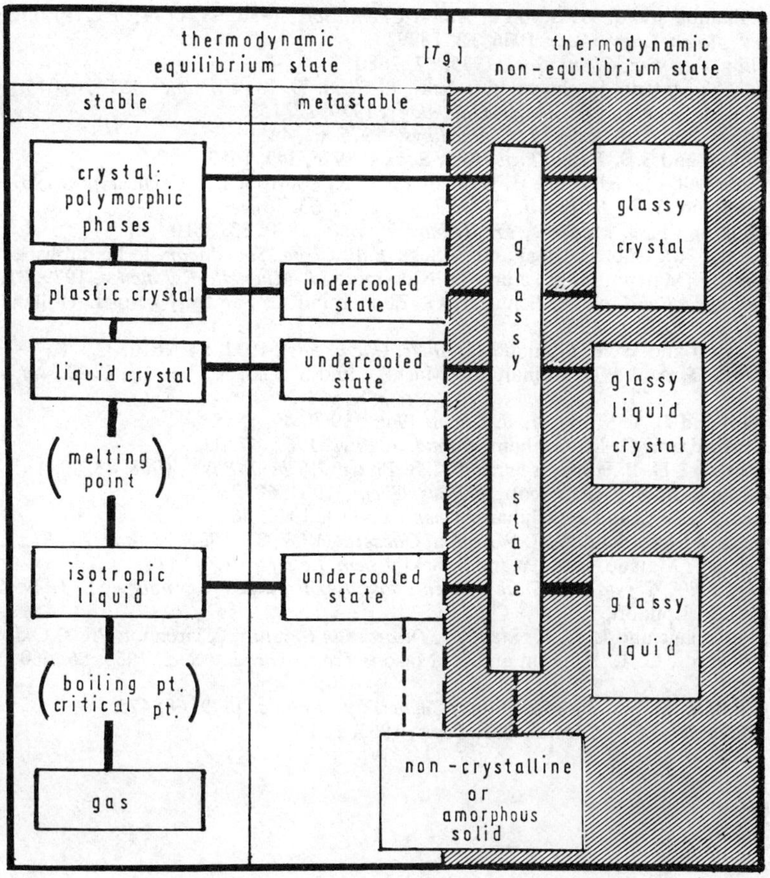

FIG. 15.—Various states of aggregation of pure substance and their inter-relationships, classified by mechanical, structural and thermodynamic features.

typically 1 ks. The method gives insight into slow molecular motions in solids for which the conventional spectroscopic techniques such as dielectric measurement are impractical because of the prolonged relaxation time. Thus, adiabatic calorimetry proved to have a function as an ultra-low-frequency spectrometer in addition to its traditional mission.

[1] W. Kauzmann, *Chem. Rev.*, 1948, **43**, 219.
[2] H. Suga and S. Seki, *J. Non-cryst. Solids*, 1974, **16**, 171.
[3] H. Suga and S. Seki, *Bull. Chem. Soc. Japan*, 1965, **38**, 1000.
[4] M. Tatsumi, T. Matsuo, H. Suga and S. Seki, *Bull. Chem. Soc. Japan*, 1975, **48**, 3060.
[5] K. Kishimoto, H. Suga and S. Seki, to be published in *Bull. Chem. Soc. Japan.*
[6] H. Kiriyama, K. Kitahama, O. Nakamura and R. Kiriyama, *Bull. Chem. Soc. Japan*, 1973, **46**, 1389; R. Kiriyama, H. Kiriyama, K. Kitahama and O. Nakamura, *Chem. Letters*, 1973, 1061; 1973, 1105; 1976, 689.
[7] M. Tatsumi, T. Matsuo, H. Suga and S. Seki, *Bull. Chem. Soc. Japan*, 1979, **52**, 716; 1979, **52**, 728.
[8] T. Matsuo, M. Oguni, H. Suga, S. Seki and J. F. Nagle, *Bull. Chem. Soc. Japan*, 1974, **47**, 57.
[9] B. M. Craven and T. M. Sabine, *Acta Cryst.*, 1966, **20**, 214.
[10] M. Oguni, T. Matsuo, H. Suga and S. Seki, *Bull. Chem. Soc. Japan*, 1977, **50**, 825.
[11] Personal communication from J. F. Nagle to one of the authors (H. S.), 1977.

[12] W. F. Giauque and J. W. Stout, *J. Amer. Chem. Soc.*, 1936, **58**, 1144; E. A. Long and J. D. Kemp, *J. Amer. Chem. Soc.*, 1936, **58**, 1829.
[13] L. Pauling, *J. Amer. Chem. Soc.*, 1935, **57**, 2680.
[14] O. Haida, T. Matsuo, H. Suga and S. Seki, *J. Chem. Thermodynamics*, 1974, **6**, 815.
[15] O. Haida, H. Suga and S. Seki, *J. Glaciology*, 1979, **22**, 155.
[16] G. P. Johari and S. J. Jones, *J. Chem. Phys.*, 1975, **62**, 4213.
[17] G. P. Johari and S. J. Jones, *Proc. Roy. Soc. A*, 1976, **349**, 1659.
[18] E. S. Campbell, G. Gelernter, H. Heinen and V. R. G. Moorti, *J. Chem. Phys.*, 1967, **46**, 2690.
[19] H. S. Kim and G. A. Jeffrey, *J. Chem. Phys.*, 1970, **53**, 3610.
[20] J. S. Cook and R. J. Meakins, *Trans. Faraday Soc.*, 1970, **53**, 3610.
[21] M. Oguni, T. Matsuo, H. Suga and S. Seki, *Bull. Chem. Soc. Japan*, to be published.
[22] M. Oguni, T. Matsuo, H. Suga and N. Nakamura, *J. Magnetic Resonance*, 1979, **31**, 51.
[23] *The Hydrogen Bond*, ed. P. Schuster, G. Zundel and C. Sandorfy (North Holland Amsterdam, 1976), vol. II, p. 420.
[24] J. O. Clayton and W. F. Giauque, *J. Amer. Chem. Soc.*, 1932, **54**, 2610.
[25] T. Shinoda, T. Atake, H. Chihara, Y. Mashiko and S. Seki, *Kogyo Kagaku Zasshi*, 1966, **69**, 1619.
[26] E. K. Gill and J. A. Morrison, *J. Chem. Phys.*, 1966, **45**, 1585.
[27] J. C. Burford and G. M. Graham, *Canad. J. Phys.* 1969, **47**, 23.
[28] R. F. Curl, Jr., H. P. Hopkins Jr, and K. S. Pitzer, *J. Chem. Phys.*, 1968, **48**, 4064.
[29] M. W. Melhuish and R. L. Scott, *J. Phys. Chem.*, 1964, **68**, 2301.
[30] T. Atake, H. Suga and H. Chihara, *Chem. Letters*, 1976, 567.
[31] L. Cavalca, M. Nardelli and I. W. Bassi, *Gazzetta*, 1957, **87**, 153.
[32] K. Moriya, T. Matsuo, H. Suga and S. Seki, *Chem. Letters*, 1977, 1427.
[33] *The Plastically Crystalline State* (*Orientationally-Disordered Crystals*), ed. J. N. Sherwood (John Wiley, London, 1979).
[34] N. G. Parsonage and L. A. K. Staveley, *Disorder in Crystals*, (Clarendon Press, Oxford, 1978).
[35] J. W. Edwards, G. L. Kington and R. Mason, *Trans. Faraday Soc.*, 1960, **56**, 660; 1962, **58**, 1334.
[36] K. Ogasahara, M. Sorai and H. Suga, *Chem. Phys. Letters*. 1979, **68**, 475.
[37] P. O. Davies and G. O. Jones, *Adv. Phys.*, 1953, **2**, 370.

Melting and the Liquid and Glassy States

By Rodney M. J. Cotterill and Jeffery L. Tallon†

Department of Structural Properties of Materials,
The Technical University of Denmark,
Building 307, DK-2800 Lyngby, Denmark

Received 3rd *December*, 1979

A unified model is presented which treats the melting, glass, and critical transitions by a single approach. It is shown that the important event in each case is the loss or gain of entropy associated with bifurcation of the on-site potential. This explains the first-order nature of the melting transition. Lindemann's melting rule is neither fundamental nor obeyed exactly. The relevance of various defects, invoked in earlier theories of the melting and glass transitions, lies in their providing favoured sites for bifurcation. Liquid helium II is believed to be unique in having bifurcated on-site potentials yet lacking the bifurcation entropy.

1. INTRODUCTION

This communication describes a microscopic model for the melting, glass and critical transitions of simple molecular substances. Its novel features are inevitable consequences of the nature of interatomic interactions, and as such are believed to be applicable, in principle at least, to all such systems. The underlying idea is that the nature of the on-site potential is of paramount importance to the manner in which a system behaves. If this is bifurcated, an extra entropy factor, $\boldsymbol{R}\ln 2$, arises. This entropy and the associated bifurcation appear to be the hallmark of the liquid state. It is lost from the liquid when it crystallizes, but it also disappears at the glass and critical transitions. Finally, there is a somewhat exotic loss of the bifurcation entropy in He^4 when, crossing the λ-line, it transforms from helium I to helium II. In the following sections we discuss, in turn, the application of the model to these four transactions.

2. MELTING

Three short papers already in the literature,[1–3] when taken together, might hold the secret of the long-standing melting problem. First, Stishov *et al.*[1] examined the melting entropy, ΔS_m, as a function of volume change on melting, ΔV_m, and noticed that on extrapolation to $\Delta V_m = 0$ one obtains $\Delta S_m = \boldsymbol{R}\ln 2$ for both sodium and argon. This indicates a fundamental choice of position between two alternatives for each atom upon melting. Secondly, Tallon *et al.*,[2] discussing Born's suggestion of vanishing shear modulus at the melting point, found that although this is not true at the melting volume, V_m, one of the shear moduli is apparently continuous through the melting expansion and falls to a value, at the freezing volume, V_f, which is sufficiently small that a mechanical instability occurs by thermal activation. They demonstrated

† Permanent address: Physics and Engineering Laboratory, Department of Scientific and Industrial Research, Lower Hutt, New Zealand.

that this is consistent with a first-order melting transition. Finally Cotterill,[3] considering the role of anharmonicity in the process, proposed that melting might be associated with bifurcation of the on-site potential. He suggested that with increasing volume at the melting point the crystal structure will be pushed progressively closer to a bifurcation instability, and that the liquid state might correspond to a situation of " cascading bifurcations ". The factor 2 found by Stishov *et al.* lends support to the bifurcation idea,[4] so the microscopic instability proposed by Cotterill might be one and the same as the macroscopic shear instability of Tallon *et al.* In this communication it is demonstrated that such an identification is substantially correct for the Lennard-Jones system. It is also shown, using a dynamic computer simulation, that the bifurcations of the on-site potential have a vital influence on the dynamics of the liquid state.

The arguments presented by Tallon and co-workers[2,5,6] show why a phase change based on a bifurcation instability has all the qualitative features of the melting process. The appearance of the entropy $\boldsymbol{R} \ln 2$ at the onset of bifurcation, somewhere near the volume V_f at the freezing point, allows the material to transform discontinuously from a lower volume, V_m, the volume at the melting point. Melting is therefore first-order and heterogeneous, with a nucleation barrier of about $\boldsymbol{R}\,T_m \ln 2$. The transition has to be nucleated and superheating is, in principle, possible. As discussed by Tammann,[7] Klæstrup Kristensen and Cotterill,[8] however, surface free-energies are such that the transition would usually nucleate at a free surface and superheating would be rare. In normal melting, because V_m is substantially less than the instability volume, V_i, fluctuations at and below V_m would be insignificant, and there would be no premelting behaviour, in agreement with experimental evidence.[9] With increasing pressure, V_m and V_f (which must bracket V_i) move closer together, however, and it is possible that the melting line may traverse part of the fluctuation regime. Premelting may therefore become significant at elevated pressures, and some substances, such as the alkali metals,[10] might already have entered this regime at atmospheric pressure.

Equating the bifurcation instability with the shear instability accounts for the fluidity of the melt, and the spontaneous symmetry breaking inherent in the bifurcation accounts for the loss of long-range molecular order.[3] In every respect, then, the qualitative features of a transition around a bifurcation instability point correspond to those of the melting transition. There are indications, indeed, that a bifurcation instability is the universal cause of melting for simple substances, because a shear instability occurs between V_m and V_f for a large number of materials including inert gases, metals, alkali halides, silver bromide and benzene.[2,5,6] Moreover, the volume-independent entropy change on melting, $\boldsymbol{R} \ln 2$, exhibited by argon and sodium is characteristic of a large number of metals[11] and also of caesium, at least in its normal melting regime well away from the melting maxima. (For the alkali halides there is a complication because an additional volume-independent entropy change, namely part or all of the communal entropy, may appear on melting.[12])

Bifurcation of the on-site potential occurs when the neighbouring atoms lie farther than a certain distance from the atom in question. Such a situation can arise as an instantaneous event resulting from the combined effects of lattice expansion and lattice vibration.[3] We thus examine the local dilation provided by specific thermal phonons. One may reduce the number of modes which are candidates for this process by first noting that for the face-centred cubic, and indeed all cubic, lattices the symmetry is such that the destructive mode is longitudinal and not transverse. For a Lennard-Jones f.c.c. crystal the on-site potential for small displacements from the ideal lattice position is softest in a $\langle 110 \rangle$ direction, and of all the longitudinal phonons directed in low index directions, the second lowest order optical phonons in $\langle 110 \rangle$ directions are the most effective in softening the local potential in the $\langle 110 \rangle$ direction. Fig. 1 shows the projection of adjacent $(1\bar{1}0)$ planes of the f.c.c. lattice, the arrows indicating the direction and magnitude of the displacements associated with this particular longitudinal optical phonon in the [110] direction at the instant of maximum dilation. The displacements instantaneously soften the potential seen by all [001] transverse phonons propagating along the dashed line and polarized in the [110] direction. The identical longitudinal phonon propagating in the $[1\bar{1}0]$

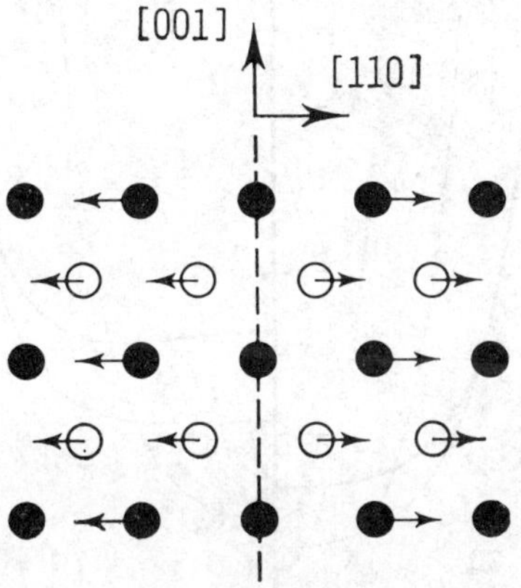

FIG. 1.—Schematic diagram of the $(1\bar{1}0)$ plane of an f.c.c. lattice, showing the displacements associated with the low-order longitudinal optical phonon in the [110] direction, as discussed in the text. Full circles denote atoms lying in the plane of the figure, and open circles denote atoms immediately above and below the plane. The arrows indicate the directions and relative magnitudes of the atomic displacements at the instant of maximum dilation along the dashed [001] line.

direction normal to the figure, *i.e.*, with polarization normal to the transverse phonon, is also effective, although to a lesser extent, in instantaneously softening both the potential seen by the transverse phonon and the [110] longitudinal phonon. Thus the two longitudinal phonons cooperatively soften each other's potential and collectively soften the potential of the transverse phonon, driving it in the limit to instability.

One must first calculate the local potential an atom would see, if displaced along the [110] direction, arising from neighbours out to three nearest-neighbour distances, when, for a given density, they are fixed in their mean positions. The particular densities investigated are those at the melting line and freezing line for argon at pressures of 0 and 1.71×10^8 Pa,[13,14] and fig. 2 and 3 show the results of calculations for each of these pressures, respectively, the melting-line and freezing-line results being shown on the same plot for each of the two pressures. In Lennard-Jones reduced units for argon, the temperatures and densities are, respectively,

$$T^* = 0.70 \quad \rho_{\mathrm{m}}^* = 0.97 \quad \rho_{\mathrm{f}}^* = 0.84$$

and

$$T^* = 1.01 \quad \rho_{\mathrm{m}}^* = 1.01 \quad \rho_{\mathrm{f}}^* = 0.92.$$

The mean on-site potential arising from atoms fixed in their mean positions is shown in the figures by the dashed curves. The displacements corresponding to energies T^* above the minima of the potential wells are read off from these curves and assigned to the phonon depicted in fig. 1. The potential seen by the parallel-polarized transverse phonon is then recalculated with atoms frozen in the maximum dilation and compression extremes. In fig. 2 and 3 these new " instantaneous " potential profiles are shown by the light full curves and the dotted curves, respectively. The marked disparity between these two curves, for any one temperature and density, exposes the gross anharmonicity present, especially at the low densities corresponding to the freezing line. Finally, allotting the same displacements to the [1$\bar{1}$0] longitudinal phonon modifies the potential profile to the heavy full curves, shown in fig. 2 and 3, for the

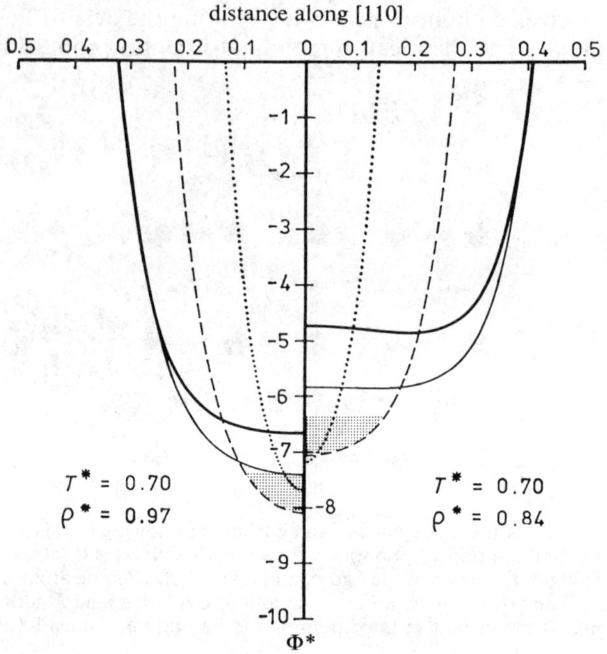

FIG. 2.—Reduced potential, φ^*, along a [110] axis for a Lennard-Jones f.c.c. lattice at the zero pressure melting density (left-hand side) and freezing density (right-hand side). Dashed curve: atoms fixed in their mean positions. Dotted curve and light solid curve: single parallel-polarized longitudinal optical phonon frozen at moments of, respectively, maximum compression and maximum dilation. The heavy solid curve describes the potential at maximum dilation due to both parallel- and perpendicularly-polarized longitudinal phonons. The shaded region denotes the region occupied by an atom with a thermal energy of T^*.

most favourable circumstance when both longitudinal phonons produce a maximum dilation along the dashed line in fig. 1. In fig. 2, the on-site potential is now markedly bifurcated for the freezing density, and presumably the co-operative mutual softening of the two longitudinal modes (which have not been included) is sufficient to guarantee the instantaneous instability of the transverse phonons. The potential curve shown in fig. 3 is only slightly bifurcated, but at this elevated pressure the anharmonicity is more significant and this first-order calculation less valid. The figure does show, however, that the potential profiles at either side of the melting point are not as dissimilar as they were at zero pressure. This illustrates the possibility of premelting becoming significant at elevated pressures.

These preliminary calculations confirm the existence of a hidden bifurcation instability located at some volume intermediate between V_m and V_f, and the instability is associated with a transverse phonon propagating in ⟨100⟩ directions for which the elasticity modulus is the shear modulus, c_{44}. This was the modulus found to be associated with the shear instability for the inert gases,[6] and one concludes that the shear and bifurcation instabilities are identical.

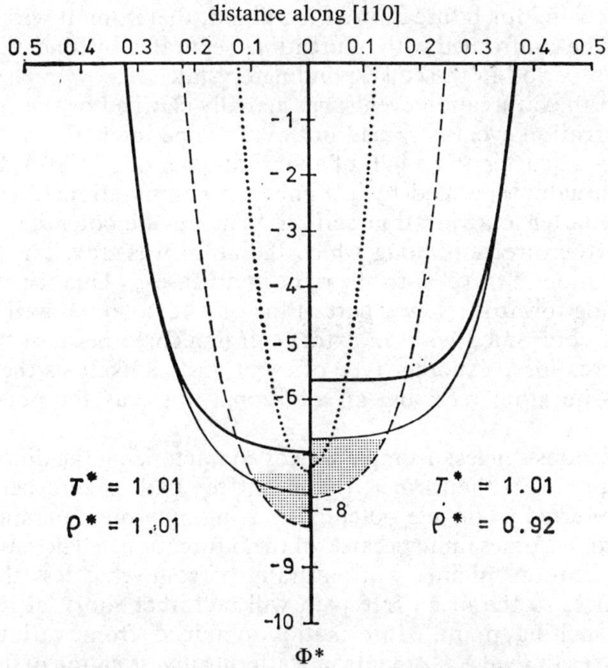

FIG. 3.—The reduced potential, Φ^*, along a [110] axis for a Lennard-Jones f.c.c. lattice at the melting density (left-hand side) and freezing density (right-hand side) for a pressure of 1.71×10^8 Pa. Dashed curve: atoms fixed in their mean positions. Dotted curve and light solid curve: single parallel-polarized longitudinal optical phonon frozen at moments of, respectively, maximum compression and maximum dilation. The heavy solid curve describes the potential at maximum dilation due to both parallel- and perpendicularly-polarized longitudinal phonons. The shaded region denotes the region occupied by an atom with a thermal energy of T^*.

3. BIFURCATION DYNAMICS AND THE NATURE OF THE LIQUID STATE

The results of the quasistatic calculations described in the preceding section are encouraging in their tentative endorsement of the bifurcation hypothesis, but it is clear that the inherently dynamic nature of the instability demands a fully dynamic computational treatment. Computer simulation is well suited to this task and has of course already been applied to the investigation of liquid structure and dynamics.[15,16] In the present study, a search was made for bifurcation of the on-site potential, using a system which was reported on recently.[17] It consisted of instantaneously fixing the positions of all the atoms in the 336-atom irreducible cell, and probing the on-site potential of randomly chosen atoms by letting these sample the potential energy at various positions within their " frozen " environment, the atomic motions being resumed after each bifurcation search.

The criterion used in the preceding section for the instability was that the on-site potential of each and every atom should develop a double-well character at some time during a vibrational period. A simple and unambiguous definition is possible for a crystal lattice: the underlying symmetry permits one to treat all crystallographically equivalent atoms in the same manner. In a liquid, one might be forced to accept a somewhat more nebulous description of bifurcation. It must be remembered, however, that the factor $\boldsymbol{R} \ln 2$ alluded to earlier shows that each atom must retain its accessibility to bifurcation in spite of the atomic rearrangements accompanying melting. In searching for bifurcation in the liquid, therefore, it was anticipated that each atom would be confronted with bifurcations separated by time intervals averaging about a vibration period in the corresponding crystal.

The nature of the bifurcation events was actually clarified by the computer simulations. The bifurcation event is spread out over a time interval that was found to lie typically between a quarter and a half of a vibration period. Fig. 4 shows a sequence of situations, mutually separated by a number of computational cycles, there being about fifty of the latter to a vibration period. The on-site potential was, in this case. being probed in the direction along which the atom was travelling at each instant, The forward direction of travel is to the right in all cases. One sees that the atom is initially descending towards a lower part of the on-site potential well, but that it suddenly finds itself confronted with an extra retarding force because the potential has developed a bifurcation. Another type of event, just as likely as the one illustrated, involves the on-site atom receiving an additional push, as the potential bifurcates behind it.

These observations suggest a simple way of characterizing the difference between a crystal and a liquid. In the former, the mean free path is zero because all atomic motion will be referred to on-site potentials having only one minimum. In a liquid the mean free path becomes finite because of the bifurcation. The separation distance between the bifurcation minima will naturally be somewhat less than the nearest-neighbour distance, so the mean free path will be rather short, at least for a liquid just above the melting point. Information obtained from various experimental techniques endorses this view. A neutron scattering investigation of liquid rubidium[18] indicated that the individual displacements were somewhat smaller than typical interatomic spacings, in agreement with what has been found in the present study. Moreover, a computer study of liquid sodium[19] showed that the atomic motion is not of the interatomic jump diffusion type; it is rather a smooth oscillatory migration involving cooperation with 10-15 surrounding atoms. This latter result can now be explained: the surrounding atoms must contrive to bifurcate the on-site potential of the moving atom in order to give it a non-zero free path.

Although these results establish the intimate connection between bifurcation and the melting transition, they provide no information on the manner in which the individual bifurcations interact with one another. In the liquid just above T_m these interactions must be very pronounced, but the elucidation of their many-body nature represents a formidable challenge. The introduction of the concept of mean free path is qualitatively useful in that it emphasizes the essential similarity between the liquid and gaseous states. For the small mean free paths in the liquid state, however, the assumption of independent collisions is hardly tenable.

In default of the necessary formalism for the interacting-bifurcation situation, one seeks simplifying principles, and it is tempting to compare this new approach with existing theories. Because some crystal defects involve a local concentration of free volume, a defect mechanism of melting suggests itself. In particular, dislocation-mediated melting has been suggested,[20–24] and dislocations have been observed in melting models studied by computer simulation.[25–27] It will now be clear, however, that earlier dislocation theories of melting were inadequate because of the omission of the volume-independent entropy, $\boldsymbol{R} \ln 2$, which must appear on melting. Using the quasistatic approach described earlier, and a previously simulated dislocation,[28] it was readily established that the atom nearest the centre of the core (for each atomic plane threaded by the dislocation) experiences bifurcation for temperatures in excess of about $0.5T_m$. Each such central atom will therefore contribute an extra entropy, $k_B \ln 2$, above this temperature, and the melting crystal will be able to progress towards

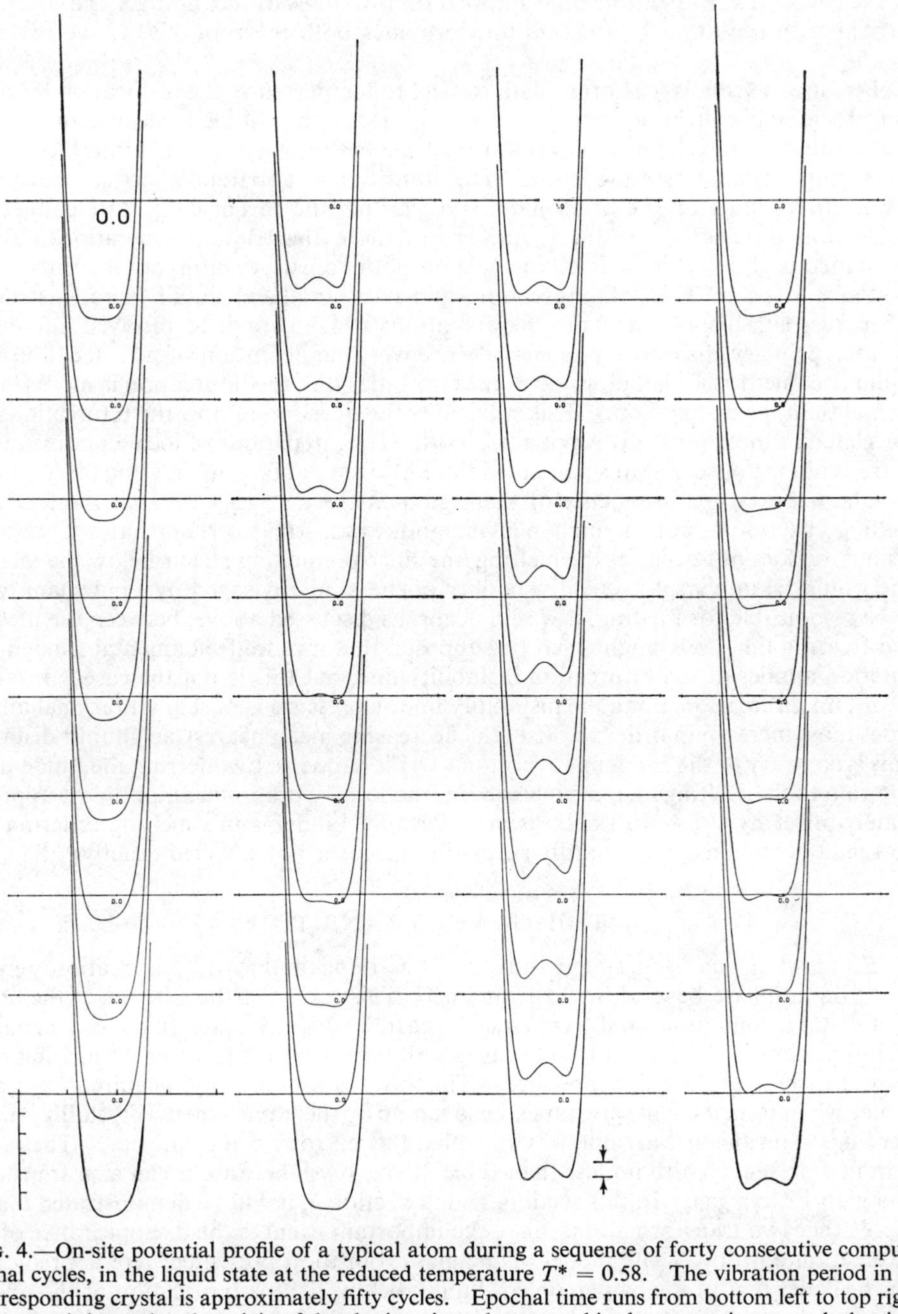

FIG. 4.—On-site potential profile of a typical atom during a sequence of forty consecutive computational cycles, in the liquid state at the reduced temperature $T^* = 0.58$. The vibration period in the corresponding crystal is approximately fifty cycles. Epochal time runs from bottom left to top right. The atom is located at the origin of the abscissa, in each case, and is always moving towards the right. The units on the abscissa are half a reduced nearest-neighbour distance. The interval given at the bottom of the figure indicates the reduced temperature interval of 0.58.

its $\boldsymbol{R}$ ln 2 state *via* the proliferation of dislocations. This suggestion is supported by the earlier demonstration[3] that bifurcation promotes dislocation generation. It is intriguing to note that bifurcation thus provides both the root of, and the route to, melting.

Because melting is first order with respect to temperature, the dislocation mechanism of melting will be a virtual transition,[29] remaining hidden because of the discontinuous volume change, but appearing locally at the crystal–liquid interface, where that volume change is established. The transition is apparently second order with respect to volume, on the other hand. Regarding the earlier conjecture connecting dislocation dipoles to melting,[22] one should note that dipole generation and the spontaneous glide of an isolated dislocation both constitute bifurcation events. But these are a special kind of bifurcation, and in a simulated liquid there exist many other free-path displacements over bifurcations which cannot be resolved into either of these primary dislocation events. Moreover, the definition of a dislocation in a liquid becomes rather nebulous, and one concludes that the bifurcation is more fundamental than the dislocation. This still leaves the question of how the bifurcations are correlated in the liquid. It is possible that the concatenations of local concentrations of free volume, observed in a simulated liquid,[17] were, in fact, interacting bifurcations.

One important consequence of the arguments set out above is that Lindemann's melting criterion is not of fundamental significance. It is certain that a vibrational instability does not occur at the melting line since premelting effects below the melting line would be universal. If an instability of the kind envisaged by Lindemann were to be responsible for melting, it would occur, as discussed above, between the melting and freezing lines. It might then be supposed that a more fundamental Lindemann criterion applies at the bifurcation instability line, but this is not the case. In order to sustain bifurcation along the instability line as pressure is raised, vibrational amplitudes must increase in order to offset the decreasing mean nearest-neighbour distance. This is contrary to the Lindemann notion. As it happens, by referring the Lindemann criterion to the melting line, the increase in fractional vibrational amplitude is approximately offset by $(V_i - V_m)$ decreasing. Because Lindemann's melting criterion has no fundamental basis, it is hardly surprising that it is not satisfied exactly.[30,31]

4. THE GLASS TRANSITION

From what has been stated thus far, it is implicit that the bifurcation persists in the liquid phase beyond the freezing line. This results in the entropy of the liquid containing a configurational term equal to $\boldsymbol{R}$ ln 2 which was identified as a signature of bifurcation and which, in fact, ensures the first-order character of melting with respect to temperature and pressure. This bifurcation entropy, acquired by a substance when it melts, will, of course, be given up by the liquid when it crystallizes upon freezing. But this is only one way in which this entropy can disappear. The liquid is in fact presented with no less than three alternatives, because it can also transform to a glass or to a gas. In this section, and in Section 6, it will be demonstrated that in these other two transformations, also, the important event is the disappearance of the bifurcation entropy. The model of the glass transition presented here is predicated on the belief that in the (supercooled) liquid, below T_m, the bifurcation of the on-site potential disappears over a fairly narrow temperature range around T_g. If bifurcation does have a pronounced influence on fluidity, this would explain the rather abrupt stiffening of the material at the glass transition. The retention of the bifurcations in the supercooled liquid state is due to this state having a volume in excess of V_m, and T_g becomes, in effect, a false and somewhat extended " melting point ".[32] Two facts lend strong support to this notion. First, it will be shown that the bifurcational entropy is indeed retained in the supercooled liquid, and that the greater part of it vanishes within a narrow temperature range around T_g. Secondly, it will be demonstrated, using a model Lennard-Jones substance, as simulated by molecular dynamics, that as the temperature approaches the glass transition from below, the on-site potential starts to bifurcate in the vicinity of T_g.

The entropy of a model glass has been derived from the molecular dynamics simulation of a quenched Lennard-Jones liquid reported by Damgaard Kristensen.[33] Using the power spectrum, obtained by cosine transformation of the velocity autocorrelation function, he calculated, assuming validity of the harmonic approximation, the vibrational component of the entropy to within an arbitrary constant. His data points are shown as squares and circles in fig. 5, and the entropy origin has been chosen

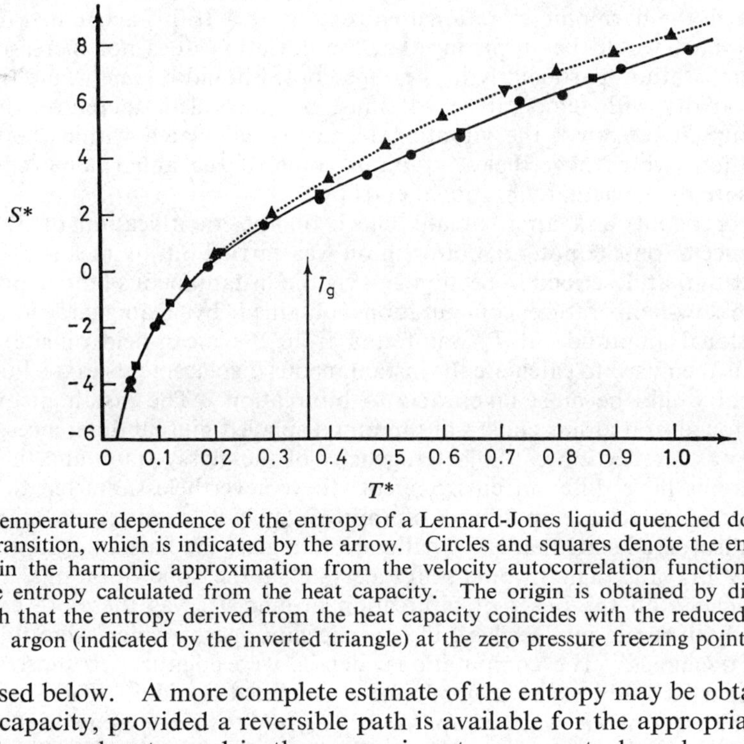

FIG. 5.—Temperature dependence of the entropy of a Lennard-Jones liquid quenched down through the glass transition, which is indicated by the arrow. Circles and squares denote the entropy calculated within the harmonic approximation from the velocity autocorrelation function. Triangles denote the entropy calculated from the heat capacity. The origin is obtained by displacing the curves such that the entropy derived from the heat capacity coincides with the reduced entropy for argon (indicated by the inverted triangle) at the zero pressure freezing point.

as discussed below. A more complete estimate of the entropy may be obtained from the heat capacity, provided a reversible path is available for the appropriate integration. The quench rate used in these experiments appears to have been sufficiently slow compared with the relaxation time for bifurcation, yet sufficiently fast compared with restructuring processes, that the quenches were pseudo-reversible, and this is supported by the minimal rate dependence found by Dnmgaard Kristensen. The entropy found in this way should contain the complete vibrational entropy plus any configurational entropy associated with the bifurcation, but no configurational terms associated with restructuring. The values obtained *via* the heat capacity, from the temperature dependence of the internal energy along the zero-pressure isobar, are indicated by the triangles in fig. 5. The origin is determined by letting the entropy coincide with the measured value,[34] shown by the inverted triangle, for liquid argon at the zero-pressure melting temperature ($T^*_{\rm m} = 0.71$, the asterisk indicating reduced units). The origin for the vibrational entropy is obtained by super-imposing both curves in the low-temperature region below $T^* = 0.2$. In the reduced-temperature range

0.5-1.0, the total entropy exceeds the vibrational entropy by a roughly constant amount lying between 0.6 and 0.8 $\boldsymbol{R}$, and this is consistent with the persistence of the postulated configurational entropy of $\boldsymbol{R}$ ln 2 arising from bifurcations of the onsite potential. The agreement is, in fact, within the limits of computational uncertainty of 0.1 k_B per atom (k_B being Boltzmann's constant). A similar result has been extracted from experimental data for liquid argon over a range of temperatures and pressures above the melting line.[35] The additional, and most significant, information provided by fig. 5 is that this entropy vanishes in the neighbourhood of the glass transition which is indicated by the arrow at $T_g = 0.36$. It is thus concluded that the bifurcations of the on-site potential vanish from the system around the glass transition and that this is indeed the cause of the fairly abrupt drop in fluidity.

An alternative interpretation of the deviation between the two curves in fig. 5 is that the entropy calculated from the heat capacity contains the anharmonic contributions which the harmonic vibrational entropy lacks. If this accounted for the entire deviation, it would be surprising that the deviation does not increase with increasing temperature, particularly in the supercooled liquid regime where the density decreases rapidly with temperature. It must be concluded, therefore, that in the density range investigated, the vibrational entropy calculated within the harmonic approximation (which nevertheless contains some of the anharmonic effects) is a good measure of the actual vibrational entropy.

An independent, and approximate, check on the identification of T_g with the disappearance of on-site potential bifurcation was carried out by means of a calculation similar to that described in Section 2 using the instantaneous atomic positions of one of the low-temperature configurations obtained by Damgaard Kristensen.[33] The vibrational amplitude at T_g was found from the mean-field on-site potential, and this was then used to calculate the instantaneous displacements in the longitudinal phonon that would be most favourable to bifurcation. The instantaneous on-site potentials for several atoms chosen at random displayed mutual differences, and this, of course, was a reflection of the heterogeneity of the glassy state, and the fact that different atoms have different environments they, nevertheless, started to bifurcate over a range of reduced temperatures of approximately 0.1, centred on T_g.

In the disordered state that is the hallmark of a glass the harmonic approximation is obviously less valid than it would be for a crystal, at the same vibrational amplitude. A further check on the onset of bifurcation around T_g was therefore carried out using a fully dynamic model glass, the atomic motions of which were followed by molecular dynamics. The computational details were identical to those used in a recently reported study of the dynamics of a simulated liquid.[17] The glass used for the static calculations was warmed up towards T_g, and the on-site potential monitored at various instants for a number of atoms chosen at random. It was found that, in the vicinity of T_g, the frequency of occurrence of bifurcations arose from essentially zero to once during a time roughly equal to the mean vibrational period.

Given this new insight into the fundamental events which occur at and around T_g, one may rationalize some of the characteristics of the glass transition. The dominant feature is of course the change in fluidity, but equally important is the abrupt decrease in thermal expansion as the liquid is cooled through T_g. In the present investigation of the density dependence of the Lennard-Jones on-site potential for both f.c.c. lattices and glasses, it was found that the local potential wells increased in diameter with decreasing density. This was initially slow, but then quite suddenly the increase became rapid in the density range where bifurcations develop. This is not a special characteristic unique to the Lennard-Jones potential, but is common to all potentials for which the attractive component has a much larger range than the repulsive part. Consequently, in the bifurcation regime, the atomic displacements which comprise a phonon increase far more rapidly with temperature than they do in the lower-temperature non-bifurcation region. The thermal expansion in the liquid thus exceeds that in the glass, and it experiences a fairly abrupt change at T_g.

5. GLASS STRUCTURE, BIFURCATION AND TWO-LEVEL SYSTEMS

One might speculate on the possible connection between bifurcation and the two-level systems[36,37] that have been suggested as the possible origin of the anomalous behaviour of the low-temperature heat capacity of glasses. The latter is characterized by a linear variation with temperature, contrasting with the third-power behaviour that is normally seen for the crystalline state. This has been seen in a variety of glasses,[38–43] and it has been suggested by Anderson[44] that the anomaly might actually be universal. In the proposed two-level systems, the states that are being excited at the low temperature are presumed to be associated with quantum-mechanical tunneling of some units of the glass which reside in double-well potentials, but the nature and origin of the latter has remained a mystery. In order to account for the magnitude of the observed anomaly, as few as 10^{-4} of such traps need be invoked. From what has been stated earlier, it is possible that the two-level systems are simply residual bifurcated on-site potentials that have become trapped in the amorphous structure.

It has been demonstrated[45] that, in the glassy state, the phonon mean free path is dominated by localized low-energy excitations. Moreoever, an approximately linear relationship exists[46] between the concentration of phonon scattering sites and the magnitude of the anomalous part of the heat capacity referred to above. Finally, it has been found that there is a pronounced correlation between the low-temperature phonon mean free path and the glass-transition temperature.[47] Taken together, these results suggest that there is an intrinsic connection between the localized low-energy excitations and those features of the amorphous structure that are responsible for determining the magnitude of T_g. This is just what one would expect if the two-level systems involved in the low-energy excitations and the structural changes occurring around T_g have their common origin in bifurcated on-site potentials.

This identification suggests that the relaxation centres proposed by Berry *et al.*[48] to explain an internal friction peak in the metallic glass Nb_3Ge may also be a manifestation of trapped-in bifurcated potentials. The relaxation centres must have an axial character which enables them to reorient with the applied stress field, and it was suggested that the relaxation is caused by the localized motion of a bound or trapped atom contained in a rather stable grouping which is itself relatively immobile. A trapped-in bifurcated potential certainly has these characteristics although the relaxation would not consist of an atom traversing the bifurcation which involves energies much lower than the observed relaxation activation energy of 0.52 eV. Rather, the relaxation must consist of a reorientation of the axis of the bifurcation by means of, for instance, a transverse displacement of an atom around the end of a bifurcated region.

The present analysis makes no predictions regarding the actual microstructure in a glass, and the validity of the model is independent of structural detail, but it is nevertheless interesting to note that the salient features of previous structural studies of the glass transition appear to be consistent with the idea presented here. Thus Wendt and Abraham[49] give an empirical criterion for locating the liquid–glass phase boundary that is based on certain features of the pair distribution function, $g(r)$. They find that when the ratio of the heights of the first maximum of $g(r)$ to that of the first minimum reaches a certain value (≈ 7.0), the glassy state is formed. This result can be interpreted in terms of the " crowding in " of a sufficient number of atoms around any given atom. This is in fact the opposite of the situation which promotes bifurcation of the on-site potential. Indeed, it would not be difficult, given a particular form of interatomic potential, to derive a criterion for the onset of the amorphous state based on features of $g(r)$.

There have been many different suggestions as to which are likely to be the most salient features of amorphous structures. Among the most familiar of these are the dense random packing,[50,51] microcrystalline[52] and dislocation[24,53–56] models. Other recent suggestions have involved mixed clusters of competing polymorphs,[57,58] disclinations[59] and mixtures of dislocations and disclinations.[60] These different possibilities are not mutually exclusive. Ninomiya[61] has, for instance, found an intimate connection between the most common holes in the dense random structure and the atomic arrangement at the cores of dislocations, and this has led to an estimate of dislocation density in a model glass.[62] Similarly, the microcrystalline model implies grain boundaries, and the latter can formally be regarded as arrays of dislocations. Indeed, it has recently been demonstrated[63] that the regions of greatest disturbance in a grain boundary have a striking similarity to the most common holes in the dense random packed model, thus establishing, in conjunction with Ninomiyas' result, a three-way connection between dense random packing, dislocations and grain boundaries. In default of evidence to the contrary, and in view of the wide variety of substances which can be prepared in the amorphous state, it seems reasonable to assume that a range of different defect structures would be identified in the atomic-scale observation of an amorphous material, if such observation were possible. Moreover, it seems likely that different amorphous materials would contain different proportions of the various defect types. From what was stated earlier, it is clear that the present approach to the glass transition would require an exhaustive analysis of each defect structure to discover at which temperature the atoms most perturbed by the defect display bifurcation of the on-site potential. This has not been attempted in the present study but, as was stated in Section 3, in a quasistatic calculation using atomic-scale data from an earlier simulation,[22] it was shown that bifurcation does set in at approximately half the melting point. This applies to the Lennard-Jones substance that is explored in the present analysis, and for which T_g is indeed approximately half of T_m. The conclusion is, therefore, that the on-set of bifurcation around T_g occurs first in the most defective part of the glass structure, irrespective of whether this resembles a grain boundary, a dislocation, or a Bernal hole.

Although there does appear to be an intimate connection between the onset of bifurcation and the location of T_g on the temperature scale, this does not provide information on the interactions between the bifurcations. These are, of course, not independent. The bifurcation of the on-site potential of a given atom is due to the movements of that atom's immediate neighbours, but these are to a certain extent influenced by the motion of the atom itself. If trapped-in bifurcations do exist at low temperatures, this would indicate the existence of a regime in which the bifurcations are isolated and independent. With increasing temperature the individual bifurcations might provoke bifurcation of the on-site potentials of neighbouring atoms, and a cascade process enabling percolation[32] might ensue around T_g. Isolated bifurcations would not be expected to influence fluidity. Conversely, the deactivation of bifurcations around T_g, with falling temperature, might ultimately cause an inverse percolation, isolation of bifurcations, and a sharp drop in fluidity.

6. THE LIQUID–GAS TRANSITION

It has been shown that the feature which distinguishes the liquid phase from both the solid and glassy phases is its possession of the entropy $\boldsymbol{R} \ln 2$. It will now be shown that the same distinction exists between the liquid and gaseous phases, namely that the liquid possesses all or part of the entropy $\boldsymbol{R} \ln 2$, while the gas does not. To understand this, one must realize that this entropy arises not just from the presence of bifurcations, but from the choice provided by the existence of a double well. In general, two distinct kinds of choice may be discerned. First, if a line of bifurcation exists in a fluid, either of the atoms at each end of the line may traverse the bifurcation which it sees along this line. Secondly, there is the choice presented to each atom as to which potential well it will occupy. In the case of a gas, the first situation does not exist and there is no choice involved in the second situation, for, following a collision between two atoms, the traversal of the potential energy barrier is inevitable and the atoms separate. One may be certain, therefore, that in spite of the ubiquity of bifurcations in the gas, this phase possesses no contributions to its configurational entropy of the form $k_B \ln 2$. As to which of the above two kinds of choice is predominant in a liquid one cannot yet say. One such choice is presented to each atom, each effective vibration period, thus resulting in a total entropy contribution of $\boldsymbol{R} \ln 2$. This appears to arise from the fact that an atom may be on either side of single-valley well when a bifurcation suddenly appears and hinders its progress across the centre of its co-ordination shell.

At the virtual bifurcation instability experienced by the superheated solid, each atom is presented with a bifurcated potential which falls away symmetrically on either side, and the configurational entropy arising from this situation is exactly $\boldsymbol{R} \ln 2$. Since the average first coordination shell for the liquid near the freezing point does not differ greatly from that of the solid at the melting point, the configurational entropy should remain somewhere near the value $\boldsymbol{R} \ln 2$. Inevitably, if the density were to be decreased, the ultimate fate of this entropy would be to diminish and ultimately disappear. Nevertheless, the experimental data for argon and the molecular dynamics data shown in fig. 5 reveal that this configurational entropy remains constant over a rather extended range of temperature and density and for the 100 K isotherm for argon, up to the boiling point.[35] Thus, at least for the 100 K isotherm, the entropy, $\boldsymbol{R} \ln 2$, is lost during the liquid to gas transition. At the same time, both the internal energy and enthalpy rise, and one might ask how it is that the transition is first order with no change in Gibbs function. The additional feature which ensures the first-order behaviour is the appearance of part if not all of the communal entropy, $\boldsymbol{R}$. Thus, in principle an entropy increase of at least $\boldsymbol{R} - \boldsymbol{R} \ln 2 = 0.3\, \boldsymbol{R}$ is achievable, and because a volume change accompanies the transition, there is a further volume-dependent increase in entropy.

The above distinction between liquid and gas is well defined in the neighbourhood of the triple point where the two fluid phases have widely different densities, but the distinction is rather less certain in the region closer to the critical point where liquid and gas merge. Nonetheless, in the belief that it is a natural consequence of the ideas developed thus far, it can be tentatively proposed that the distinction extends to the critical point. With decreasing liquid density, the traverse of bifurcated potentials becomes increasingly inevitable, and the bifurcation entropy must diminish. There exists, therefore, a line running over the pVT fluid surface at which the bifurcation entropy just vanishes and along which liquid and gas are indistinguishable. It is consequently suggested that this line cuts the liquid–gas coexistence curve at the critical point, and it will be demonstrated below that such an interpretation may be placed on the entropy–volume isotherms near the critical point for fluid argon.

Fig. 6 shows entropy as a function of volume along the 150.7 K isotherms for solid,[6] liquid[64] and gaseous[65] argon. The isotherms are the heavy lines denoted S, L and G, respectively. The isotherm lies extremely close to the critical isotherm of 150.9 K, and consequently, the liquid gas coexistence region is very narrow. It is vital for the construction described below to note that the extrapolated liquid and gas isotherms meet with a discontinuous change of slope. At the extremes of low and high volume, the curves have a well defined asymptotic behaviour, which in the solid

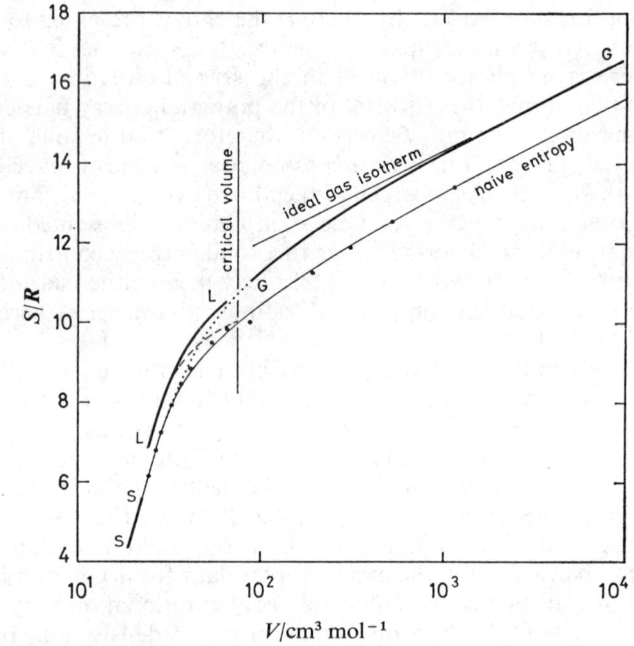

FIG. 6.—Entropy (heavy lines) of solid (S–S), liquid (L–L), and gaseous (G–G) argon as a function of molar volume along the 150.7 K isotherm. The naive entropy is obtained by subtracting **R** from the entropy for the gas and **R** ln 2 from the liquid entropy. Some of the displaced data points are shown. The naive entropy is the single smooth curve which just osculates these new data points and the solid curve. The divergence between the extrapolated entropy of the gas and the naive entropy describes the acquisition of communal entropy. Subtracting this acquired entropy from the entropy of the liquid yields the dashed curve, and the divergence between this and the entropy of the liquid describes the depletion of bifurcation entropy which is complete at the critical volume.

and high density liquid is governed by a linear dependence of entropy on volume[35] with gradient equal to the product of the volume thermal expansion coefficient, α, and the isothermal bulk modulus, B_T. Thus

$$S - S_0 = \alpha B_T(V - V_0).$$

The other asymptote is the perfect-gas behaviour which is delineated in the figure and satisfies

$$S - S_0 = \boldsymbol{R} \ln (V/V_0).$$

It will be assumed that the entropy may comprise certain volume-independent configurational components and a well defined volume-dependent part which we will call the naive entropy and which has essentially the same form irrespective of the phase. To obtain the naive entropy, **R** is subtracted from the entropy for the gas and **R** ln 2 from the entropy for the liquid, while the solid entropy is left unchanged, and one then constructs a single smooth curve which osculates the three resultant curves. This construction is shown in the figure. The displaced data points are shown as dots, and the naive entropy depicted by the dashed curve follows a smooth course from the linear dependence on volume at high densities, to a logarithmic dependence on volume at low densities.

A smooth interpolation between the actual isotherm for the gas and the high-density naive isotherm results in the dotted line shown, and the divergence between these two curves describes the progressive acquisition of the communal entropy over an approximately eight-fold increase in volume. If the proportion of communal entropy is now subtracted from the actual liquid isotherm, the dashed line results, and its convergence on the naive isotherm describes the depletion of the bifurcation entropy, which is complete at the critical volume shown by the vertical line. This is not a proof (as the construction is subject to uncertainty), but it lends credence to the notion that the bifurcation entropy vanishes at the critical point. The depletion of bifurcation entropy and acquisition of communal entropy as depicted in fig. 6 is entirely reasonable. The two processes commence at about the same density, and this is consistent with the expectation that when confinement to bifurcations begins to reduce, so accessibility to volumes outside the coordination shell begins to increase. On the other hand, at lower density, the former process must be complete before the latter process is, and indeed these two events occur at densities which differ by a factor of four.

7. QUANTUM MELTING AND THE SUPERFLUID PHASE OF He^4

In the previous section it has been shown that bifurcation is the hallmark of the liquid state, and that it is responsible both for the volume-independent entropy, $\boldsymbol{R} \ln 2$, which appears on melting and for a liquid's fluidity. In simple systems that behave in a classical (*i.e.*, Boltzmann) manner, the entropy $\boldsymbol{R} \ln 2$ can be lost by the liquid in any of three ways: by crystallization upon freezing; by becoming a glass at sufficient supercooling; or by evaporation to the gaseous state. For the quantum liquid, He^4 below a pressure of 25 atm, there is a fourth way of losing the bifurcation entropy: by transforming to the superfluid state below the λ-point.

Evidence that the entropies of the crystal and normal liquid forms of He^4 differ by about $\boldsymbol{R} \ln 2$ has, in fact, long been available. Dugdale[66] found that the difference was 0.7 $\boldsymbol{R}$ ($=\boldsymbol{R} \ln 2$) at 5 K, while just above the λ-line it is approximately 0.6 $\boldsymbol{R}$. Below the λ-line, the entropy difference rapidly falls to zero.

As shown earlier for argon,[35] the excess configurational entropy of the liquid phase over that of the solid phase may be obtained by subtracting out the volume-dependent entropy along isotherms which traverse the melting line. In fig. 7 entropy is plotted as a function of molar volume (full data points) for various isotherms between 5 and 20 K for both solid and liquid He^4. Anticipating that the liquid phase for He^4 will possess the bifurcation entropy while the solid does not, $\boldsymbol{R} \ln 2$ is now subtracted from the liquid isotherms to obtain the open data points which now extrapolate for 9, 15 and 20 K precisely to the solid data points. These new curves are the naive entropy curves described earlier with local gradient approximately equal to the product of the volume thermal expansion coefficient and the isothermal bulk modulus, and the construction confirms the presence of the bifurcation entropy in liquid He^4 I. For isotherms below 7 K the extrapolation indicates that the bifurcation entropy is progressively surrendered. At 7 K it is 0.6 $\boldsymbol{R}$, at 5 K it is 0.5 $\boldsymbol{R}$, and below the λ-line it rapidly falls to zero.

Further support for these ideas is presented in fig. 8 which shows three isobars (full lines) of entropy as a function of temperature which pass through the λ transition, located at the knees in these curves. The dotted curves are obtained by subtracting $R \ln 2$ from the full curves above the λ points, and they are seen to fall to zero smoothly. Thus, commencing from just above the λ transition, the bifurcation entropy is progressively lost as the proportion of normal fluid decreases. He^4 II is therefore anomalous in that it is a liquid but lacks this entropy. In view of the earlier identification of bifurcation entropy with a liquid, and of bifurcation as the cause of its fluidity,

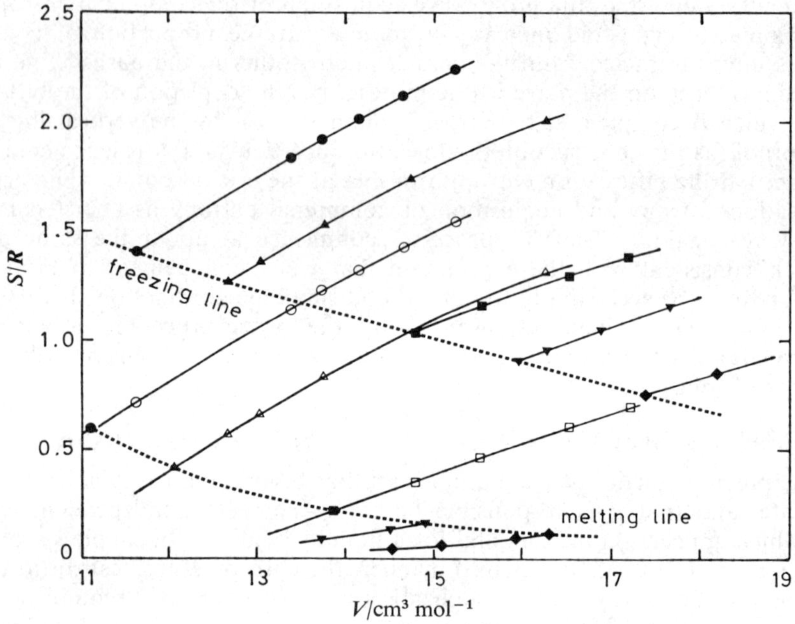

FIG. 7.—Isotherms of molar entropy as a function of molar volume (full data points) for solid and liquid He^4 I. The data are taken from ref. (67)-(70). For the 9, 15 and 20 K isotherms the entropy, $R \ln 2$, is subtracted from the actual entropy data giving the open symbols which extrapolate almost exactly to the data points for the solid: ●, 20; ▲, 15; ■, 9; ▼, 7 and ♦, 5 K.

the anomalous behaviour of He^4 II raises a fundamental question: is it a non-bifurcated liquid or is it bifurcated yet for some reason lacking bifurcation entropy? It will be argued that it is the latter and that this is important for the understanding of superfluidity.

Helium is a liquid at atmospheric pressure because the zero-point energy is so large[72] ($\approx$70 cal mol^{-1}) compared with the enthalpy of melting[72] ($\approx$0.5 cal mol^{-1}). It was earlier shown that a short-wavelength longitudinal optical phonon provides the bifurcation of the on-site potential, in argon, and this provokes shear instability of a transverse phonon. At such a low temperature compared with the Debye temperature[72] (24 K), only long-wavelength phonons can be excited in crystalline He^4 below the λ-point,[73,74] so thermal vibrations cannot contribute appreciable local instantaneous dilation. Instead the dilation is provided by the zero-point positional uncertainty of the neighbours which for crystalline He^4 at the melting point is of the order of 0.3 nearest-neighbour distances.[75] This will result in a roughly spherically-symmetric broadening of the potential wells, but provided the wells contain no negative curvature, $\Psi\Psi^*$ (where Ψ is the ground-state wave function for an atom), maximizes at the mean position. This is the situation which in all likelihood obtains

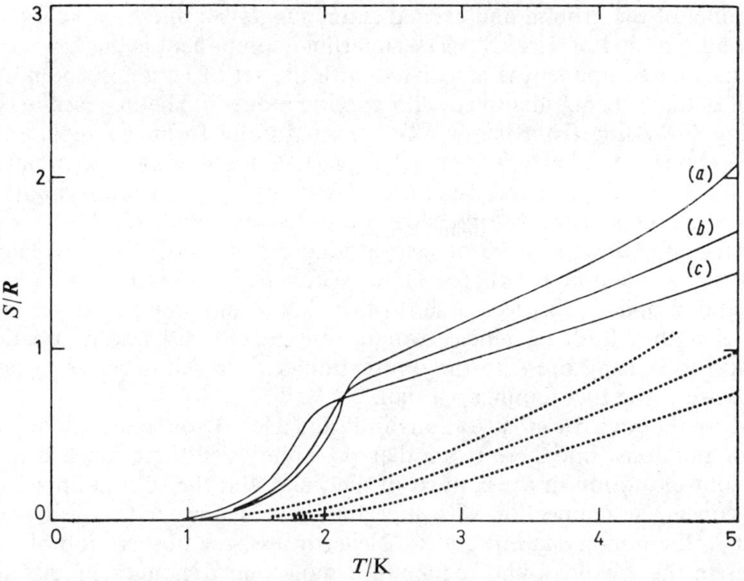

FIG. 8.—Isobars of molar entropy as a function of temperature (full lines) for liquid He^4 through the λ-transition. The dashed lines are obtained by subtracting $\boldsymbol{R}$ ln 2 from the full curves, and these fall smoothly to zero without anomaly. The data are taken from ref. (67) and (71). (*a*) 2.5×10^5 (*b*) 10×10^5 and (*c*) 25×10^5 Pa.

in the crystal. With decreasing density the potential will ultimately bifurcate, but with spherical symmetry, and $\Psi\Psi^*$ no longer maximizes at the mean position. It is suggested that this condition is appropriate to the superfluid state. For a two-dimensional system the bifurcated on-site potential will have circular symmetry and resemble a Mexican hat, while the ground-state $\Psi\Psi^*$ is toroidal. These geometries have obvious counterparts in three dimensions; and we denote by the symbol Σ such ground-state wave functions which are circularly or spherically symmetric and for which $\Psi\Psi^*$ does not maximize at the centre.

If superfluid He^4 is a bifurcated liquid, its lack of entropy would seem to raise doubt about the universal validity of the bifurcation theory. However, even for classical systems, a bifurcated on-site potential will automatically imply the entropy $\boldsymbol{R}$ ln 2 only if that bifurcation is axial and thus temporarily confines individual atoms to one of the two potential wells. If the bifurcation were spherically symmetric, there would be no confinement since the bifurcation barrier could be circumvented, and the bifurcation entropy would be absent. Returning to the quantum domain, this argument indicates that, because $\Psi\Psi^*$ corresponding to the ground-state Σ wave function for liquid He^4 below the λ-line is a three-dimensional torus, it possesses no bifurcation entropy. Because it seems reasonable to assume that the radial part of the wave function will not be substantially altered, excitations will influence the angular component and will locally break the spherical Σ symmetry, forcing upon the potential an axial bifurcation for which the wave function is polarized, having a form which, using the convention for the hydrogen-like electron wave function, is *p*-like. Such excited states will be called Π states. The polarization of this wave function gives an entropy k_B ln 2 for each such excitation, and the wave function possesses angular momentum. According to the arguments of Feynman,[73] for a Bose–Einstein system, this angular momentum is indistinguishable from the angular momentum associated with long wavelength rotation, and is therefore dissipative.

This model of the ground and excited states suggests a microscopic interpretation of the two-fluid model of Tisza.[76] The superfluid component is the Σ wave function, while the normal component is associated with the set of Π states, the normal component not actually being identified with specific atoms. The increase in density on approaching the λ-line from below[77] follows naturally from the replacement of Σ states by Π states. Similarly, the model suggests a microscopic interpretation of the excitations observed, for wavenumber ≈ 2 Å^{-1}, by neutron scattering.[78–81] They have been attributed to the roton modes proposed by Landau,[82–84] and Feynman[85] has suggested that they must be of near-atomic dimensions. It is evident that the rotons are to be identified with the Π wave functions. With increasing pressure, Henshaw and Woods[78] observed a shift of the roton minimum to lower energy and lower wavelength. Both tendencies would be expected with the bifurcation model because with increasing pressure the bifurcation barrier will decrease in height, and distance between the two minima will fall.

Regarding the non-roton excitations in liquid He^4 II, these are usually described in terms of phonons, but there is another possibility. Bifurcation has earlier been linked to soliton motion in the context of melting,[3] and the soliton appears to be an attractive concept in connection with superfluidity. These excitations have not been looked for in the non-crystalline state. Nevertheless, the observation of soliton-like heat pulses in the second-sound regime of a molecular dynamics simulated classical system[86] is suggestive of a connection with superfluidity. Equally interesting, in this respect, is the discovery of soliton-like loss-free motion of excitations in a linear chain.[87,88]

The nature of the bifurcated state contains the essential distinction between superfluid He^4 and both the ideal Bose–Einstein gas and crystalline He^4. The Bose–Einstein gas has a degenerate condensate phase which is not superfluid, and this is a natural consequence of the absence of interactions and hence of bifurcations. In crystalline He^4, the non-isotropic symmetry precludes the existence of a spherically symmetric Σ wave function, and the ground-state wave function both possesses angular momentum and cannot extend through the entire medium, as it does in the superfluid. It seems likely that at low temperatures, crystalline He^4 possesses some localized bifurcations. This would explain the apparent decrease in bifurcation entropy for liquid He^4 I which, as shown in fig. 7, commences at about 7 K, well above the λ-transition. The interpretation of this is that the bifurcation entropy for liquid He^4 I is not being depleted, but rather is being acquired by the crystal. This is consistent with the observation of Minkiewicz *et al.*[89] of single-particle excitations in crystalline He^4 which are identical in kind to those found in liquid He^4 II. The " well-defined groups " for which the energy transfer is very high may be identified with excitations over bifurcation barriers.

We are grateful to the Danish Natural Science Research Council (R. M. J. C.), the New Zealand Department of Scientific and Industrial Research, and the Technical University of Denmark (J. L. T.) for financial support.

1 S. M. Stishov, I. N. Makarenko, V. A. Ivanov and A. M. Nikolaenko, *Phys. Letters*, 1973, **45A**, 18.
2 J. L. Tallon, W. H. Robinson and S. I. Smedley, *Nature*, 1977, **266**, 337.
3 R. M. J. Cotterill, *Physica Scripta*, 1978, **18**, 37.
4 R. M. J. Cotterill, *J. Cryst. Growth*, 1980, **48**, 582.
5 J. L. Tallon, W. H. Robinson and S. I. Smedley, *Phil. Mag.*, 1977, **36**, 741.
6 J. L. Tallon, *Phil. Mag.*, 1979, **39**, 151.
7 G. Tammann, *Z. phys. Chem.*, 1910, **68**, 205.

[8] J. Klæstrup Kristensen and R. M. J. Cotterill, *Phil. Mag.*, 1977, **36**, 437.
[9] G. Mair, K. Hamacher and H. Wenzl, *Z. Phys.*, 1976, **B24**, 301.
[10] G. Fritsch, in *Anharmonic Lattices, Structural Transitions and Melting*, ed. T. Riste (Noordhoff, Leiden, Netherlands, 1974), p. 441.
[11] M. Lasocka, *Phys. Letters*, 1975, **51A**, 137.
[12] L. V. Woodcock and K. Singer, *Trans. Faraday Soc.*, 1971, **67**, 12.
[13] G. L. Pollack, *Rev. Mod. Phys.*, 1964, **36**, 748.
[14] R. K. Crawford and W. B. Daniels, *Phys. Rev. Letters*, 1968, **21**, 367.
[15] B. J. Alder and T. Wainwright, *J. Chem. Phys.*, 1959, **31**, 459.
[16] A. Rahman, *Phys. Rev.*, 1964, **136**, A405.
[17] R. M. J. Cotterill, *Phys. Rev. Letters*, 1979, **42**, 1541.
[18] K. Sköld, J. M. Rowe, G. Ostrowski and P. D. Randolph, *Phys. Rev.*, 1971, **A6**, 1107.
[19] M. Tanaka, Y. Fukui, and S. Takeuchi, *J. Jap. Inst. Metals*, 1973, **37**, 907.
[20] S. Mizushima, *J. Phys. Soc. Japan*, 1960, **15**, 70.
[21] A. Ookawa, *J. Phys. Soc. Japan*, 1960, **15**, 2191.
[22] D. Kuhlmann-Wilsdorf, *Phys. Rev. A*, 1965, **140**, 1599.
[23] J. M. Kosterlitz and D. J. Thouless, *J. Phys. C*, 1973, **6**, 118.
[24] R. M. J. Cotterill, E. J. Jensen, W. Damgaard Kristensen, R. Paetsch and P. O. Esbjørn, *J. Physique*, 1975, **36**, C2–35.
[25] R. M. J. Cotterill and L. B. Pedersen, *Solid State Comm.*, 1972, **10**, 439.
[26] R. M. J. Cotterill, E. J. Jensen and W. Damgaard Kristensen, *Phil. Mag.*, 1974, **30**, 245.
[27] R. M. J. Cotterill, *Phil. Mag.*, 1975, **32**, 1283.
[28] R. M. J. Cotterill and M. Doyama, *Phys. Rev.*, 1966, **145**, 465.
[29] J. L. Tallon, *Phys. Rev. B*, to be published.
[30] M. Ross, *Phys. Rev.*, 1969, **184**, 233.
[31] S. Toxværd, *J. Chem. Phys.*, 1978, **69**, 4750.
[32] R. M. J. Cotterill, *Proc. NATO Conf. Liquid and Amorphous Metals*, Zwiesel, 1979, ed. E. Lüscher (to be published).
[33] W. Damgaard Kristensen, *J. Non-cryst. Solids*, 1976, **21**, 303.
[34] R. K. Crawford, W. F. Lewis and W. B. Daniels, *J. Phys. C*, 1976, **9**, 1381.
[35] J. L. Tallon, *Phys. Letters*, in press.
[36] W. A. Phillips, *J. Low Temp. Phys.*, 1972, **7**, 351.
[37] P. W. Anderson, B. I. Halperin and C. M. Varma, *Phil. Mag.*, 1972, **25**, 1.
[38] O. L. Anderson, *J. Phys. Chem. Solids*, 1959, **12**, 41.
[39] R. S. Craig, C. W. Massena and R. M. Mallya, *J. Appl. Phys.*, 1965, **36**, 108.
[40] C. L. Choy, R. B. Hunt and G. L. Salinger, *J. Chem. Phys.*, 1970, **52**, 3629.
[41] R. C. Zeller and R. O. Pohl, *Phys. Rev.*, 1971, **B4**, 2029.
[42] R. B. Stephens, G. S. Cieloszyk and G. L. Salinger, *Phys. Letters*, 1972, **38A**, 215.
[43] R. B. Stephens, *Phys. Rev. B*, 1973, **8**, 2896.
[44] A. C. Anderson, *Proc. NATO Conf. Liquid and Amorphous Metals*, Zwiesel, 1979, ed. E. Lüscher (to be published).
[45] T. L. Smith, P. J. Anthony and A. C. Anderson, *Phys. Rev.*, 1978, **B17**, 4997.
[46] J. R. Matey and A. C. Anderson, in *Thermal Conductivity*, ed. P. G. Klemens and T. K. Chu (Plenum, New York, 1976), vol. 14, p. 31.
[47] C. L. Reynolds, *J. Non-cryst. Solids*, 1979, **30**, 371.
[48] B. S. Berry, W. C. Pritchet and C. C. Tsuei, *Phys. Rev. Letters*, 1978, **41**, 410.
[49] H. R. Wendt and F. F. Abraham, *Phys. Rev. Letters*, 1978, **41**, 1244.
[50] J. D. Bernal, *Proc. Roy. Soc. A.*, 1964, **280**, 299.
[51] J. L. Finney, *Nature*, 1977, **266**, 309.
[52] D. Turnbull and D. E. Polk, *J. Non-cryst. Solids*, 1972, **8**, 19.
[53] J. J. Gilman, *J. Appl. Phys.*, 1973, **44**, 675.
[54] M. F. Ashby and J. Logan, *Scripta Metallurg.*, 1973, **7**, 513.
[55] J. C. M. Li, in *Frontiers in Materials Science*, ed. L. E. Murr and C. Stein (Marcel Dekker, New York, 1976), p. 527.
[56] S. F. Edwards and M. Warner, *Phil. Mag.*, in press.
[57] C. H. L. Goodman, *Nature*, 1975, **257**, 370.
[58] R. Wang and M. D. Mertz, *Nature*, 1976, **260**, 35.
[59] N. Rivier, *Phil. Mag.*, in press.
[60] R. C. Morris, *J. Appl. Phys.*, 1979, **50**, 3250.
[61] T. Ninomiya, in *The Structure of Non-Crystalline Materials*, ed. P. H. Gaskell (London, Taylor and Francis, 1977), p. 45.
[62] R. Yamamoto and M. Doyama, *J. Phys. F.*, 1979, **9**, 617.
[63] M. F. Ashby, F. Spaepen and S. Williams, *Acta Metallurg.*, 1978, **26**, 1647.

[64] R. K. Crawford and W. B. Daniels, *J. Chem. Phys.*, 1969, **50**, 3171.
[65] A. Michels, J. M. Levels and G. J. Wolkers, *Physica*, 1958, **24**, 769.
[66] J. S. Dugdale, in *Physics of High Pressure and the Condensed Phase*, ed. G. Itterbeck (North Holland, Amsterdam, 1965), p. 382.
[67] R. D. McCarty, *J. Phys. Chem. Ref. Data*, 1973, **2**, 923.
[68] G. Ahlers, *Phys. Rev. A*, 1970, **2**, 1505.
[69] J. S. Dugdale and J. P. Franck, *Phil. Trans. A*, 1974, **257**, 1076.
[70] E. R. Grilly and R. L. Mills, *Ann. Phys.*, 1959, **8**, 1.
[71] J. S. Brooks and R. J. Donnelly, *J. Phys. Chem. Ref. Data*, 1977, **6**, 51.
[72] J. Wilks, *The Properties of Liquid and Solid Helium* (Clarendon Press, Oxford, 1967), p. 267.
[73] R. P. Feynman, *Phys. Rev.*, 1953, **91**, 1301.
[74] R. P. Feynman, *Phys. Rev.*, 1954, **94**, 262.
[75] C. Domb and J. S. Dugdale, *Progr. Low Temp. Phys.*, 1957, **2**, 338.
[76] L. Tisza, *Phys. Rev.*, 1947, **72**, 838.
[77] H. Kamerlingh Onnes and J. D. A. Boks, Rept. Comm. 4th International Cong. Refrig. London, 1924; Leiden Communication Number 170a.
[78] D. G. Henshaw and A. D. B. Woods, *Proc. 7th Int. Conf. Low Temp. Phys.*, ed. G. M. Graham and A. C. Hollis Hallet (Toronto University Press, Toronto, 1961), p. 539.
[79] O. W. Dietrich, E. H. Graf, C. H. Huang and L. Passell, *Phys. Rev. A*, 1972, **5**, 1377.
[80] A. D. B. Woods and R. A. Cowley, *Rep. Progr. Phys.*, 1973, **36**, 1135.
[81] E. H. Graf, V. J. Miniewicz, H. Bjerrum Møller and L. Passell, *Phys. Rev. A*, 1974, **10**, 1748.
[82] L. D. Landau, *J. Phys. U.S.S.R.*, 1941, **5**, 71.
[83] L. D. Landau, *J. Phys. U.S.S.R.*, 1944, **8**, 1.
[84] L. D. Landau, *J. Phys. U.S.S.R.*, 1947, **11**, 91.
[85] R. P. Feynman, *Progr. Low Temp. Phys.*, 1957, **1**, 17.
[86] T. Schneider, E. Stoll and Y. Hiwatari, *Phys. Rev. Letters*, 1977, **39**, 1382.
[87] Y. Y. Earmme and J. H. Weiner, *Phys. Rev. Letters*, 1974, **33**, 1550.
[88] N. Flytzanis, S. Crowley and V. Celli, *Phys. Rev. Letters*, 1977, **39**, 891.
[89] V. J. Minkiewicz, T. A. Kitchens, G. Shirane and E. B. Osgood, *Phys. Rev. A*, 1973, **8**, 1513.

Theory of Phase Transitions in Diluted Systems with Competing Interactions

By Kurt Binder, Wolfgang Kinzel and Stephane Sarbach

IFF, KFA Jülich, D-5170 Jülich, Postfach 1913, W. Germany

and Dietrich Stauffer

Theoretische Physik, Universität Köln, D-5000 Köln 41,
Zülpicher Str. 77, W. Germany

Received 3*rd December*, 1979

As a simple model of a diluted system with competing interactions, we first consider an Ising spin system on the square lattice with nearest-neighbour interaction J_1 positive, next-nearest-neighbour interaction J_2 negative, and where a fraction $1-x$ of the spins is removed at random. Systematic expansions in x or $1-x$ as well as Monte Carlo calculations are used to investigate the ordering of this model system. It is shown that for broad ranges of x and J_2/J_1 a spin-glass phase occurs, which consists of (anti-) ferromagnetic clusters coupled together by partially " frustrated " bonds, giving rise to a high ground-state degeneracy. Then we briefly discuss various analytical methods which can be used to calculate spin-glass properties, and compare them with our numerical results. Also other related models are mentioned. Finally we discuss to what extent these models can be applied to randomly mixed molecular crystals.

---

1. INTRODUCTION AND OVERVIEW

This paper considers the effects of dilution on order–disorder phase transitions in solids. We first identify a degree of freedom of atoms or (atomic groups) which plays the role of a local order parameter φ_i of the transition: in a magnet, φ_i is the magnetic moment μ_i of the atom at lattice site i; in a ferroelectric or a uniaxial structural transition, φ_i is a pseudospin σ_i describing the orientation of the ith molecular group (this also applies to molecular crystals like NH_4Cl);[1] in molecular crystals like KCN[2] or solid H_2[3] φ_i has tensor character. Keeping the set $\{\varphi_i\}$ fixed and integrating out all other (lattice-dynamical) degrees of freedom, one is left with some effective interaction V_{ij} between φ_i, φ_j, which is responsible for the ordering. In our study of dilution effects, we only consider a very simplified case: (i) suppose atoms (or molecules) with an order parameter φ_i are *randomly* replaced by other ones without local order parameter; (ii) suppose that $V_{ij} = 0$ if a site $i(j)$ is such a " defect site "; (iii) suppose that other V_{ij} (in the vicinity of defect sites) remain unchanged (*i.e.*, one ignores the change of V_{ij} due to local lattice distortions around defect sites, *etc.*). These assumptions are reasonable for ferromagnets like EuS diluted with Sr, or molecular crystals like $pH_2 + oH_2$ mixtures;[3,4] it is not so clear whether they also hold for KCN diluted with Br,[5] or NH_4Cl diluted with Cs.[6] However, this simple case must first be understood before more complicated ones can be tackled.

Effects of dilution are then very different depending on whether the interactions V_{ij} are " *competing* " or not: in the latter case, the mutual " orientation " of any pair,

φ_i, φ_j in the classical ground state (quantum effects are ignored here throughout) of the ordered phase is the *same* as that which would be produced by V_{ij} *alone*, while in the former case for *some* pairs the orientation would be *different*. An example for non-competing interactions is a ferromagnet with only nearest-neighbour exchange $J_2 > 0$. Dilution effects then are simple: the ordered state remains ferromagnetic for concentrations x of the magnetic atoms exceeding the nearest-neighbour percolation threshold x_p^{nn}, at which also the critical temperature $T_c(x)$ vanishes.[7] If there is also a next-nearest neighbour exchange $J_2 > 0$ (like in EuO[8]), this occurs at the (smaller) next-nearest neighbour percolation threshold x_p^{nnn},[9] *etc.* In any case for $x < x_p$ one has only finite clusters of the φ_i linked together by the V_{ij}, while for $x > x_p$ an infinite network appears which hence can carry an ordered phase.

A simple example for competing interactions is a magnet with $J_1 > 0$, $J_2 < 0$ [like EuS[8]]. Then dilution destroys the ordering much more rapidly: although on the f.c.c. lattice $x_p^{nnn} = 0.136$, $x_p^{nn} = 0.198$,[7] ferromagnetism in $Eu_xSr_{1-x}S$ breaks down even at $x_c = 0.51$.[4] In $oH_2 + pH_2$ mixtures ordering breaks down at $x_p \cong 0.55$.[3,10] In the regime $x_p < x < x_c$ one observes at low temperatures a " spin-glass " phase[4] or a " molecular-glass " phase,[3] respectively. It is the aim of the present paper to study these phases and the transitions between them. Attention will be paid to understanding the structure of the glass-like phases as well as the diluted but ordinarily ordered phases (Section 2). It will be shown (Section 3) that even in the simplest cases analytical techniques (based on the replica method[11] together with mean-field or Bethe methods)[12] are unsatisfactory, and one has to resort to complicated numerical techniques (Monte Carlo simulations, high-temperature and low-concentration expansions, *etc.*[13]), in order to obtain reliable results. Even then the questions concerning the description of the glass-like " phases " in terms of order parameters and their correlations need further clarification (Section 4).

While in spin-glasses[14] competing interactions often arise because the sign of the exchange oscillates with distance (*i.e.*, competition between ferro- and antiferromagnetic spin alignment), we emphasize that one may also have competing interactions of *uniform sign*: *e.g.*, in magnets with $J < 0$ on certain lattices part of the interactions remain " frustrated "[15] (*i.e.*, unfavourable) even in the ground state. This happens with only nearest-neighbour interactions J_p for the triangular and f.c.c. lattices; but on d-dimensional hypercubic (square, s.c., *etc.*) lattices $J_2 < 0$ is also " frustrated " in the antiferromagnetic structure, and for $J_2 = J_2^m = J_1(d-1)$ this ground state becomes degenerate with a layered antiferromagnetic structure. Although for $J_2^m < J_2 < 0$ the energy of this structure is higher in the pure case, random impurities may locally stabilize very small irregular domains of this structure (see Section 2). Similar competition effects may occur in molecular crystals, too: *e.g.*, in KCN the CN^- molecules order below $T = 168$ K (where the elastic constant c_{44} softens) along the b-axis of the orthorhombic low-temperature phase, while other orderings would have only slightly higher ground-state energies [as seen from the components of the " susceptibility tensor " derived in ref. (2)], and act as " competing orderings " in a similar sense as in the simpler cases considered here.

2. A " SIMPLE " MODEL: THE DILUTED ISING SQUARE LATTICE

We consider a square lattice of Ising spins ($S_i = \pm 1$) and interactions $J_1 > 0$, $J_2 < 0$ with $J_2 > J_2^m$ $(= -\frac{1}{2})$, *i.e.*, the pure system has a ferromagnetic ground state. When it is randomly diluted, it happens that some spins (fig. 1(*a*)] or cluster of spins [fig. 1(*b*)] close to dilution sites must align antiparallel to the ferromagnetic network, because they experience many more negative than positive bonds. Also degenerate configurations of spins result [fig. 1 (c_1) and (c_2)]. We may consider the latter case as an antiferromagnetic cluster of two spins, which due to frustration and disorder is effectively split off from the ferromagnetic network (*i.e.*, magnetic correlations involving one spin of the cluster and otherwise spins of the network vanish). As dilution proceeds, more and more clusters are turned antiparallel or are split off from the ferromagnetic network, until utimately there are no longer infinite ferromagnetic correlations (although still most of the spins belong to an " infinite cluster " of spins linked together by interactions). This breakdown of order depends sensitively on $R = J_2/J_1$, since one must count numbers of positive and negative bonds at the surface of clusters to decide whether energy is gained by overturning them.[13] Hence also for

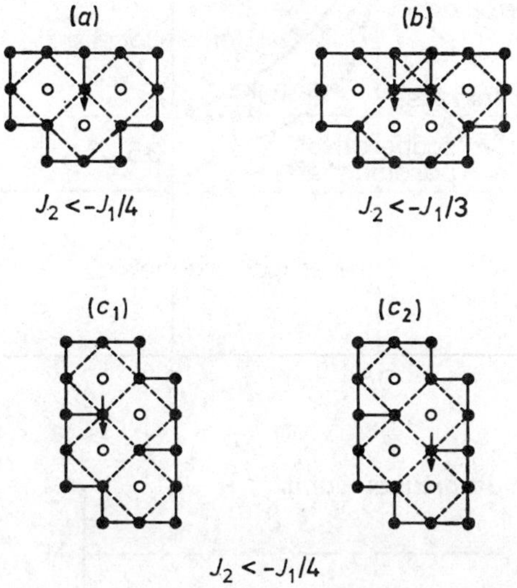

FIG. 1.—Reversed spins in the vicinity of nonmagnetic atoms (open circles) occurring in the ferromagnetic region of a diluted system of magnetic atoms (black dots). Nearest-neighbour bonds $J_1 > 0$ are indicated by full and next-nearest neighbour bonds $J_2 < 0$ by broken bonds.[13]

rational values of R, additional degeneracies occur due to " loose " clusters in the network. By systematic counting of all possibilities of vacancy clusters of 3 sites [fig. 1(*a*)], 4 sites [fig. 1(*b*) and (*c*)], *etc.*, one obtains a series expansion for the magnetization in powers of the vacancy concentration $1 - x$. This series is expected to become singular at the concentration $x_c''(R)$ where the ferromagnetic long-range order breaks down. Unfortunately the coefficients of this series depend on R in a highly nonanalytic manner, due to the degeneracies mentioned above, and hence the series is hard to analyse. The same holds for the complementary approach starting in the dilute regime (small x), where one derives a systematic expansion for the ferromagnetic susceptibility $k_B T \chi_F = \sum_{i,j} \langle S_i S_j \rangle_T / N$ in powers of x: for the kth power of x in this series one has to collect the contribution of all clusters containing precisely k magnetic atoms. Similarly, one can define a susceptibility $k_B T \chi_{SG}$ sensitive to spin-glass ordering,

$$k_B T \chi_{SG} = \sum_{ij} \langle S_i S_j \rangle_0 \langle S_i S_j \rangle_T / N, \tag{1}$$

and again derive a series expansion in powers of x for [details see ref. (13)]. Since χ_{SG} is sensitive to any kind of correlation between the spins, not only to ferromagnetic alignment, one expects that χ_{SG} diverges at a low concentration $x_c'(R) < x_c''(R)$, and hence the ordering in the regime $x_c'(R) < x < x_c''(R)$ is identified as spin-glass ordering. Fig. 2(*a*) shows qualitatively the ground-state phase diagram expected from these considerations, and fig. 2(*b*) shows that our series estimates confirm this expectation. These estimates were obtained from Padé approximants to series containing terms up

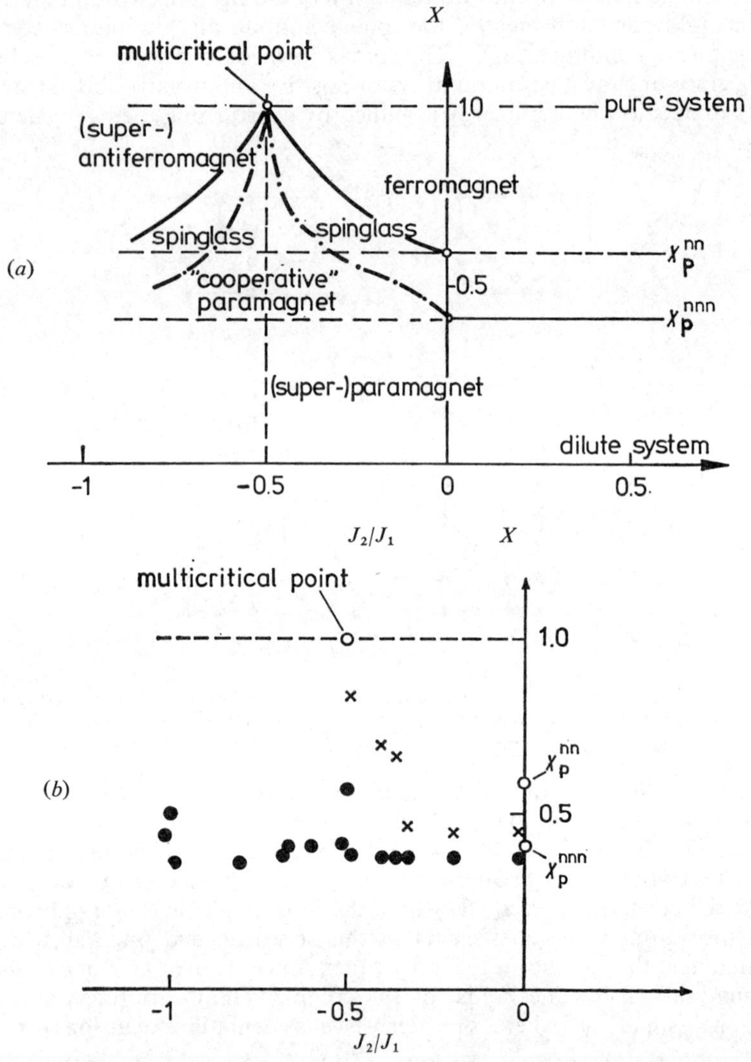

FIG. 2.—Ground-state phase diagram of the diluted Ising square, schematic (*a*) and as obtained from Padé approximants (*b*). Full curves (and crosses) denote x_c'' while dash-dotted curves (and dots) denote x_c'.[13]

to x^9.[13] Since for each configuration of magnetic atoms one has to identify all ground states of the cluster and then calculate the average spin correlation for each pair of spins to obtain the weight of the cluster for the series expansion of eqn (1), time consuming computer programs are needed and an extension to higher order is not straightforward. It would also be hard to extend this method to 3-dimensional systems, because then many more cluster configurations occur. In the 3-dimensional case we found it more convenient to apply Monte Carlo simulation methods. Fig. 3 shows ground-state results for the case of the classical Heisenberg f.c.c. lattice (which models the $Eu_xSr_{1-x}S$ system, as mentioned above). In this case $J_2^m = -J_1$. From fig. 3(*a*) we note that the ground-state magnetization steadily decreases with increasing dilution [since only a finite system of about 10^3 spins is treated, there is some residual contribution to the root-mean-square magnetization also in the paramagnetic and spin-glass phases. The breakdown of ferromagnetism at $x_c''(R)$ is hence identified from

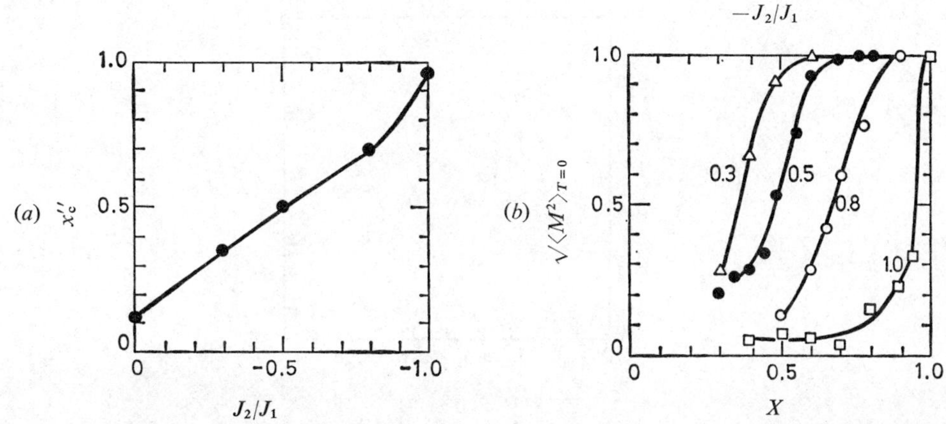

FIG. 3.—(*a*) Ground-state magnetization of classical f.c.c. Heisenberg ferromagnets plotted against concentration for various values of J_2/J_1. (*b*) Resulting variation of x_c' with J_2/J_1.[13]

the inflection point of the M against x curves]. For $J_2/J_1 = -0.5$ the magnetization breaks down at about $x_c'' \approx 0.5$, in agreement with experiment.[4] Note also that at the multicritical point J_2^m arbitrarily small dilution suffices to destroy any of the (degenerate) homogenous phases, as in the case of the square lattice. Note that at J_2^m not only the bulk energies of the two phases become equal, but also their interface energy vanishes. Therefore very large domains can be produced by very few dilution sites at the domain boundary, since there is no large cost of domain wall energy.

The Monte Carlo method works equally well for non-zero temperature, and hence one can study the transition from the ferromagnetic to the disordered phase as a function of dilution, fig. 4. The results for $R = -\frac{1}{2}$ agree nicely with the data for $Eu_xSr_{1-x}S$, and hence we can explain the early breakdown of order due to dilution in this material quantitatively. For $x \gtrsim 0.7$ the Monte Carlo data are also confirmed by extrapolation of high-temperature series [see ref. (13)], while for $x \lesssim 0.7$ the available (short) series are too irregular to allow meaningful extrapolation. However, the series was worked out for the quantum-mechanical $S = 7/2$ Heisenberg system as well as for the classical ($S \rightarrow \infty$) limit. The differences were shown to be numerically insignificant.

From the above considerations and " data " [fig. 3(*a*)] we conjecture that the spin-glass to ferromagnetic transition is of second order. This implies a ferromagnetic correlation length to appear in the spin-glass phase, which diverges at the transition, as does the susceptibility χ_F [since the root-mean-square magnetization of a finite lattice is just $\sqrt{\chi_F/N}$, the enhancement of χ_F as $x \rightarrow x_c''(R)$ is consistent with fig. 3(*a*)]. Equivalently one can express the same fact in other terms by noting that there are large (ferromagnetic) clusters in the spin glass phase, which should show up in neutron scattering (and do so[4]). In fact, the microscopic states of the spin-glass " phase " can be considered as an arrangement of irregular (ferromagnetic) clusters of spins (with strong and predominantly unfrustrated bonds inside the clusters) coupled together by partially frustrated (and hence in their net effect, weaker) bonds, such that no unique ground-state configuration can be found [see ref. (13) for a direct bond-counting analysis of computer-generated defect configurations which confirm this picture].

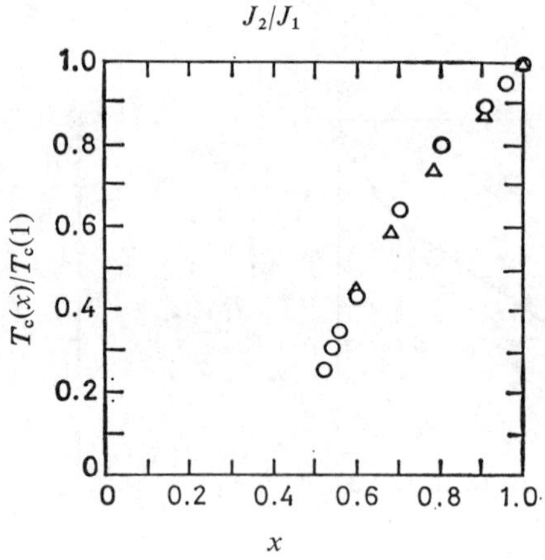

FIG. 4.—Monte Carlo estimates for $T_c(x)$ plotted against x for $J_2 = -J_2/2$. Data for $Eu_xSr_{1-x}S$ are included.[4,13] □, Monte Carlo; ○, experimental.

Here we assumed that the pure system orders ferromagnetically. But from the (qualitative, not quantitative!) symmetry of fig. 2 around the line $J_2 = J_m^2$ it is clear that similar considerations apply if one dilutes an antiferromagnet with competing interactions rather than a ferromagnet (χ_F is replaced by the appropriate staggered susceptibility, *etc*). Qualititatively similar considerations apply to molecular crystals as well (but due to the tensor character of the order parameter conclusions about the order of transitions may differ).

3. "REPLICA THEORY" OF DISORDER EFFECTS

The above discussion relied heavily on computer simulations (which are time-consuming and not always quantitative) and systematic series (which are tedious to perform and often ambiguous to analyse). Clearly, it would be better to have analytical techniques.

One method is to treat the statistical mechanics of the pure system in mean-field approximations or generalizations thereof like variants of the constant coupling approximation or Green's function decoupling. Frozen-in disorder is then treated by the average **T**-matrix approximation, coherent potential approximation or related techniques [*e.g.*, ref. (16) and (17)]. However, many claims of success for such treatments can often be shown to have no foundation [see, *e.g.*, ref. (17) and (18)]. In our opinion, the accuracy of such treatments is quite uncertain.

We follow a more ambitious programme, where frozen-in disorder is treated by the " replica method ",[11] which is expected to be exact in the paramagnetic phase up to the transition to the respective ordered phases.

Let us denote by $\{c_j\}$ the variables describing the frozen-in disorder. (In a site-dilution problem, $c_j = 1$ if the jth site is occupied and $c_j = 0$ if it is empty; for bond disorder, c_j stands symbolically for a random bond V_{jk} between two sites j, k; *etc.*). One has to calculate

$$F_{\{c_i\}} = -k_B T \ln Z_{\{c_i\}} = -k_B T \ln \mathrm{Tr} \exp(-\mathscr{H}\{c_i\}/k_B T) \tag{2}$$

for a fixed set $\{c_i\}$, and then one uses the appropriate probability distribution $P(\{c_j\})$ to compute

$$F = [F_{\{c_i\}}]_{av} = \int d\{c_i\} P(\{c_i\}) F_{\{ci\}}. \tag{3}$$

This average can be done by the " replica trick ".[11] Writing

$$F = k_B T [\ln Z_{\{c_i\}}]_{av} = -k_B T \lim_{n\to 0} \frac{1}{n}([Z^n{}_{\{c_i\}}]_{av} - 1), \tag{4}$$

where

$$Z^n_{\{ci\}} = \mathrm{Tr} \prod_{j,k} \prod_{\alpha=1}^{n} \exp[V_{jk}(\{c_i\})\varphi_j^\alpha \varphi_k^\alpha / k_B T], \tag{5}$$

n being a positive integer and φ_j^α the local order parameter at site j of the αth replica of the system. Performing the average now yields[11,12]

$$[Z^n_{\{ci\}}]_{av} = \mathrm{Tr} \exp(-\mathscr{H}_n^{eff}/k_B T), \tag{6}$$

$$-\mathscr{H}_n^{eff}/k_B T = \tilde{V}_1 \sum_{j,k} \sum_{\alpha} \varphi_j^\alpha \varphi_k^\alpha + \tilde{V}_2 \sum_{j,k} \sum_{\alpha,\beta} \varphi_j^\alpha \varphi_k^\alpha \varphi_j^\beta \varphi_k^\beta + \tilde{V}_3 \sum_{,k} \sum_{\alpha,\beta,\gamma} \varphi_j^\alpha \varphi_k^\alpha \varphi_j^\beta \varphi_k^\beta \varphi_j^\gamma \varphi_k^\gamma + \ldots, \tag{7}$$

where the $\tilde{V}_\nu$ are the cumulants

$$\tilde{V}_1 = [V_{jk}/k_B T]_{av} = \int d\{c_i\} p(\{c_i\}) \frac{1}{N} \sum_{j,k} V_{jk}(\{c_i\})/k_B T,$$

$$V_2 = [(V_{jk}/k_B T)^2]_{av} - [V_{jk}/k_B T]^2_{av}, \ldots. \tag{8}$$

Note that in the standard case of gaussian bond disorder only $\tilde{V}_1$ and $\tilde{V}_2$ are non-zero, while otherwise the set of interaction parameters $\tilde{V}_\nu$ is infinite. For the effective hamiltonian, eqn (7), we may define the following order parameters $Q^{(\nu)}$

$$Q^{(1)} = m_\alpha = \langle \varphi_j^\alpha \rangle, Q^{(2)} = Q_{\alpha\beta} = \langle \varphi_j^\alpha \varphi_j^\beta \rangle (\alpha \neq \beta), Q^{(3)} = \langle \varphi_j^\alpha \varphi_j^\beta \varphi_j^\gamma \rangle (\alpha \neq \beta \neq \gamma), \ldots. \tag{9}$$

While $Q^{(1)}$ is just the standard order parameter of the pure system (where only $\tilde{V}_1 \neq 0$) the $Q^{(\nu)}$ with $\nu \gtrsim 2$ are spin-glass order parameters (usually only the Edwards–Anderson order parameter[11] $Q^{(2)}$ is considered).

Eqn (4), (6) and (7) show that the problem of averaging over the disorder has been reduced to the consideration of an equivalent translationally invariant hamiltonian. It is however, an open problem to establish the proper meaning of $Q_{\alpha\beta}$ (as well as $Q^{(3)}$, $Q^{(4)}$, *etc.*) in the limit of $n \to 0$, since $Q_{\alpha\beta}$ then becomes a " zero-dimensional matrix ", and it is unclear how to relate physical order parameters $q^{(\nu)} = [\langle \varphi_j \rangle^\nu]_{av}$ to the $Q^{(\nu)}$.[19] The general belief, however, is that the $q^{(\nu)}$ become non-zero at those temperatures where the $Q^{(\nu)}$ become non-zero (and the associated response functions diverge), *i.e.*, one can locate the paramagnetic-spin-glass transition without answering the problems which appear in the spin-glass phase.

Having averaged over the disorder in eqn (4)-(7) exactly, the thermal averaging can be done by standard approximations (which are more complicated than usual, since we have infinitely many order parameters). However, for site disorder the meanfield approximation (which for gaussian bond-disorder reduces to Sherrington–Kirkpatrick theory)[20] yields for the critical temperature $T_c^{(\nu)}$ where the νth order parameter Q_ν starts to be non-zero $T_c^{(\nu)} \gtrsim T_c^{(\nu+1)}$ [the equality sign holds only on the percolation threshold x_p which in this approximation is estimated as[12] $x_p = 1 - \exp(-1/z)$ where z is the coordination number]. Applying the Bethe approximation yields similar results, the estimate for the percolation concentration is slightly improved $x_p^{nn} = 1/(z-1)$, in agreement with other formulations of the Bethe method for impure systems.[21]

For competing interactions $J_1 > 0$, $J_2 < 0$ one finds (in the Bethe approximation) $T_c^{(\nu)} \gtrsim T_c^{(\nu+2)}$ and hence only two order parameters $Q^{(1)}$, $Q^{(2)}$ need to be considered. However, both the mean-field and Bethe approximations are in trouble, since already their description of the pure case is too inaccurate. *E.g.*, mean-field approximations imply for $J_2 = J_2^m$ a coexistence of ferro- and antiferro-magnetic order up to a bicritical

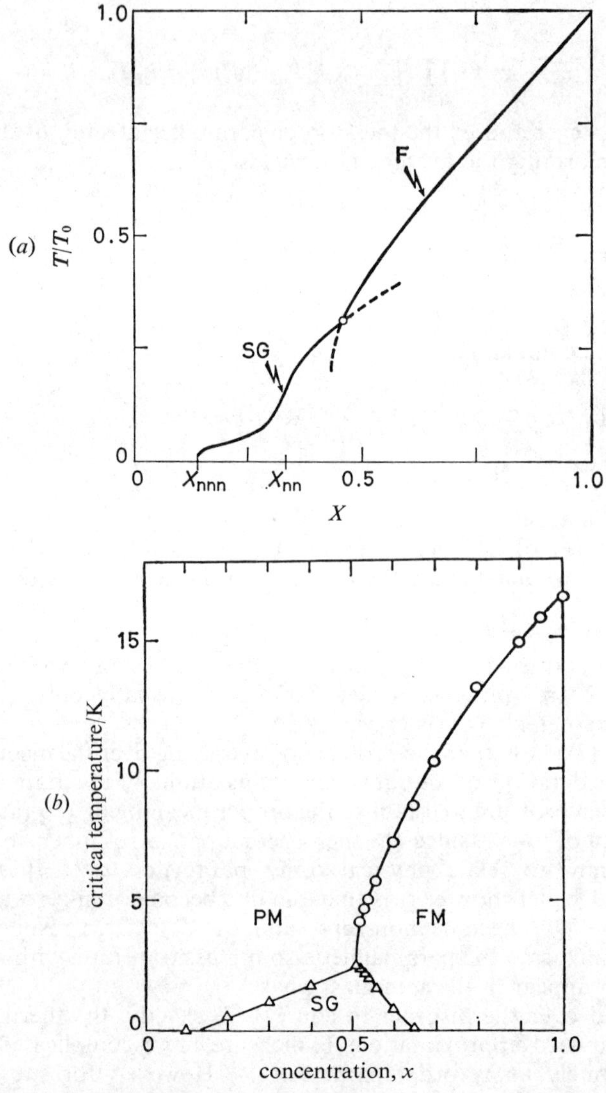

FIG. 5.—Phase diagram of a diluted competing system ($J_2/J_1 = -0.05$) according to replica theory and Bethe approximation (*a*),[12] compared to $Eu_xSr_{1-x}S$ data[4] (*b*).

point T_b, while actually $T_b = 0$.[13,14] This feature is correctly given by the Bethe approximation. In fact, for $|R| = |J_2/J_1|$ small enough a reasonable phase diagram results [fig. 5(*a*)], qualitatively similar to experimental ones [fig. 5(*b*)]. For larger $|R|$, however, the Bethe method also gets into trouble: a spin-glass ordering is predicted to occur even for $x \to 1$ (fig. 6), *i.e.*, in the pure system! This result again makes no sense.

It might be hoped that one could obtain better results by more sophisticated approximations like those of Kikuchi[22] or by real-space renormalization methods.[23] So far we have not done either of them: the numerical effort would be quite large and the reliability of the results quite uncertain in both cases.

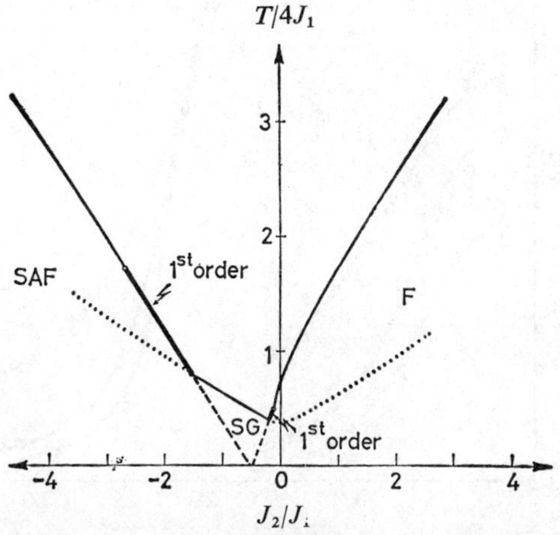

FIG. 6.—Phase diagram of a diluted competing Ising system according to replica theory and Bethe approximation in the limit of no dilution.[12]

4. DOES A SPIN-GLASS PHASE EXIST AT ALL?

There are many arguments that the freezing of spin glasses is not a phase transition but rather a non-equilibrium phenomenon [relaxation times of the " co-operative paramagnet "[24] become large in comparison to characteristic times of the measurement; see ref. (14) for a review]. We summarize here some of the most recent results relevant to this question.

Fig. 7 shows the " order parameter " $\psi^2 \equiv [k_B T \chi_{SG}/N]_{av}$ as obtained from a recursive transfer matrix approach for a square Ising lattice with nearest-neighbour interaction of random spin ($\pm J$ model)[25] [Note that for $T \to 0$ and $N \to \infty$, $\psi^2 = (q^{(2)})^2$]. Also included are results for a Mattis spin glass,[26] which is related to an Ising ferromagnet *via* gauge transformations[15] and hence $q^{(2)} = (q^{(1)})^2 = \psi^2$, *i.e.*, it has both an order parameter and an equilibrium phase transition. In this latter case, $\psi = q^{(2)} \equiv 1$ at $T = 0$, and this is unambiguously seen from the calculation of quite small lattices. Also the transition temperature T_c can be identified within an accuracy of a few percent from the inflection point of these curves. For the non-trivial $\pm J$ model, however, we conclude that $\psi \to 0$ for $N \to 0$: there is no order parameter, not even in the ground state! Rather the data are consistent with a power-law decay of correlations, $\langle \varphi_j \varphi_k \rangle^2 \propto 1/|r_j - r_k|^p$, with $p \approx 1$. Note, however, that then χ_{SG} is infinite for $N \to \infty$, and hence one has a phase transition (which is believed to occur at $T = 0$ in that case).[25] A divergence of χ_ψ has also been found from series extrapolation methods for the same model.[27] Monte Carlo simulations, however, indicated a transition to an apparently non-zero order parameter at $k_B T/J \approx 1$: in this case the above interpretation in terms of long relaxation times is hence established to be correct.

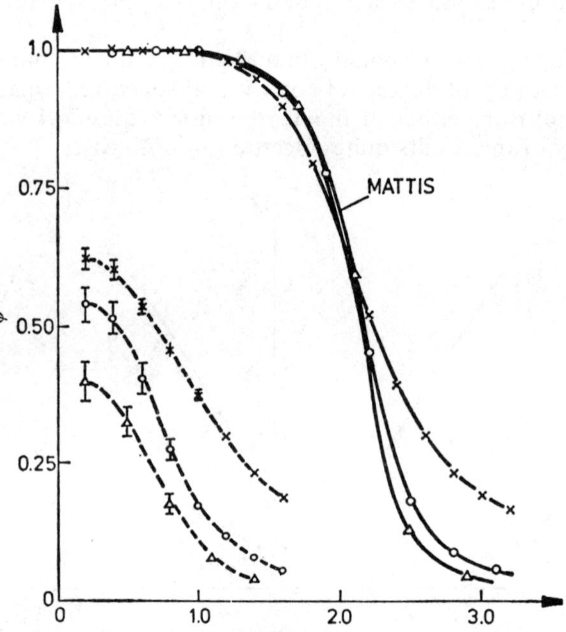

FIG. 7.—ψ^2 plotted against temperature for square lattices of various linear dimensions L for both the $\pm J$ model (broken curves) and the Mattis model (full curves).[25] ×, 6; ○, 12; △, 16.

For a square Ising lattice with symmetric gaussian distribution the recursive transfer matrix approach yields $\psi = q^2 \equiv 1$ at $T = 0$ as in the Mattis case, however.[25] But the decrease of the location of the inflection points with increasing N suggests $T_c = 0$ again.[25]

In all other cases the situation seems as yet unclear. Monte Carlo simulations, which have proven very useful for the study of diluted ferromagnets (Section 2), yield transitions to a spin-glass phase with non-zero order parameter (for all dimensionalities) in the case of both Ising[28] and Heisenberg[29] spins, as well as a molecular glass phase[30] in the case of tensor order parameters, if the averaging time is not too long. However, long runs establish that the order parameter

$$q(t) = \sum_i \left[\int_0^t S_i(t')\mathrm{d}t'/t \right]^2 \Big/ N$$

still relaxes with increasing observation time t: While in the Heisenberg case the simulations suggest $q(t \to \infty) \to 0$ [fig. 8(*c*)], it is not clear whether this also holds in the Ising case [fig. 8(*a*) and (*b*)]. It would be interesting to study this long-time behaviour for molecular glasses also.

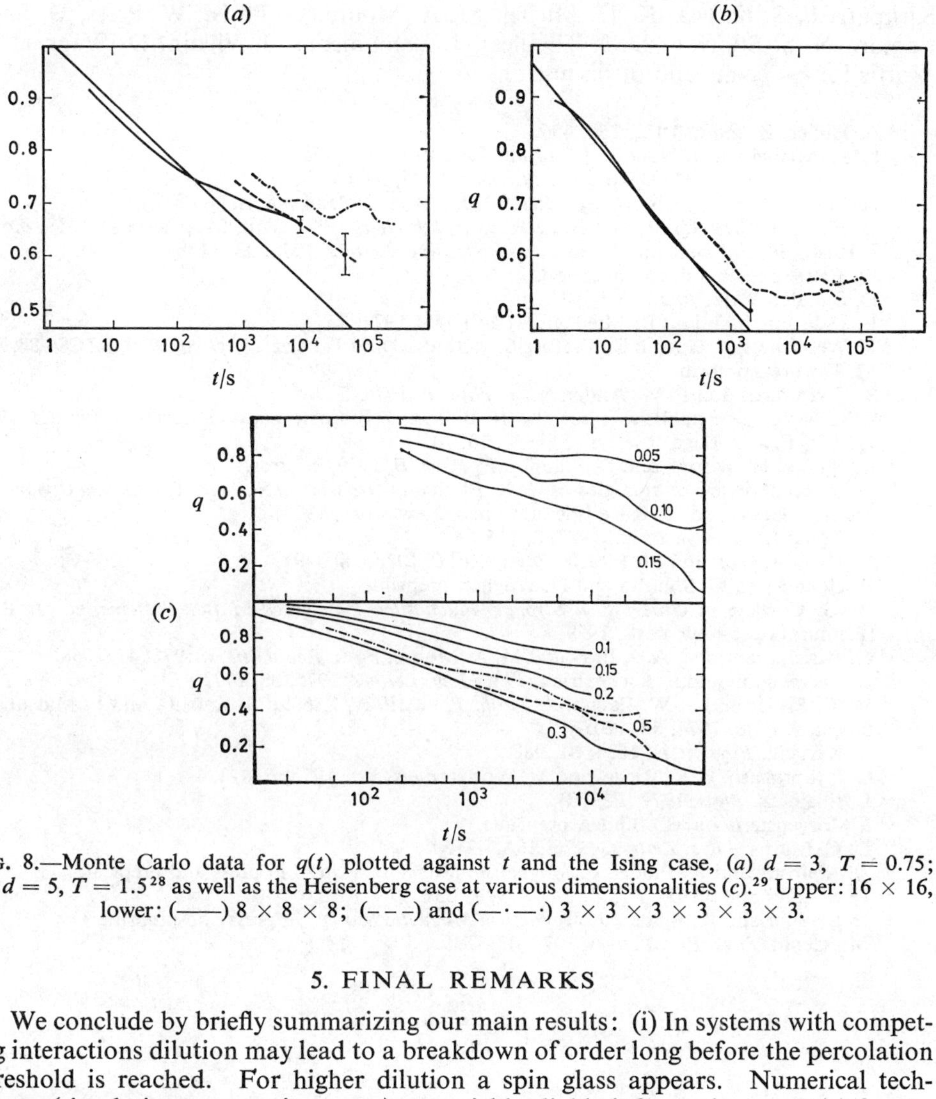

FIG. 8.—Monte Carlo data for $q(t)$ plotted against t and the Ising case, (*a*) $d = 3$, $T = 0.75$; (*b*) $d = 5$, $T = 1.5$[28] as well as the Heisenberg case at various dimensionalities (*c*).[29] Upper: 16×16, lower: (——) $8 \times 8 \times 8$; (- - -) and (—·—·) $3 \times 3 \times 3 \times 3 \times 3 \times 3$.

5. FINAL REMARKS

We conclude by briefly summarizing our main results: (i) In systems with competing interactions dilution may lead to a breakdown of order long before the percolation threshold is reached. For higher dilution a spin glass appears. Numerical techniques (simulations, expansions, *etc.*) may yield reliable information on $T_c(x)$ for the ordered phase, but it is not clear for which systems (order parameter and spatial dimensionalities) a spin-glass *phase* exists, and whether it is characterized by a non-zero order parameter or only a power-law decay of correlations. (ii) A characteristic feature of these dilution systems seems to be the inhomogeneity of the ordering. Even within the ferromagnetic phase one has antiparallel spins as well as antiferromagnetic or decoupled clusters. These clusters show up in many physical properties. Clearly, an accurate description of such an inhomogeneously ordered system in terms of an equivalent homogeneous hamiltonian, eqn (7), should be tricky, and in fact is: simple approximations for its statistical mechanics fail dramatically.

We thank H. Maletta and I. Morgenstern for allowing us to include their results [fig. 5(*b*) and 7]. We are grateful to them as well as to A. Aharony, M. Devoret, S. F. Edwards, S. Fishman, G. S. Grest, B. I. Halperin, A. Hüller, J. Kötzler, S.

Kirkpatrick, S.-K. Ma, K. H. Michel, M. A. Moore, G. Parisi, W. Press, D. Sherrington, N. S. Sullivan, D. J. Thouless, J. Vannimenus, J. Villain, D. Wagner, M. Wortis for preprints and/or discussions.

[1] A. Hüller, *Z. Phys.*, 1972, **254**, 456.
[2] K. H. Michel and J. Naudts, *J. Chem. Phys.*, 1977, **67**, 547.
[3] N. S. Sullivan and M. Devoret, *J. Physique*, 1978, **39**, C6-72.
[4] H. Maletta and W. Felsch, *Phys. Rev*, 1979, *B***20**, 1245, and preprints.
[5] K. H. Michel, J. Naudts and B. de Raedt, *Phys. Rev.*, 1978, **B18**, 648; see also J. M. Rowe, J. Rush, D. G. Hinks and S. Susman, *Phys. Rev. Letters*, 1979, **43**, 1158.
[6] W. Press, personal communication.
[7] D. Stauffer, *Phys. Rep.*, 1979, **54**, 1.
[8] L. J. de Jongh and A. R. Miedema, *Adv. Phys.*, 1974, **23**, 1.
[9] K. Westerholt, B. Ghosh, K. Siratori, S. Methfessel and T. Petzel, *Physica*, 1979, **BZC86-88**, 740.
[10] M. Devoret, preprint.
[11] S. F. Edwards and P. W. Anderson, *J. Phys. F*, 1975, **5**, 965.
[12] S. Sarbach, to be published; see also R. Bidaux, J. P. Carton and G. Sarma, *J. Phys. A*, 1976, **9**, L87; E. J. S. Lage, *J. Phys. C*, 1977, **10**, 701.
[13] K. Binder, W. Kinzel, and D. Stauffer, *Z. Phys. B*, 1979, in press.
[14] For a recent review on spin glasses see K. Binder, in *Ordering in Strongly-Fluctuating Condensed-Matter Systems*, ed. T. Riste (Plenum Press, New York, 1979).
[15] G. Toulouse, *Comm. Phys.*, 1977, **2**, 115.
[16] H. Dvey-Aharon and M. Fibich, *Phys. Rev. B*, 1978, **18**, 3491.
[17] B. Hoeppener, G. Sobotta and D. Wagner, preprint.
[18] R. G. Cowley, in *Ordering in Strongly-Fluctuating Condensed-Matter Systems*, ed. T. Riste (Plenum Press, New York, 1979).
[19] G. Parisi, preprints; A. J. Bray and M. A. Moore, *Phys. Rev. Letters*, 1978, **41**, 1068.
[20] D. Sherrington and S. Kirkpatrick, *Phys. Rev. Letters*, 1975, **35**, 1972.
[21] M. E. Fisher and J. W. Essam, *J. Math. Phys.*, 1961, **2**, 609; S. Katsura and F. Matsubara, *Canad. J. Phys.* 1974, **52**, 120.
[22] R. Kikuchi, *Phys. Rev.*, 1951, **81**, 988.
[23] C. Jayaprakash, E. K. Riedel and M. Wortis, *Phys. Rev*, 1978, *B***18**, 2244.
[24] J. Villain, *Z. Phys*, 1979, *B***33**, 31.
[25] I. Morgenstern and K. Binder, preprints.
[26] D. C. Mattis, *Phys. Letters*, 1976, **56A**, 421.
[27] A. Aharony and K. Binder, *J. Magnetism Magnetic Mater.*, in press and preprint.
[28] D. Stauffer and K. Binder, *Z. Phys*, 1979, *B***34**, 97.
[29] D. Stauffer and K. Binder, *J. Magnetism Magnetic Mater.*, in press and preprint.
[30] M. Klenin, *Phys. Rev. Letters*, 1979, **42**, 1549.

Computer Simulation of Glass Formation from Simple Model Liquids

BY JULIAN H. R. CLARKE

Chemistry Department,
University of Manchester Institute of Science and Technology,
Manchester M60 1QD

AND JOHN F. MAGUIRE AND LESLIE V. WOODCOCK

Laboratory for Physical Chemistry,
University of Amsterdam,
Amsterdam, The Netherlands

Received 4*th January*, 1980

Computer simulation techniques have been applied to the process of glass formation from simple model atomic and molecular liquids. Owing to the economic restrictions of ultrafast quench rates the " computer glass transition " as characterised by the change in apparent heat capacity is seen to be so diffuse that applications to quasi-thermodynamic transitions as observed in the laboratory are severely limited. Preliminary results for a simulation of a small periodic diatomic molecular liquid (modelling chlorine) are compared with data on model atomic liquids. The influence on glass formation of the form of the interaction potential, the periodic sample size and the quench rate are discussed and the structural properties of the quenched states are compared.

1. INTRODUCTION

The transition from supercooled liquid to amorphous solid has the form of a diffuse transition of second order but questions regarding a true underlying thermodynamic transition remain matters of uncertainty.[1,2] Computer simulation can extend the range of " experimental " investigation since there is virtually no restriction in the choice of intermolecular potential. Thus, glass-forming properties of liquids with simple effective two-body potentials can be investigated whereas in the laboratory the low barrier to nucleation precludes this possibility for simple systems which may be amenable to theoretical study.

Heretofore, molecular dynamics (MD) studies of the glass-forming properties of simple repulsive potential models following the inverse power law

$$\varphi(r, n) = \varepsilon(r_0/r)^n \tag{1}$$

have been reported for hard spheres[3-5] ($n = \infty$) and for soft spheres[6,7] ($n = 12$) and also for the Lennard-Jones (LJ) model at low temperatures.[8-11] The latter can be regarded as a perturbation of the soft-sphere model at low temperatures by adding an attractive component to the potential

$$\varphi_{LJ}(r) = \varphi(r, n = 12) - 2\varepsilon(r_0/r)^6. \tag{2}$$

where r_0 is the minimum potential-energy separation in the LJ case.

In the present work we extend these investigations by considering specifically the limitations in these simulation experiments in relation to laboratory glass formation.

The first question concerns the extent to which these simple models exhibit the thermodynamic and transport behaviour in the vicinity of the glass transition characteristic of laboratory molecular liquids. We report preliminary data for glass formation using a simple LJ diatomic model for chlorine.[12] Secondly we investigate how the two serious restrictions of using small-periodic samples and ultrafast quench rates affect the apparent thermodynamic properties in the glass-transition range and the structure of the amorphous solids thus produced.

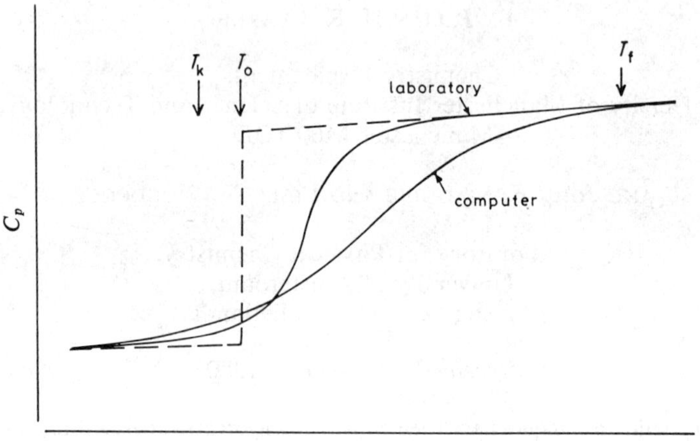

FIG. 1.—Schematic showing the expected temperature variation of the heat capacity (C_p) during glass formation using quench rates typical of computer and laboratory experiments and also for an infinitely slow quench assuming that crystallisation can be avoided and assuming an underlying second order transition at T_0. T_f and T_K are, respectively, the fusion temperature and the temperature, of zero excess entropy.

The situation is illustrated schematically in fig. 1 where is sketched the behaviour of an instantaneous " heat capacity " defined in terms of the enthalpy H and the time t

$$C_p(t) = \mathrm{d}H(t)/\mathrm{d}t. \tag{3}$$

If, as has been speculated,[1,2] there were to be an underlying second-order transition then there would be a sharp break in this plot. When the system has degrees of freedom which significantly contribute to H and with relaxation times comparable with $\mathrm{d}H(t)/\mathrm{d}t$, then a diffuse transition is observed. In computer experiments quenches are performed in the same order of time as the characteristic relaxation times for self diffusion in the stable liquid range, so that the " glass transition " is so diffuse that it cannot be characterised as a transition in the way that we ordinarily understand the term. Whereas the transformation range extends over only a few degrees in laboratory glass transitions, in computer simulations it can extend from above the freezing transition to temperatures well below the point of zero excess entropy.[1] Consequently it is meaningless to speak of a single operational glass-transition temperature in the computer models since the point of initial kinetic retardation will occur at temperatures well above any underlying singularity.

In Section 2 we draw up a comparison of previous results and assess the role of the potential in determining (*a*) the pVT dependence of the transition range and further (*b*) discuss the effects on the properties of the glasses so produced by the drastic quench rates which have been used. In Section 3 we present some new data on relaxation effects and discuss sample-size and quench-rate effects on the glass transformation. In Section 4 we discuss new data on the structural aspects of glass formation in atomic and molecular systems.

2. *pVT* PROPERTIES

There is in computer simulation experiments the distinct advantage that virtually no restrictions are placed upon the choice of the intermolecular potential defining the system. The hard-sphere model represents probably the simplest that might reasonably be expected to exhibit glass formation reminiscent of atomic or molecular laboratory glasses. The glass-forming properties of this system were first examined by Alder and Wainwright.[3] A close examination of their results for 32 spheres however, shows that the system actually nearly crystallised. It seems that for very small systems the periodic boundary induces the formation of phases that resemble more highly defective crystals rather than glasses. For larger systems ($\gtrsim$500 particles) the periodic boundary has less influence and the distinction between amorphous and crystalline configurations is clearer since the existence of the latter requires a rather sharply defined nucleation event. A 500-sphere model was subsequently found to compress to a truly amorphous solid[4] with a glass transition in the density (ρ) range below $N\sigma^3/V = 0.99$, where σ is the hard-sphere diameter and V is the volume of the N-particle system. This density is close to that for which the diffusion and viscosity extrapolate to zero.[4,13] The general conclusion was that the glass transition is instigated by the restrictive interplay of repulsive forces whereas the apparent discontinuities in the second-order derivatives of the free energy are largely *incidental manifestations* of attractive components of the pair potential where they exist. If this is true then we expect that the glass transition should occur at the same molecular-reduced volume for many real and model systems and that it will be largely insensitive to the strength of any attractive forces.

It has been argued[14] that the glass transition should occur at approximately the same " effective " hard-sphere density in several different systems and that this density is close to that found for the hard-sphere glass transition. Whilst this is, or could be, a useful generalisation, there are at present difficulties in testing the postulate accurately. The first arises in the precise definition of a glass-transition point in computer experiments. The second difficulty rests in the definition of the effective hard-sphere diameter for realistic potentials. Thus in the zeroth order of perturbation it is always possible to vary this parameter to fit the data for a single system. What we seek is a universal criterion that is approximately applicable for many, including molecular, systems.

A pronounced discontinuity in the heat capacity is the foremost manifestation of experimental transitions as reported from laboratory studies. Unfortunately there is still a dearth of accurate simulation data for derivative properties of simple models; the results to date for hard spheres,[4] soft spheres[7] and the LJ model[11] are collected in fig. 2. These data serve to emphasize some important points in our understanding of glass formation. The isobaric heat capacity is compared as a function of density scanning the entire amorphous range from gas to the high density glass. To make the comparison we choose the LJ effective-hard-sphere diameter to correspond to the potential minimum. The value is somewhat higher than as predicted using alternative criteria.[11] The soft-sphere density is then the appropriate high-temperature limit and we see that when $\varepsilon = kT$ the density of the glass transition is more or less the same as for the LJ model but substantially greater than that for the hard-sphere model. This indicates that soft sphere-behaviour (with $\varepsilon = kT$) provides a better zeroth-order representation of the glass transition for LJ and possibly other more realistic interactions.

It is interesting to observe the manner in which the change in C_p upon glass formation develops on going from the hard- and soft-sphere models to the zero-pressure low-temperature LJ case. In the hard spheres the only contribution to C_p arises from density fluctuations and these can be regarded above the glass transition as involving vibrational and diffusional contributions. When the density is increased through the transition only the vibrational modes remain so that there is a discontinuity in the derivative of C_p but no pronounced change in C_p itself. The absence of potential

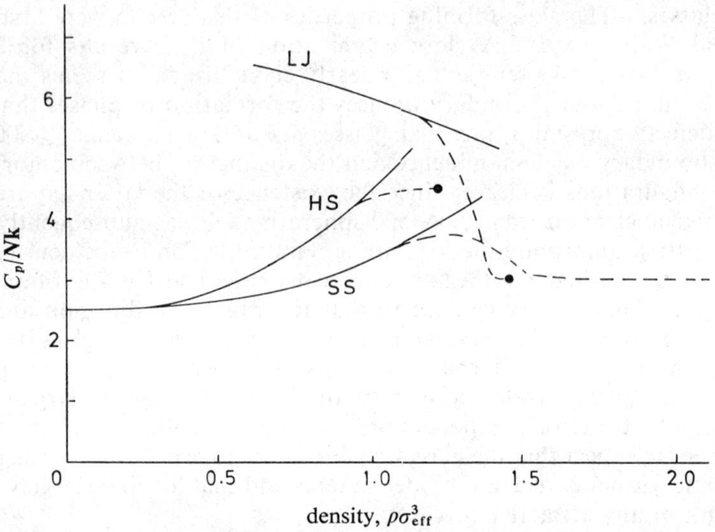

FIG. 2.—Heat capacity plotted as a function of reduced density for computer glass formation experiments on hard spheres (HS), soft spheres (SS) and Lennard-Jones particles (LJ), the latter results being those at zero pressure. The broken lines show the non-equilibrium glass formation. ● indicates the limits of densification for HS and for LJ at zero pressure. r_0 is taken as the effective HS diameter $\sigma_{\rm eff}$ for LJ particles and SS. For SS ε is set equal to kT [see eqn (1) and (2) in text].

energy fluctuations leads to a glass transition without the usual second-order discontinuity.

In the case of the LJ system there is a large contribution to the potential-energy fluctuations as a result of the attractive forces. As the glass transition is traversed there is $\approx 40\%$ change in C_p although this is not as abrupt as in laboratory experiments, for the reasons discussed previously. The soft-sphere data show a situation more akin to hard spheres; despite the finite repulsive potential there is still only a weak second-order discontinuity. Since the soft spheres represent the high-temperature and high-pressure limit of the LJ model (when attractive forces no longer are important) we can see how ΔC_p through the glass transition is expected to behave with increasing temperature and pressure.

The hard-sphere and LJ amorphous solids are classical with heat capacities of 4.5 Nk and 3 Nk, respectively, at the 0 K limit. The soft-sphere model compresses indefinitely as the pressure goes to infinity.

3. RELAXATION EFFECTS

Since the glass transition is determined by the relaxational properties of the sample, in this section we discuss the influence of quench rate and sample size in determining the nature of the transition in computer experiments. Sample size is important because at high densities one might expect that density fluctuations would tend to be correlated over large distances even for simple models and that these fluctuations would be slow to decay. Use of small samples would eliminate such effects and might lead to faster configurational relaxation. For systems of particles with rotational degrees of freedom the sample-size dependence of relaxation times would also be expected if long-range orientational correlations were important. The question is whether or not this would significantly affect the final state achieved using a particular quench rate.

It is essential in these studies of course that structural rearrangements leading to crystal nucleation should not occur. A combination of fast quench rates and small sample size renders these events improbable although care must be taken to avoid the periodic-boundary-induced effects for very small samples which have previously been mentioned.

In fig. 3 are shown the results of two MD experiments on LJ argon in which samples

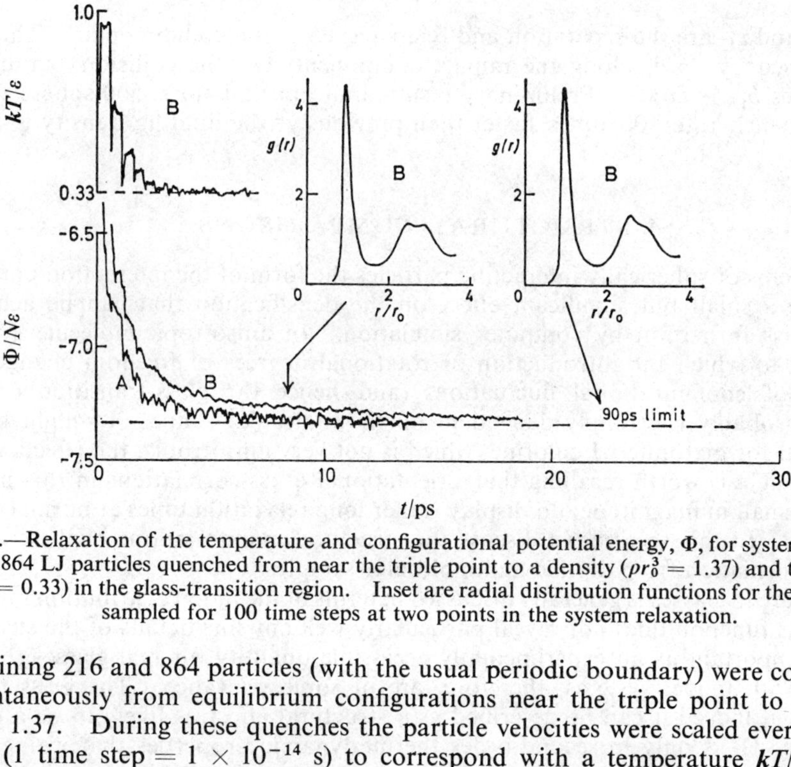

FIG. 3.—Relaxation of the temperature and configurational potential energy, Φ, for systems of A-216 and B-864 LJ particles quenched from near the triple point to a density ($\rho r_0^3 = 1.37$) and temperature ($kT/\varepsilon = 0.33$) in the glass-transition region. Inset are radial distribution functions for the 864 system sampled for 100 time steps at two points in the system relaxation.

containing 216 and 864 particles (with the usual periodic boundary) were compressed instantaneously from equilibrium configurations near the triple point to a density $\rho r_0^3 = 1.37$. During these quenches the particle velocities were scaled every 50 time steps (1 time step $= 1 \times 10^{-14}$ s) to correspond with a temperature $kT/\varepsilon = 0.33$. This point is predicted to be in the glass-transformation range for the zero-pressure isobar of the LJ system.[11] Since MD is normally a constant-energy technique, this scaling of the velocities is essential in order to simulate the temperature decrease and to avoid annealing of the sample as potential energy is converted into kinetic energy.

It is seen that thermal equilibration is essentially complete within 2-3 ps. Of particular interest in the present context, however, is the time taken for configurational properties (here the potential energy) to achieve equilibrium values. The statistical fluctuations are quite large as expected for small samples. Nevertheless there is clear evidence that the smaller sample has a faster relaxation to equilibrium. All though at very long times (80-100 ps) the final states seem indistinguishable within the statistical uncertainties, these times are much longer than economically feasible " sampling " times during simulation quenches.

The above data also illustrate that it should be possible to observe a quench rate dependence of the glass transformation in computer experiments. The most promising system for investigating this phenomenon is the hard spheres since a wide range of quench rates can be achieved without the annealing effects that occur in, for instance, the LJ model. An earlier exploratory study was, however, not conclusive.[5] The statistical errors were large and the comparison of different sample sizes further increased the uncertainties. New experiments currently are in progress in an attempt to clarify the situation. In these simulations quenching is achieved by allowing the hard-sphere diameters (σ) to swell at a fixed rate ($\mathrm{d}\sigma/\mathrm{d}t$). In the molecular dynamics the quadratic equation used to predict the time of the next collision[3] becomes

$$r_{ij} - \sigma^2 + 2t\left\{r_{ij}v_{ij} - \frac{\sigma\mathrm{d}\sigma}{\mathrm{d}t}\right\} + t^2\left\{v^2{}_{ij} - \left(\frac{\mathrm{d}\sigma}{\mathrm{d}t}\right)^2\right\} = 0 \tag{4}$$

where r_{ij} and v_{ij} are the separation and relative speeds of a colliding pair. The sizes are increased by $\mathrm{d}\sigma/\mathrm{d}_t$ along the impact component, *i.e.*, the collision parameter[3] b_{ij} becomes $b_{ij} - \mathrm{d}\sigma/\mathrm{d}t$. Preliminary results indicate that for a 500-sphere model using a quench rate 100 times faster than previously[4] the limiting density is $\approx 5\%$ smaller.

4. STRUCTURAL COMPARISONS

In systems of spherically interacting particles the form of the interaction potential does have a small but significant effect on the densification that can be achieved during glass formation by computer simulation. In anisotropic-molecule systems the extent to which the introduction of rotational degrees of freedom changes the spectrum of configurational fluctuations (and hence the glass-transition characteristics) probably is strongly dependent on the molecular shape. It might be expected that for diatomic LJ chlorine, which is not very anisotropic, the effects would be small but it is worth recalling that orientational cross-correlations in this model, although small in magnitude, do display rather long relaxation times at normal liquid densities.[15] In this section we discuss the structural manifestations of these effects for glass formation from atomic and molecular samples.

Structure in glasses is generally discussed in terms of the radial distribution function $g(r)$. This function does not reveal particularly well any fine details of the structure but it is important as an experimentally accessible quantity for real glasses. Comparisons with model systems, therefore, are of some relevance. The glass transition itself, if indeed it can be described as a structural effect, is likely to be a rather subtle one. It is only in second-order thermodynamic properties that we perceive any change at the glass transition. This suggests that perhaps an examination of structural fluctuations might be a fruitful approach although we postpone further discussion on this point until later.

As the density is increased beyond the melting density in any of the spherically interacting systems[16] there are three significant effects in $g(r)$. First, the peaks become narrower and more well defined as a result of restricted particle displacements. Secondly, discernible " structure " begins to extend to larger distances. Thirdly the second-nearest-neighbour peak begins to split into two components.

In regard to the nearest-neighbour geometry, the narrowing of the first peak in $g(r)$ is indicative of a narrower range of co-ordination numbers. Detailed analysis shows, however, that the most probable number up to the first minimum in $g(r)$ is always 13 and, as should be expected, there is no distinction between the metastable liquid and the glass. It is more fruitful to examine the second-nearest-neighbour packing.

The doublet structure of the second peak in $g(r)$, which has been discussed at length in connection with the geometry of hard-sphere amorphous close packing, is common to all densely packed systems of spherical particles, both model and experimental.[17] It is important to realise, however, that this feature in no way *distinguishes* the glassy and metastable liquid states since the splitting appears gradually as the normal liquid is cooled or compressed. The longer of the two preferred lengths corresponds to colinear arrangements of atom triplets whereas the shorter corresponds to the distance of closest approach of a second-nearest-neighbour across an intervening pair of atoms in the nearest-neighbour shell.

It is interesting to observe the important role played by the second-nearest-neighbour packing in densification effects during glass formation. Illustrated on fig. 3 are two plots of $g(r)$ for a LJ system at $\rho r_0^3 = 1.37$ in the glass-transition region but with different allowed relaxation times for structural rearrangement. As more time is allowed for relaxation the shorter distance component of the doublet grows at the expense of the other. The second-nearest-neighbour distribution achieves a more compact arrangement which, of course, results in a lower pressure at constant density. This emphasises the entirely operational definition of the glass transition. It is only reasonable therefore to compare structural features in different glasses when they have been prepared by comparable quench rates. When this is done from the available data on hard spheres,[16] soft spheres[7] and the LJ model[11] it is apparent that the relative magnitudes of the constituent peaks does vary with interaction potential. For the latter two models the inner peak is much more prominent than in the hard-sphere case showing that a more compact second-nearest-neighbour structure can be achieved. " Softness " of the interaction potential does appear to have an important influence on relaxational behaviour in the glass-transition region. Densification relative to the hard-sphere model is to be expected for a given quench rate and this is largely to be associated with structural relaxations involving *second*-nearest neighbours.

Since the glass transition is characterised by discontinuous changes in second (and higher) order thermodynamic properties it would seem relevant to give some attention to what might be termed " second-order " structural properties. Just as $n(r)$, from which $g(r)$ is derived, is the average number of particles within a sphere radius r around a reference particle so we can define moments about this mean distribution. Here we shall focus attention on the second moment, termed the radial fluctuation function, $W(r)$, and defined as follows

$$W(r) = [\langle n(r)^2\rangle - \langle n(r)\rangle^2][\tfrac{4}{3}\pi r^3\rho]^{-1}. \tag{5}$$

One significance of $W(r)$ is that its limiting value gives the isothermal compressibility

$$\chi = \left[\frac{\mathrm{d}(\ln V)}{\mathrm{d}p}\right]_T = W(r \to \infty)V/NkT. \tag{6}$$

A small discontinuous change in χ at the isobaric glass transition has been observed for the LJ liquid[11] so that it is perhaps appropriate to examine the behaviour of $W(r)$ in this system, although of course only the *short-range* behaviour can be determined. Fig. 4 displays $W(r)$ for the LJ liquid and glass at various densities along the $p = 0$ isobar. The form of this function follows very closely the form of $g(r)$ indicating quite sensibly that the fluctuations are largest where the local density of particles is largest. The significant narrowing of the first peak with increasing density is again indicative of a consequent decrease in the distribution of co-ordination numbers but there is no apparent discontinuity through the glass transition. As for $g(r)$ a doublet structure appears in the peak corresponding to second-nearest-neighbour distances as the density is increased. One interesting point to note, however, is the change in relative intensities of these two components between the metastable liquid and the glass. The small-r component in fact changes little but the larger-r component decreases significantly in height. Fluctuations at distances corresponding to this peak appear to be suppressed in the glass.

One can distinguish two contributions to $W(r)$ in a dense monatomic liquid. The

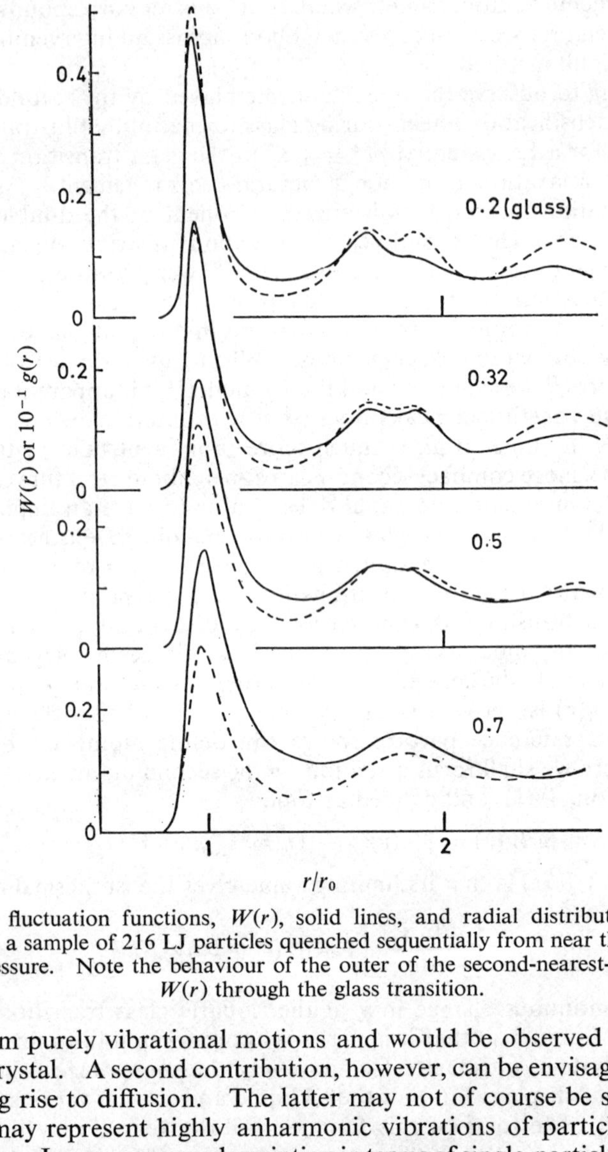

FIG. 4.—Radial fluctuation functions, $W(r)$, solid lines, and radial distribution functions, $g(r)$, broken lines, for a sample of 216 LJ particles quenched sequentially from near the triple point to the glass at zero pressure. Note the behaviour of the outer of the second-nearest-neighbour peaks in $W(r)$ through the glass transition.

first arises from purely vibrational motions and would be observed even for a finite-temperature crystal. A second contribution, however, can be envisaged as arising from motions giving rise to diffusion. The latter may not of course be simple translation motions but may represent highly anharmonic vibrations of particles between local energy minima. In any case any description in terms of single-particle motions is likely to be inadequate. Since a glass is characterised by the absence of diffusion on the experimental time-scale then $W(r)$ here will be dominated by purely vibrational effects. Perhaps the intensity increase in the large r component of the second peak can, therefore, be associated with the onset of diffusional motions. The " local compressibility " for spheres defined by this radial distance (which depends on the available distribution of configurations) increases in going from the glass to liquid. The fact that this has more to do with the dynamics of second-nearest-neighbour packing than nearest-neighbour effects emphasises the co-operative nature of particle motions at high densities.

It is interesting to compare some of the above results with preliminary structural data that we have obtained for the diatomic LJ model of chlorine. For densities up

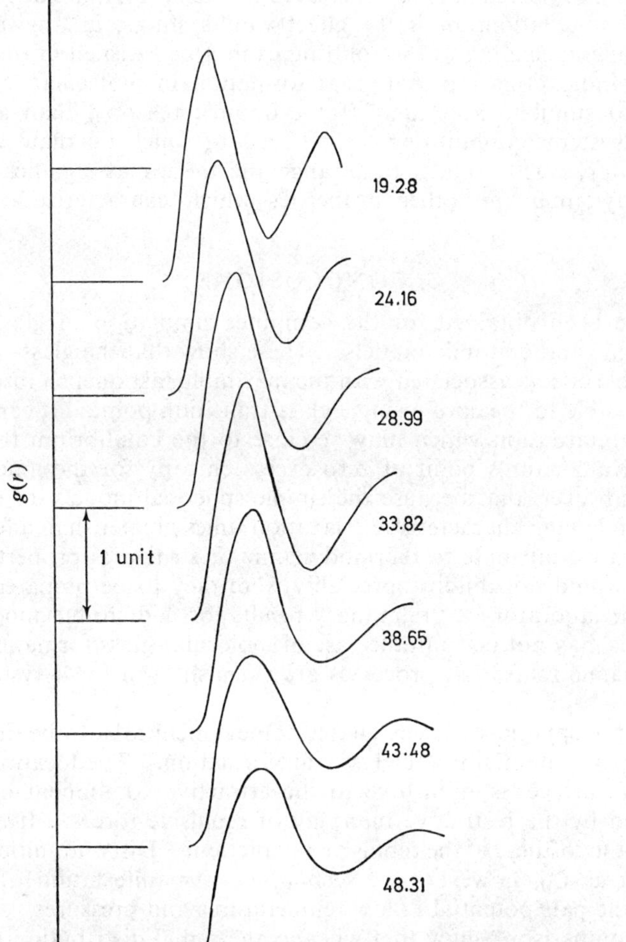

FIG. 5.—Centre of mass radial distribution functions, $g(r)$, for a 64 molecule sample of the diatomic LJ model of chlorine at 273 K with the usual periodic boundary conditions. The molar volumes are indicated in $cm^3\ mol^{-1}$. The bottom curve is for a normal liquid density at near zero pressure. All the results are from Monte Carlo simulations. The range of $g(r)$ is limited to half the length of the primary sample cell. Referring to eqn (2), $\varepsilon/k = 174$ K and $r_0 = 0.376$ nm and the bond length $= 0.542\ r_0$.

to the normal liquid range this system has been investigated extensively.[12] In fig. 5 is shown the centre of mass $g(r)$ for six high density states examined by the Monte Carlo method following instantaneous compressions from a normal liquid density. In each case the system (64 molecules) was relaxed over 200 000 moves and the properties sampled for a further 100 000 moves. Inspection of the configurational averages suggests that the time required for configurational relaxation is substantially longer than for spherical models at comparable densities. The distribution functions for the diatomic model are less structured and more diffuse at short range than for the atomic models. There is no indication of restructuring of the second peak at high densities. It is not yet certain whether this is to be expected for a centre of mass function which takes no account of the relative orientations of the molecule pairs, whether it is a consequence of the lengthening of structural relaxation times due to the presence of orientational fluctuations or if the effect would not occur anyway because the packing requirements leading to the splitting cannot be satisfied in this system.

Preliminary indications are that glass formation in molecular systems may be more difficult to simulate accurately (for economic reasons) than atomic models. The molecular systems contain degrees of freedom which fluctuate with relaxation times which span a much broader time range and these may significantly contribute to the thermodynamic and other properties which characterize the liquid–glass transformation.

5. CONCLUSIONS

Results have been obtained for the computer simulation of glass formation in simple monatomic and diatomic models. These show that the glass-transition range due to irreversible effects associated with the inevitable fast quench rates is broad and that it is not possible to speak of a single glass-transition point. It can vary from the point of initial retardation which may be close to the equilibrium freezing density, to well below Kauzmann's point of zero excess entropy for the amorphous phase. We conclude, however, that because the simple spherical models do not possess degrees of freedom having characteristic relaxation times greater than a few picoseconds which significantly contribute to thermodynamic or structural properties, the glasses thus produced would not differ appreciably were they to be prepared under conditions akin to the laboratory. Preliminary results for a diatomic model of chlorine indicate that this may not be so in the case of molecular glass formation. Structural and thermodynamic relaxation processes are sluggish even for a system as small as 64 molecules.

Results for the spherical models enable some conclusions to be drawn regarding the role of the pair potential in the glass transformation. The location of the transition on the pVT surfaces is insensitive to the attractive component of the potential and is instigated by the restrictive interplay of repulsive forces. It does depend to some extent on the softness of the repulsive interaction. Discontinuities in the derived properties such as C_p, however, are seen to be a manifestation of the attractive component of the pair potential at low temperatures and pressures.

Structural comparisons show that whereas the radial distribution functions show no significant change through the glass transition, the radial fluctuation functions do show a discontinuity in the region of second-nearest-neighbour distances and we tentatively associate this effect with a decrease in a " local compressibility " at the glass transition.

Further investigations to determine the quench rate dependence of the properties of both atomic and molecular glasses are in progress and we hope to be able to report on the results shortly.

J. H. R. C. and L. V. W. acknowledge the support of the Scientific Affairs Division of the North Atlantic Treaty Organisation. We thank the Netherlands Science Foundation (Z.W.O.) for financial support and for the award of a research fellowship (to J. F. M.) and are grateful to Miss Louise Copeland for obtaining data quoted in Section 3. It is a pleasure to acknowledge many helpful and stimulating discussions with Prof. C. A. Angell.

[1] W. Kauzmann, *Chem. Rev.*, 1948, **43**, 219.
[2] C. A. Angell, *J. Chem. Educ.*, 1970, **47**, 583.
[3] B. J. Alder and T. E. Wainwright, *J. Chem. Phys.*, 1960, **33**, 1439.
[4] L. V. Woodcock, *J.C.S. Faraday II*, 1976, **72**, 1667.
[5] L. V. Woodcock, *J.C.S. Faraday II*, 1977, **73**, 11.
[6] Y. Hiwatari, H. Matsuda, T. Ogawa, N. Ogita and A. Ueda, *Prog. Theor. Phys.*, 1974, **52**, 1105.
[7] J. N. Cape and L. V. Woodcock, *J. Chem. Phys.*, in press.
[8] W. Damgaard Kristensen, *J. Non-cryst. Solids*, 1976, **21**, 303.
[9] A. Rahman, M. J. Mandell and J. P. McTague, *J. Chem. Phys.*, 1976, **64**, 1564.
[10] H. R. Wendt and F. F. Abraham, *Phys. Rev. Letters*, 1978, **41**, 1244.
[11] J. H. R. Clarke, *J.C.S. Faraday II*, 1979, **75**, 1371.
[12] K. Singer, A. Taylor and J. V. L. Singer, *Mol. Phys.*, 1977, **33**, 1757.
[13] B. J. Alder, D. M. Gass and T. E. Wainwright, *J. Chem. Phys.*, 1970, **53**, 3813.
[14] S. Hudson and H. C. Anderson, *J. Chem. Phys.*, 1978, **69**, 2323.
[15] K. Singer, J. V. L. Singer and A. Taylor, *Mol. Phys.*, 1979, **37**, 1239.
[16] C. A. Angell, P. A. Cheeseman, J. H. R. Clarke and L. V. Woodcock, in *The Structure of Non-Crystalline Materials*, ed. P. H. Gaskell (Taylor and Francis, London, 1977).
[17] G. S. Cargill, in *Solid State Phys.*, 1975, **30**, 227; J. L. Finney, *Proc. Roy. Soc. A*, 1970, **319**, 479; C. H. Bennett, *J. Appl. Phys.*, 1972, **43**, 2727.

GENERAL DISCUSSION

Dr. L. L. Boyle (*University of Kent*) said: With reference to table 1 of his paper, could Dr. Barnes please indicate how the quadrupole moment tensor is defined since it appears not to be traceless and further, since water is a dipolar molecule, could he please let us know the origin of the co-ordinate system to which the values refer.

Dr. A. I. M. Rae (*University of Birmingham*) said: It is not clear from the paper whether the hypothetical ordered form of ice Ih is polar or non-polar. In either case it is not clear how convergence of the dipole–dipole summation was ensured when it is remembered that this interaction falls off with distance as r^{-3} while the number of molecules in a shell of radius r increases as r^2 and the cut-off distance (11 Å) is less than the dimension of the basic cell.

If the disordered form of ice Ih indeed has a lower internal energy than that of the hypothetical ordered form, then the third law of thermodynamics requires either that some other ordered structure has an even lower energy or that the equilibrium zero-temperature structure involves some form of quantum-mechanical degeneracy and an ordering in momentum space such as occurs in a high-density electron gas.

Prof. R. M. Pick (*Université Pierre et Marie Curie Paris*) said: The paper by Barnes *et al.* seems to suggest that the absence of order at 0 K in ice I is not related to a quantum-mechanical effect (where the gain of energy in ordering the protons is lower than the loss of energy necessary to localize a proton, *i.e.*, lower than the tunnel splitting between the various proton states), but to a spin glass or frustration effect. This point of view can certainly not be ruled out immediately, but is also very provocative. Indeed, in a spin glass, the very weak magneto-elastic coupling and the disorder in the interaction energies are responsible for the fact that no ordered state for the magnetic spins can be obtained even if a lattice distorsion takes place. In a purely electrostatic and short-range interaction lattice as the one considered here, a weak lattice deformation and/or the formation of a (possibly incommensurate) superstructure is usually sufficient to relax the extra energy due to competing interactions of opposite signs. Further calculations seem necessary before ruling out, for ice I, the quantum-mechanical effect, and the orientational glass type of transition suggested by Suga and Seki in their paper.

Dr. A. Hüller (*KFA Jülich*) said: Dr. Rae has asked if the long-range dipole interactions have been taken into account correctly by means of an Ewald summation. I do not think that such a calculation is necessary for a structure without a net polarization.

Prof. Pick has suggested that the disorder in ice at the lowest temperatures is due to proton tunnelling. It has been objected that the barrier a proton has to pass when it jumps from one oxygen to a neighbouring one is extremely high. Tunnelling therefore should be unimportant. I want to ask if the transitions between different proton configurations in ice are achieved by a motion of the protons from one oxygen to another *via* a high barrier or by a rotation of water molecules across much lower barriers.

Dr. A. I. M. Rae (*University of Birmingham*) said: The problem of the convergence of dipole–dipole lattice sums can exist even in the absence of a net polarization unless the limits of the summation are carefully chosen to ensure that the actual unit summed over (in this case apparently a sphere) is also non-polar.

Transitions between the different proton configurations in ice are indeed believed to take place by molecular rotation, but the barriers to this are generally thought to be too high for quantum effects to be important except possibly at very low temperatures.

Prof. H. Suga (*Osaka University*) said: Barnes *et al.* calculated the energy distribution curve for many disordered proton configurations by randomizing the hypothetical ordered structure for ordinary ice. Their conclusion is based on a finding that the average energy of disordered configurations is lower than that of the starting structure. We will consider now the entropic aspect of this result. The total number of the disordered configurations that the computer realized would be approximately $(3/2)^N$, if we consider the system composed of N water molecules. If the thermal energy is larger than the spread of the energy distribution, all the configurations are regarded as degenerate. We can remove the entropy of the system by lowering the temperature, and the configuration having the least energy will be ultimately realized at the lowest temperature.

Actually, however, we have a finite amount of residual entropy in the crystal, which indicates that we are concerned with non-equilibrium state. This may only be accounted for by the existence of freezing phenomena that occur on account of the prolonged relaxation time for reorientational motion of water molecules at the very initial stage of the ordering process. Therefore, the failure of observation of hypothetical order–disorder transition in hexagonal ice is not exclusively due to the energetic problem but also to the kinetics.

Dr. R. W. Munn (*UMIST, Manchester*) (*communicated*): Because Dr. Barnes and co-workers use periodic boundary conditions, their calculation of the electrostatic energy for quasidisordered crystals can be carried out using an extension of methods developed for ordered crystals.[1,2] This obviates the need for iteration and a distance cut-off.

The repeating unit is treated as a crystal unit cell containing Z molecules. Lattice multipole sums relating the various different pairs of molecules are then calculated; with point dipoles and quadrupoles and translational order, all the sums can be related to those for the underlying lattice.[3] The dipole moments in the crystal are obtained by inverting a $3Z \times 3Z$ matrix, and the electrostatic energy then follows from a sum over distinct pairs of molecules in the repeating unit.[1,2] Use of the Ewald method to separate off the non-regular long-range part of the dipole lattice sum has the advantage of displaying part of the energy as $-\boldsymbol{E}\cdot\boldsymbol{P}$, where $\boldsymbol{E}$ and $\boldsymbol{P}$ are the macroscopic electric field and polarization. If the repeating unit has no net dipole, $\boldsymbol{P}$ is zero; otherwise one would normally expect surface charge to make $\boldsymbol{E}$ zero. In either case, the macroscopic energy contribution is zero and the energy is obtained from Lorentz-factor tensors[4] rather than Coulomb coefficients.

The method outlined here can be applied to repeating units containing orientational and translational disorder, though for large Z the amount of computation required is large. By giving explicit results, the method allows the effect of disorder to be studied starting from small repeating units. It might also allow random matrix techniques to be applied for large repeating units.

[1] D. A. Dunmur and R. W. Munn, *Chem. Phys.*, 1975, **11**, 297.
[2] R. W. Munn and R. J. Newham, *Chem. Phys.*, 1976, **14**, 309.
[3] R. W. Munn and T. Luty, *Chem. Phys.*, 1979, **38**, 413.
[4] P. G. Cummins, D. A. Dunmur, R. W. Munn and R. J. Newham, *Acta Cryst.*, 1976, **A32**, 847.

Dr. P. Barnes (*Birkbeck College, London*) said: The convergence of dipolar energies does fall off as $1/r$, which is indeed a slow convergence rate. All the Ice I structures considered were, however, non-polar by design and this coupled with the normal periodic boundary conditions ensures ultimate convergence. It is true that the 11 Å cut-off is just smaller than half the smallest unit cell dimension (22.08 Å) but one is limited here by available computer resources (which must increase as r^3!). However, a surprising consequence of non-additive systems is that the *effective* cut-off can paradoxically greatly exceed the computed cut-off as a result of the non-additive (polarization) interactions of intervening molecules. The convergence aspects of these indirect interactions is not yet clear but tests[1] have indicated that indirect interactions in condensed systems (such as bulk water at 0 °C) can swamp the direct dipole–dipole interactions for separations *ca.* 4 Å.

The question of whether low-temperature disordered Ice phases contravene the third law of thermodynamics is problematical. Further, regardless of our calculations on Ice I, there is no reason why Ice I should be unique in this aspect or that model systems cannot be contrived so as to produce no driving force to an ordered structure. One way round the paradox might be to argue in favour of a super-large unit cell of Ice I, the required ordering then being a kind of super-lattice ordering with a very large unit cell. Accessing this state could well require very long relaxation times as argued by Prof. Suga. Alternatively ordering in momentum space as suggested by Dr. Rae might be feasible. We would welcome further thought on these lines.

Concerning the comments of Dr. Munn, we are grateful for his suggestions and we are looking into the possibility of this application to our quasidisordered crystalline system.

[1] P. Barnes, J. L. Finney, J. D. Nicholas and J. E. Quinn, *Nature*, 1979, **282**, 459.

Dr. M. Eldrup (*Risø National Laboratory, Roskilde*) said: Could Profs. Suga and Seki comment on the possible role of molecular vacancies for the effects they observe in ice at *ca.* 100 K and which they ascribe to proton ordering. We have studied ice with the positron annihilation technique and found a much higher vacancy concentration (*ca.* 10^{-6}-10^{-5} at the melting point) than usually assumed.[1] Furthermore we have found that long-range vacancy migration takes place at 100 K and above over periods of hours.[1] Is it just a coincidence that these two temperatures are the same or does it have some significance? Does the reorientation of molecules perhaps take place at vacancies and therefore stop when vacancy migration becomes very slow below 100 K? In fact, it seems likely that a certain binding exists between vacancies and Bjerrum defects[1] and therefore that the migration of vacancies will contribute to the reorientation of molecules. This also means that the models used so far to obtain, *e.g.*, formation energies for Bjerrum defects, without taking binding to vacancies into account, probably do not give correct values for these formation energies.

[1] M. Eldrup, *J. Chem. Phys.*, 1976, **64**, 5283; O. E. Mogensen and M. Eldrup, *Phys. Letters*, 1977, **60A**, 325; O. E. Mogensen and M. Eldrup, *J. Glaciology*, 1978, **21**, 85; M. Eldrup, O. E. Mogensen and J. H. Bilgram, *J. Glaciology*, 1978, **21**, 101.

Prof. H. Suga (*Osaka University*) said: Our interpretation is based on the experimental findings that the calorimetric relaxation time coincides with the dielectric relaxation time for orientational polarization of water molecules and also the activation enthalpy for the relaxational process agrees with that for migration of the Bjerrum defect. If there exists a certain binding between vacancies and Bjerrum defect, *i.e.*, the movement of the Bjerrum defect is intimately associated with that of the vacancies, then we can say that both of our experimental observations have a common basis.

Prof. J. N. Sherwood (*University of Strathclyde*) said: In their summary on glass-forming materials Profs. Seki and Suga indicate that in some cases there is a need for impurities in the material in order to stabilise the glassy state, whereas in others a high purity is desirable to inhibit crystallization. Could they explain their comment and define the purity of their present materials?

Prof. S. Seki (*Kwansei Gakuin University*) said: As to the impurity introduction to the glass, as you know very well, glass manufacturers of window glass or other glass products pay special attention to preventing crystallization. That means the know-how to increase the glass-transition temperature and to avoid the devitrification is crucial in the case of industry. In our case, however, from the scientific point of view, to clarify the intrinsic and the essential thermodynamical or other properties, all the samples used for investigation were purified as well as we could possibly manage and we employed only the samples of which the absolute purity determined calorimetrically in our laboratory exceeds 99 mol % or more. That is the reason why we purified and re-purified the samples, since the realization of the glassy state of substances is very sensitive to impurities, as you know well from your experience on diffusional processes in solid materials.

Dr. A. Fuchs (*Université Paris Sud, Orsay*) said: As remarked by Prof. Seki, conventional dielectric techniques cannot be used to study very long relaxation times. There is, however, a dielectric technique termed *depolarization thermo-currents* (d.t.c.) which can be used for this purpose.

This technique was used for studying relaxation phenomena in some polar organic glasses, and a computational method was designed for the mathematical analysis of the d.t.c.: assuming that the activation energy remains constant over a small range of temperature, it is possible to obtain a relaxation time distribution function so as to get very long relaxation times (up to centuries!).[1] The experiment consists of polarizing the sample while cooling it. When the lowest temperature is reached, the electric field is switched off and the depolarization current is recorded as the temperature is slowly raised. D.t.c. can also be obtained from the glassy crystalline state of cyclohexanol, as we have shown recently. The so-called α and β domains were observed at temperatures of *ca.* 90 and 150 K, the relaxation times being of the order of 10 Hz in this experiment. A systematic investigation of polar glassy crystalline materials by d.t.c. will be performed shortly.

[1] R. Heitz and H. Szwarc, *Rev. Appl. Phys.*, 1980, **15**, 687.

Prof. H. Suga (*Osaka University*) said: I agree with Dr. Fuchs' opinion. Actually Drs. Johari and Jones[1] applied this technique to ordinary ice. They concluded that the current peak due to thermostimulated depolarization observed around 100 K arose from the relaxation of frozen-in orientation polarization of water molecules in ice. From an analysis of the current–temperature plots, they obtained the relaxation time data up to 10^3 s. We pursued the spontaneous temperature drift due to enthalpy relaxation under good adiabatic conditions for about one month and obtained relaxation time data up to 10^6 s.

[1] G. P. Johari and G. P. Jones, *J. Chem. Phys.*, 1975, **62**, 4313.

Dr. A. Dworkin (*Université Paris Sud, Orsay*) said: Johari[1] has shown that the glass-like transition T_g' in a glassy crystal can be detected by dielectric measurements. Among the substances described by Johari, two had not been studied by thermal methods and therefore I wish to report preliminary results on neohexanol (2,2-dimethyl-butan-l-ol) and neopentanol (2,2-dimethyl-propan-l-ol). Measurements were performed on a Perkin Elmer differential scanning calorimeter DSC-2* with samples of a few milligrams of Aldrich products used without purification. These results are shown in fig. 1 and 2.

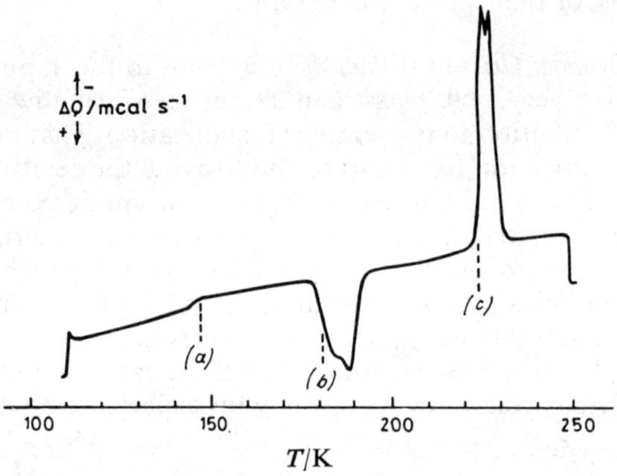

FIG. 1.—D.s.c. results for 2,2-dimethyl-butan-l-ol (2.75 mg). (*a*) T_g = 147 K; (*b*) T_r = 180 K; (*c*) T_m = 224.5 K.

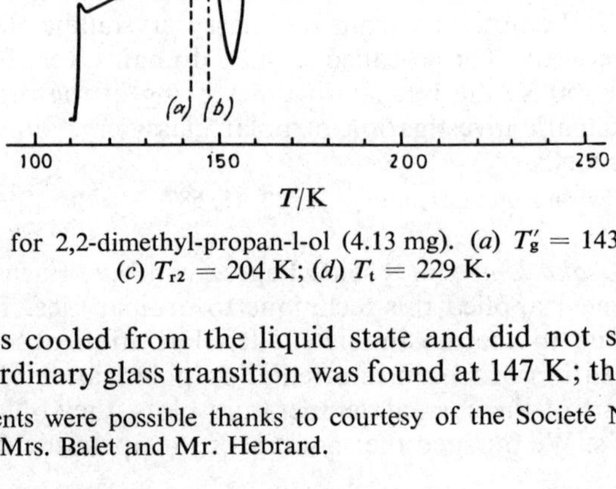

FIG. 2.—D.s.c. results for 2,2-dimethyl-propan-l-ol (4.13 mg). (*a*) T_g' = 143 K; (*b*) T_{r1} = 149 K; (*c*) T_{r2} = 204 K; (*d*) T_t = 229 K.

Neohexanol was cooled from the liquid state and did not show evidence of a glassy crystal. An ordinary glass transition was found at 147 K; the liquid crystallized

* These measurements were possible thanks to courtesy of the Societé Nationale Aerospatiale (Les Mureaux) and of Mrs. Balet and Mr. Hebrard.

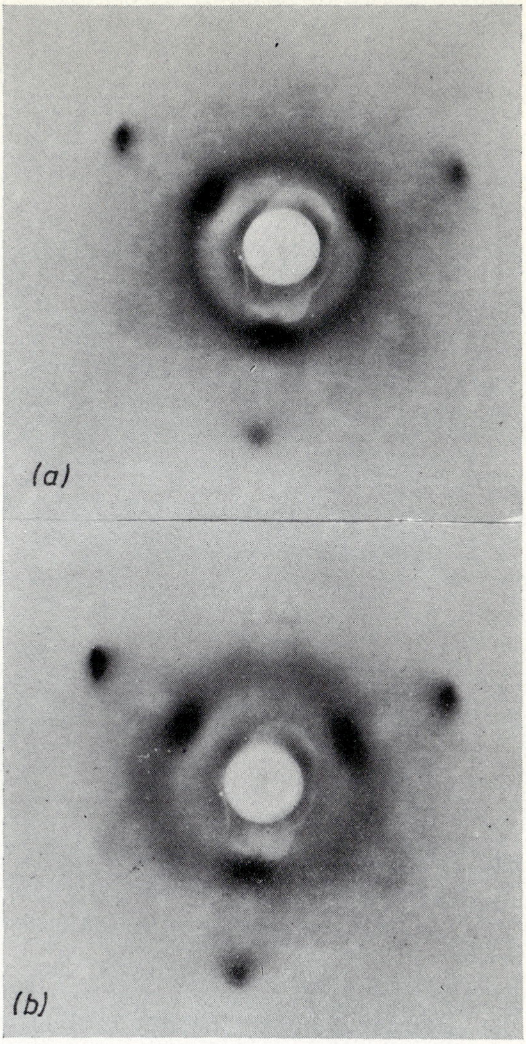

FIG. 3.—Diffuse X-ray scattering by a cyclohexanol single crystal with the incident beam perpendicular to plane (111): (*a*) plastic phase at 289 K; (*b*) glassy crystalline phase at 120 K.

[*To face page* 288

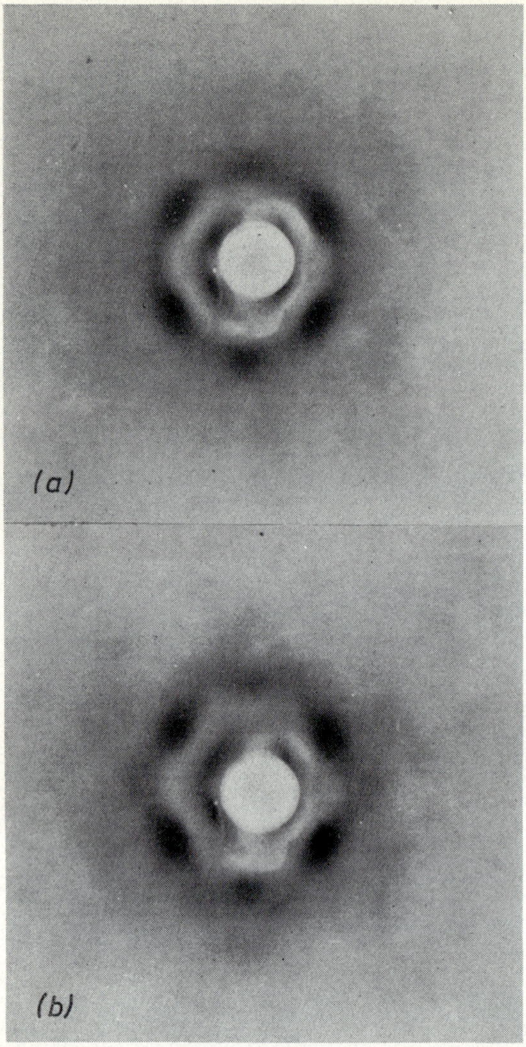

FIG. 4.—Diffuse X-ray scattering by a cyclohexanol single crystal with the incident beam perpendicular to plane (110): (*a*) plastic phase at 289 K; (*b*) glassy crystalline phase at 120 K.

[*To face page* 289

at 180 K and the crystal melted at 225 K (*lit*:213 K). No vitreous state could be obtained from liquid neopentanol. When crystalline form I is cooled down to 110 K and subsequently heated, a glass-like transition is noted at 143 K (133 K in Johari's paper) followed immediately by a recrystallization into a metastable form which recrystallizes in turn at 204 K just before the transition at 229 K. Another T'_g seems to appear just before this second recrystallization. Melting occurs at 318 K (*lit*:325 K).

This difference in behaviour between two closely related substances suggests that minor changes of symmetry may be of key importance in obtaining glassy crystals instead of ordinary glasses.

[1] G. P. Johari, *An. N.Y. Acad. Sci.*, 1976, **279**, 117.

Dr. H. Szwarc (*Université Paris Sud, Orsay*) said: In connection with Prof. Seki's contributions, I wish to report some diffuse X-ray scattering experiments on the glassy crystalline state of cyclohexanol. In collaboration with Dr. Dénoyer and Prof. Lambert at Orsay, Dr. Ceccaldi and I have undertaken this study in order to add structural criteria to the thermodynamic ones that allowed Prof. Seki and his collaborators to prove that glassy crystals really represent a new state of matter aggregation.[1]

Fixed single crystals were studied by X-ray monochromatic Laue technique (Cu $K\alpha$, 1.54 Å) and the scattering was recorded on fixed photographic plates. For each crystal orientation, photographs were taken at *ca*. 289, 273 and 120 K. The first two temperatures correspond to the plastic phase, the third one to the glassy crystal.

Fig. 3 shows two photographs recorded at 289 K (*a*) and 120 K (*b*) with the incident X-ray beam perpendicular to the plane (111). Fig. 4 represents similar photographs with the incident beam perpendicular to plane (110).

Otsubo and Sugawara[2] had already shown that the two phases have the same lattices, so the observation of the same Bragg reflexions at our three experimental temperatures was no surprise.

For each orientation, the diffuse scattering intensity drops when the temperature is lowered from 289 to 273 K and does not seem to decrease any more from 273 to 120 K. One can see on fig. 3(*b*) and 4(*b*) that it is still important at 120 K: it can be concluded that at this temperature the diffuse scattering does not rise from thermal agitation but is probably related to orientational disorder, whether static or dynamic.

But the most striking feature lies in the structure of the diffusion patterns themselves. Whereas the inner limit of the diffusion ring is almost circular in the plastic phase at 289 K [fig. 3(*a*) and 4(*a*)], and also at 273 K, it is no longer true at 120 K in the glassy crystal. On fig. 3(*b*), the former diffusion ring can be described as a truncated equilateral triangle and fig. 4(*b*) exhibits an almost hexagonal diffusion pattern.

From these results, it can be inferred that strongly anisotropic short-range intermolecular correlations characterize the orientational disorder of the glassy crystalline state of cyclohexanol.

Obviously, much computional work will be necessary to find a model of the disorder involved, but this preliminary study has disclosed definite structural characters for the cyclohexanol glassy crystalline state.

[1] K. Adachi, H. Suga and S. Seki, *Bull. Chem. Soc. Jpn*, 1968, **41**, 1073.
[2] A. Otsubo and T. Sugawara, *Sci. Rep. Res. Inst. Tohoku Univ.*, 1970, **A7**, 2372.

Prof. A. J. Leadbetter (*University of Exeter*) said: Prof. Cotterill states that the constant volume entropy difference between liquid and crystal and between liquid and glass is in each case $R\ln2$. This implies that crystal and glass have the same entropy (at identical volume). I find this hard to believe, would he care to comment?

Prof. R. M. J. Cotterill (*Technical University of Denmark, Lyngby*) said: The loss of the entropy $R\ln2$ during the transition from liquid to glass occurs gradually over a range of volume, not at constant volume, and the volume of the glass when the loss is complete is not the same as that of the corresponding crystal. Further compression of the glass to the same volume as the crystal would be accompanied by an additional change of entropy, so glass and crystal at the same density and temperature would not have the same entropy.

Dr. P. Barnes (*Birkbeck College, London*) said: I would like to pose two questions in relation to the fascinating theory of the ln2 entropy factor in the liquid state:

(i) There must be higher-order divisions of the interatomic potential-well (*e.g.*, triple, quadruple minima) for 3-dimensional liquids. How would this affect the ln2 factor in the entropy which is based on bifurcation alone?

(ii) How does your theory relate to hard-sphere systems which clearly cannot show bifurcation in the energy-well?

Prof. R. M. J. Cotterill (*Technical University of Denmark, Lyngby*) said: (1) I cannot see any obvious reason why higher-order multi-furcations should not be possible and believe that some must indeed occur. They would probably be associated with higher excitation energies, however, and the experimental result, $\Delta S_m = R\ln2$ for $\Delta V_m = 0$, indicates that they have negligible statistical weight.

(2) The theory leads to the firm conclusion that the hard-sphere system, lacking bifurcation, has no liquid state!

Dr. J. F. Maguire (*University of Amsterdam*) said: I would like to make a few comments on the melting transition which bear on the thought-provoking paper presented by Cotterill and Tallon. Following Stishov,[1] these authors have noted that on plotting the fractional volume change on melting $\Delta v_f/v_s$ against the entropy of melting $\Delta S_f/Nk$ one obtains an apparently universal (at least for argon and sodium) straight line, which, on extrapolation to zero ordinate, yields a number close to ln2 (0.69). Cotterill and Tallon interpret Stishov's result as " indicating a fundamental choice of position between two alternatives for each atom, upon melting " and use this idea as the basis of their bifurcation hypothesis. I should like to point out that Stishov's data are amenable to an alternative interpretation in which the notion of bifurcation plays no part.

Fig. 5 shows a plot of the experimental fractional volume change on melting against the entropy of melting for argon (open circles) and sodium (filled triangles). The closed circles in the figure correspond to computer simulation results[2] for six soft-sphere systems of varying softness parameter n. In these systems the pair potential energy, φ, at separation, r, is defined by the relation

$$\varphi(r) = \varepsilon\left(\frac{\sigma}{r}\right)^n, \tag{1}$$

where σ is the soft-sphere diameter and ε is a constant of proportionality. The virial equation of state for a soft sphere is given by

$$\frac{pV}{NkT} = 1 + \frac{n\langle\varphi\rangle}{dNkT}, \tag{2}$$

where p is pressure, V is volume, N is the number of particles, k is Boltzmann's constant, T is temperature, d is the dimensionality and the pointed brackets represent

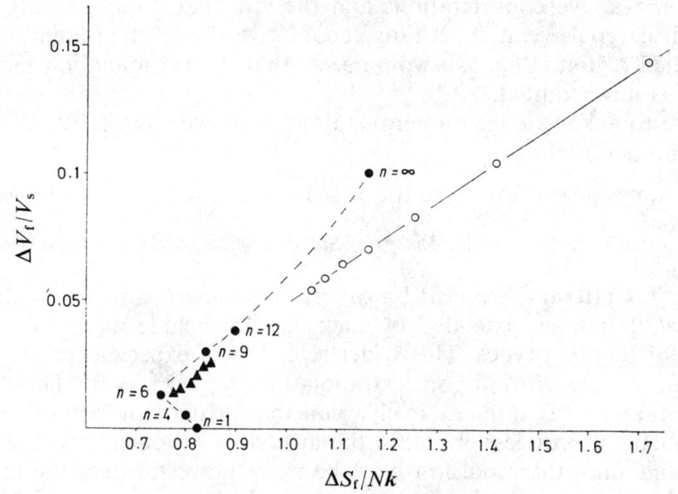

FIG. 5.—Plot of the fractional volume change on melting against the entropy of melting for argon (○), sodium (▲), and soft/hard sphere systems (●) with softness parameter n.

an ensemble average. Using the fact that along the melting line the Gibbs free energies of the solid and liquid phases are equal it is easy to show from the virial equation ($d = 3$) that

$$\frac{\Delta V_f}{V_s} = \frac{\Delta S_f}{Nk}\left[\frac{pV_s}{NkT}\left(\frac{3}{n}+1\right)\right]^{-1}. \tag{3}$$

Three points should be noted in relation to eqn (3):

(i) For a given softness parameter the compressibility factor, pV_s/NkT, is a constant hence $\Delta V/V_s$ and $\Delta S/Nk$ are both constant along the entire melting line, *i.e.*, each soft-sphere system may be represented by a single point in fig. 5.

(ii) $pV_s/NkT > 0$ for all n and only in the limit $n \to 0$ does $\Delta V/V_s \to 0$ at finite $\Delta S/Nk$. This situation is approximated in the one component plasma ($n = 1$).

(iii) For large n,

$$p\Delta V_f = T\Delta S_f,$$

i.e., the hard-sphere limit with $n = \infty$.

In the limit that $T \to \infty$ the attractive part of the intermolecular potential has no influence on the thermodynamic properties of the system, hence the soft-sphere data represent the high-temperature limiting behaviour of real systems with some, not necessarily integral, value of n. It should be realized that a reduction in $\Delta V_f/V_s$ and $\Delta S/Nk$ corresponds to ascending the melting line to higher temperatures and pressures and that in the infinite-temperature limit an appropriate soft-sphere result should be recovered. Examination of fig. 5 shows that Stishov's data extrapolate and touch the soft-sphere curve (dashed line) at $n \approx 12$ for argon and at $n \approx 6$ for sodium. The Lennard-Jones potential with repulsive parameter $n = 12$ is well known to provide a reasonable effective two-body potential for argon. Similarly recent theoretical and simulation work indicates a softness parameter $n \approx 5.6$ for the alkali metals and it is encouraging to note that the experimental data for sodium are consistent with this. The " dog-leg " behaviour of the soft-sphere data at $n \approx 6$ may reflect a change in crystal structure from f.c.c. ($n > 6$) to b.c.c. ($n < 6$).

In view of the above considerations and the fact that no spherically symmetric potential which is repulsive at short range could give rise to state points lying to the left of the dashed region in fig. 5, it would seem that the extrapolation mentioned by Prof. Cotterill is unwarranted.

I would like to acknowledge numerous discussions with Dr. L. V. Woodcock on many of the points raised above.

[1] S. M. Stishov, I. N. Makarenko, V. A. Ivanov and A. M. Nikolaenko, *Phys. Letters*, 1973, **45A**, 18.
[2] J. P. Hansen and I. R. McDonald, *Theory of Simple Liquids* (Academic Press, London, 1976).

Prof. R. M. J. Cotterill (*Technical University of Denmark, Lyngby*) said: The work of Stishov *et al.*[1] was later extended by Lasocka[2] to include many other elements, with a variety of bonding types. This wider selection of experimental data was also consistent with $\Delta S_m = R\ln 2$ upon extrapolation to $\Delta V_m = 0$. The " dog-leg " behaviour appears to be a theoretically plausible artifact of potentials having a repulsive exponent of 6 or less. Because the attractive exponent is usually taken to lie around 6, and since the repulsion must be more powerful than the attraction, it seems doubtful that this anomaly is of any practical consequence.

[1] S. M. Stishov, I. N. Makarenko, V. A. Ivanov and A. M. Nikolaenko, *Phys. Letters*, 1973, **45A**, 18.
[2] M. Lasocka, *Phys. Letters*, 1975, **51A**, 137.

Dr. D. N. Batchelder (*Queen Mary College, London*) said: (1) Prof. Cotterill has shown that a plot of ΔV_m against ΔS_m for a wide variety of materials gives a reasonable straight-line graph. A plot of ΔV_m against the equilibrium concentration of vacancies at the triple point for just as wide a range of materials gives an equally impressive straight-line graph. If this is coincidence then it is hard to see why one plot is more fundamental to melting than the other. On the other hand, if these two are strongly correlated then they both may be the result of a single underlying phenomena.

(2) In solid helium all the atoms experience a bifurcated interatomic potential, at least if that potential has been estimated from the properties of helium gas such as diffusion, viscosity and thermal conductivity. The stability of solid helium has been explained using the self-consistent phonon theory which shows that the thermal and zero-point motion of the atom produces a quantum mechanically averaged interatomic potential with a single minimum. Does one obtain $\Delta S_m = R\ln 2$ for the melting of solid helium and, if so, what are the implications for the bifurcation model of melting?

Prof. R. M. J. Cotterill (*Technical University of Denmark, Lyngby*) said: (1) I certainly agree that the linear correlation of ΔV_m with vacancy concentration at the triple point is impressive, as are several other correlations, variously involving energies, entropies and the Debye temperature. Whether such correlations lead one closer to understanding melting is more debatable. I believe that their true significance stems from the indication of a unity lying deeper than we have yet been able to probe, as Dr. Batchelder suggests. It is clear that all these parameters are inevitably linked to the interatomic potential, but there seems to be some other principle which remains stubbornly obscure. In the context of our present work it is not the linear correlation of ΔV_m and ΔS_m that is important, but rather the fact that the extrapolation to $\Delta V_m = 0$ gives the provocative result $\Delta S_m = R\ln 2$.

(2) The analysis given in our paper is more in line with the self-consistent phonon theory, wiht its quantum-mechanically averaged single-well potential in the crystalline

state. One obtains $\Delta S = \boldsymbol{R}\ln 2$ both for the transition from crystalline helium to normal fluid helium and the transition from superfluid helium to normal fluid helium. For the transition from the crystal to the superfluid, $\Delta S = 0$, and this implies a choice, for the superfluid, between either non-bifurcated on-site potentials or bifurcated on-site potentials that do not produce entropy. Because bifurcation seems to be a signature of the liquid state, the latter alternative seemed most likely, and it led to the proposal that bifurcation in the superfluid is of a special type: spherically symmetric rather than axial. If this is the case, there nevertheless remains the question of how the individual bifurcations interact.

Dr. A. Hüller (*KFA Jülich*) said: As Dr. Sarbach has said the forces leading to orientational order in molecular crystals show tensor character. There are attractive contributions (favouring ferro-ordering) in some directions and repulsive contributions (favouring antiferro-ordering) in other directions. Furthermore, the range of the interactions, being of electrostatic origin, is generally much longer than the range of exchange constants in magnetic systems. In the field of phase transitions in molecular solids one is thus always confronted with competing interactions which lead to difficulties even in pure systems. In dilute systems Dr. Sarbach's advanced methods are crucial for a description of the systems.

Can he extend his methods to cases where the interactions go beyond the first and second shell of neighbours? Is it possible to go beyond the Ising model description and to include the true tensor character of the interactions?

Dr. S. R. Sarbach (*KFA Jülich*) said: The extension of our methods to interactions with range longer than the first- or second-neighbour separation presents no difficulty in principle. Of course, the coefficients of the series expansions will prove more and more difficult to evaluate and the series will be harder to analyse. Also, the computer simulations will become even more time-consuming. The application of the replica method together with mean-field theory might prove useful in producing a qualitative phase diagram.

The real problem in applying our methods to molecular crystals is the tensor character of the interactions. Including the true tensor character of the order parameter is certainly not a straightforward extension of our calculations and obviously require more work. We hope to make progress in this direction in the near future.

ADDITIONAL REMARKS

The following late remarks were received on October 28th, 1980, and have not been checked by the respective authors.

Dr. B. Ewen (*University of Mainz*) said: (in reply to Prof. Chapuis' comment on p. 122). The occurrence of 90° rotational jumps in the modifications B and C can be excluded unambiguously from the results of the wide-line n.m.r. measurements on the uniaxially oriented n-$C_{33}H_{68}$ specimens. This is due to the fact that the decrease of the second moment d $<\Delta H^2>(\gamma)$ as a function of the alignment angle γ (γ angle between the magnetic field and the axis of orientation) has its maximum at $\gamma = \pi/4$ [see ref. (8) of our paper], which is specific for 180° rotational jumps [see ref. (14)]. In contrast, for 90° rotational jumps this maximum is predicted to appear at $\gamma = \pi/2$ [see ref. (14)].

Prof. Dr. G. Strobl (*University of Mainz*) said: (in reply to Prof. Leadbetter's question on p. 122). As long as the n-alkanes remain in the all-*trans* conformation, they behave like stiff rod-like molecules and show the typical rotational and longitudinal motions. With regard to the degree of disorder as indicated by the general appearance of the X-ray diffraction pattern, phase D can be compared to a smectic mesophase. The difference in the melting behaviour results from the internal flexibility of the n-alkanes; the larger melting entropy is associated with a coiling of the chains.

INDEX OF NAMES*

* The page numbers in heavy type indicate papers submitted for discussion.

GENERAL DISCUSSIONS OF THE FARADAY SOCIETY

Date	*Subject*	*Volume*
1907	Osmotic Pressure	Trans. 3*
1907	Hydrates in Solution	3*
1910	The Constitution of Water	6*
1911	High Temperature Work	7*
1912	Magnetic Properties of Alloys	8*
1913	Colloids and their Viscosity	9*
1913	The Corrosion of Iron and Steel	9*
1913	The Passivity of Metals	9*
1914	Optical Rotatory Power	10*
1914	The Hardening of Metals	10*
1915	The Transformation of Pure Iron	11*
1916	Methods and Appliances for the Attainment of High Temperatures in a Laboratory	12*
1916	Refractory Materials	12*
1917	Training and Work of the Chemical Engineer	13*
1917	Osmotic Pressure	13*
1917	Pyrometers and Pyrometry	13*
1918	The Setting of Cements and Plasters	14*
1918	Electrical Furnaces	14*
1918	Co-ordination of Scientific Publication	14*
1918	The Occlusion of Gases by Metals	14*
1919	The Present Position of the Theory of Ionization	15*
1919	The Examination of Materials by X-Rays	15*
1920	The Microscope: Its Design, Construction and Applications	16*
1920	Basic Slags: Their Production and Utilization in Agriculture	16*
1920	Physics and Chemistry of Colloids	16*
1920	Electrodeposition and Electroplating	16*
1921	Capillarity	17*
1921	The Failure of Metals under Internal and Prolonged Stress	17*
1921	Physico-Chemical Problems Relating to the Soil	17*
1921	Catalysis with special reference to Newer Theories of Chemical Action	17*
1922	Some Properties of Powders with special reference to Grading by Elutriation	18*
1922	The Generation and Utilization of Cold	18
1923	Alloys Resistant to Corrosion	19*
1923	The Physical Chemistry of the Photographic Process	19
1923	The Electronic Theory of Valency	19*
1923	Electrode Reactions and Equilibria	19
1923	Atmospheric Corrosion. First Report	19*
1924	Investigation on Oppau Ammonium Sulphate-Nitrate	20*
1924	Fluxes and Slags in Metal Melting and Working	20*
1924	Physical and Physico-Chemical Problems relating to Textile Fibres	20*
1924	The Physical Chemistry of Igneous Rock Formation	20*
1924	Base Exchange in Soils	20*
1925	The Physical Chemistry of Steel-Making Processes	21*
1925	Photochemical Reactions in Liquids and Gases	21
1926	Explosive Reactions in Gaseous Media	22
1926	Physical Phenomena at Interfaces, with special reference to Molecular Orientation	22
1927	Atmospheric Corrosion. Second Report	23*
1927	The Theory of Strong Electrolytes	23*
1927	Cohesion and Related Problems	24
1928	Homogeneous Catalysis	24
1929	Crystal Structure and Chemical Constitution	25*
1929	Atmospheric Corrosion of Metals. Third Report	25*
1929	Molecular Spectra and Molecular Structure	26*
1930	Colloid Science Applied to Biology	26
1931	Photochemical Processes	27
1932	The Adsorption of Gases by Solids	28
1932	The Colloid Aspect of Textile Materials	29
1933	Liquid Crystals and Anisotropic Melts	29*
1933	Free Radicals	30

Date	*Subject*	*Volume*
1934	Dipole Moments	30
1934	Colloidal Electrolytes	31*
1935	The Structure of Metallic Coatings, Films and Surfaces	31*
1935	The Phenomena of Polymerization and Condensation	32*
1936	Disperse Systems in Gases: Dust, Smoke and Fog	32*
1936	Structure and Molecular Forces in (*a*) Pure Liquids, and (*b*) Solutions	33*
1937	The Properties and Functions of Membranes, Natural and Artificial	33*
1937	Reaction Kinetics	34*
1938	Chemical Reactions Involving Solids	34*
1938	Luminescence	35*
1939	Hydrocarbon Chemistry	35*
1939	The Electrical Double Layer (owing to the outbreak of war the meeting was abandoned, but the papers were printed in the *Transactions*)	35*
1940	The Hydrogen Bond	36*
1941	The Oil–Water Interface	37*
1941	The Mechanism and Chemical Kinetics of Organic Reactions in Liquid Systems	37*
1942	The Structure and Reactions of Rubber	38
1943	Modes of Drug Action	39
1944	Molecular Weight and Molecular Weight Distribution in High Polymers (Joint Meeting with the Plastics Group, Society of Chemical Industry)	40*
1945	The Application of Infra-red Spectra to Chemical Problems	41*
1945	Oxidation	42*
1946	Dielectrics	42 A
1946	Swelling and Shrinking	42 B
1947	Electrode Processes	Disc. 1*
1947	The Labile Molecule	2
1947	Surface Chemistry (Jointly with the Société de Chimie Physique at Bordeaux) Published by Butterworths Scientific Publications, Ltd	
1947	Colloidal Electrolytes and Solutions	Trans. 43*
1948	The Interaction of Water and Porous Materials	Disc. 3
1948	The Physical Chemistry of Process Metallurgy	4
1949	Crystal Growth	5
1949	Lipo-Proteins	6
1949	Chromatographic Analysis	7
1950	Heterogeneous Catalysis	8
1950	Physico-chemical Properties and Behaviour of Nuclear Acids	Trans. 46*
1950	Spectroscopy and Molecular Structure and Optical Methods of Investigating Cell Structure	Disc. 9
1950	Electrical Double Layer	Trans. 47
1951	Hydrocarbons	Disc. 10
1951	The size and shape Factor in Colloidal Systems	11
1952	Radiation Chemistry	12
1952	The Physical Chemistry of Proteins	13
1952	The Reactivity of Free Radicals	14
1953	The Equilibrium Properties of Solutions on Non-Electrolytes	15
1953	The Physical Chemistry of Dyeing and Tanning	16*
1954	The Study of Fast Reactions	17
1954	Coagulation and Flocculation	18
1955	Microwave and Radio-Frequency Spectroscopy	19
1955	Physical Chemistry of Enzymes	20
1956	Membrane Phenomena	21
1956	Physical Chemistry of Processes at High Pressures	22
1957	Molecular Mechanism of Rate Processes in Solids	23
1958	Interactions in Ionic Solutions	24
1957	Configurations and Interactions of Macromolecules and Liquid Crystals	25
1958	Ions of the Transition Elements	26
1959	Energy Transfer with special reference to Biological Systems	27
1959	Crystal Imperfections and the Chemical Reactivity of Solids	28
1960	Oxidation-Reduction Reactions in Ionizing Solvents	29
1960	The Physical Chemistry of Aerosols	30
1961	Radiation Effects in Inorganic Solids	31
1961	The Structure and Properties of Ionic Melts	32
1962	Inelastic Collisions of Atoms and Simple Molecules	33*
1962	High Resolution Nuclear Magnetic Resonance	34
1963	The Structure of Electronically-Excited Species in the Gas-Phase	35
1963	Fundamental Processes in Radiation Chemistry	36
1964	Chemical Reactions in the Atmosphere	37
1964	Dislocations in Solids	38

Date	*Subject*	*Volume*
1965	The Kinetics of Proton Transfer Processes	39
1965	Intermolecular Forces	40
1966	The Role of the Adsorbed State in Heterogeneous Catalysis	41
1966	Colloid Stability in Aqueous and Non-Aqueous Media	42
1967	The Structure and Properties of Liquids	43
1967	Molecular Dynamics of the Chemical Reactions of Gases	44
1968	Electrode Reactions of Organic Compounds	45
1968	Homogeneous Catalysis with Special Reference to Hydrogenation and Oxidation	46
1969	Bonding in Metallo-Organic Compounds	47
1969	Motions in Molecular Crystals	48
1970	Polymer Solutions	49
1970	The Vitreous State	50
1971	Electrical Conduction in Organic Solids	51
1971	Surface Chemistry of Oxides	52
1972	Reactions of Small Molecules in Excited States	53
1972	The Photoelectron Spectroscopy of Molecules	54
1973	Molecular Beam Scattering	55
1973	Intermediates in Electrochemical Reactions	56
1974	Gels and Gelling Processes	57
1974	Photo-effects in Adsorbed Species	58
1975	Physical Adsorption in Condensed Phases	59
1975	Electron Spectroscopy of Solids and Surfaces	60
1976	Precipitation	61
1977	Potential Energy Surfaces	62
1977	Radiation Effects in Liquids and Solids	63
1977	Ion–Ion and Ion–Solvent Interactions	64
1978	Colloid Stability	65*
1978	Structure and Motion in Molecular Liquids	66
1979	Kinetics of State Selected Species	67
1979	Organization of Macromolecules in the Condensed Phase	68
1980	Phase Transitions in Molecular Solids	69

* *Not available; for current information on prices, etc, of available volumes, please contact the Marketing Officer, Royal Society of Chemistry, Burlington House, London W1V 0BN stating whether or not you are a member of the Society.*